Springer Series in Computational Mathematics

Richard S. Varga

Geršgorin and His Circles

Springer

Richard S. Varga
Kent State University
Institute for Computational Mathematics
Kent, OH 44242, USA
e-mail: varga@mcs.kent.edu

Cover jacket figure:
Wassily Kandinsky, *Heavy Circles*, 1927. Oil on canvas.
Norton Simon Museum,
The Blue Four Galka Scheyer Collection

Mathematics Subject Classification (2000): 15A18, 15A42, 15A48, 15A60, 65F15, 65F35

ISSN 0179-3632
ISBN 978-3-642-05928-5

Springer is a part of Springer Science+Business Media
springeronline.com

Printed in Germany

Cover design: *design&production*, Heidelberg

Printed on acid-free paper 46/3142LK-543210

Dedicated to: Olga Taussky-Todd† and John Todd
Department of Mathematics
California Institute of Technology
Pasadena, California

† Olga died on October 7, 1995, in Pasadena, California.

I. Preface

The **Geršgorin Circle Theorem**, a very well-known result in linear algebra today, stems from the paper of S. Geršgorin in 1931 (which is reproduced in Appendix D) where, given an arbitrary $n \times n$ complex matrix, easy arithmetic operations on the entries of the matrix produce n disks, in the complex plane, whose union contains all eigenvalues of the given matrix. The beauty and simplicity of Geršgorin's Theorem has undoubtedly inspired further research in this area, resulting in hundreds of papers in which the name "Geršgorin" appears. The goal of this book is to give a careful and up-to-date treatment of various aspects of this topic.

The author first learned of Geršgorin's results from friendly conversations with Olga Taussky-Todd and John Todd, which inspired me to work in this area. Olga was clearly passionate about linear algebra and matrix theory, and her path-finding results in these areas were like a magnet to many, including this author! It is the author's hope that the results, presented here on topics related to Geršgorin's Theorem, will be of interest to many. This book is affectionately dedicated to my mentors, Olga Taussky-Todd and John Todd.

There are two main **recurring themes** which the reader will see in this book. The first recurring theme is that a nonsingularity theorem for a matrices gives rise to an equivalent eigenvalue inclusion set in the complex plane for matrices, and conversely. Though common knowledge today, this was not widely recognized until many years after Geršgorin's paper appeared. That these two items, nonsingularity theorems and eigenvalue inclusion sets, go hand-in-hand, will be often seen in this book.

The second recurring theme in this book is the decisive role that M-matrices and H-matrices, and the related Perron-Frobenius theory of nonnegative matrices, play throughout this book. A much lesser recurring theme in this book is the observation that there have been surprisingly many published results in this area which contain major errors. Most of these errors have fortunately been corrected in subsequent papers.

As the reader will see, there are many new and unpublished items in the six chapters of this book. This is the result of trying to collect and unify many Geršgorin-type results in the literature, and it has been truly exciting to work in this area. The material in this book is almost entirely self-contained, with exercises for those who wish to test their skills. The material in this book should be within the grasp of upper-class undergraduates and graduate students in mathematics, as well as physicists, engineers, and computer scientists.

But there are a number of related topics which interested readers may find missing in these chapters. These topics include infinite dimensional extensions of Geršgorin's theory, connections with pseudo-spectra and singular values of matrices, real eigenvalues of real matrices, and generalized matrix eigenvalue problems. The results in these areas are currently, in the author's opinion, more fragmented and less unified, so that their complete coverage is intended for a future second volume of this book, with the same title. To this end, the author warmly solicits suggestions and references for this second volume, as well as comments, and corrections on this present volume. (Please send to varga@math.kent.edu.)

A large number of people have read parts of this manuscript and have offered many worthwhile suggestions to the author; I owe them my heartfelt thanks. But, there are a few whose contributions here must be fully recognized. My sincere thanks are due to my colleagues, Arden Ruttan and Laura Smithies. Arden assembled the Matlab programs in Appendix D for many of the pictures in this book, and his careful work identified some "missing eigenvalues" in Fig. 6.2, 6.3, 6.5, 6.6. Laura read the entire manuscript and set for herself the daunting task of working through all the exercises given in this book! My greatest thanks go to experts in this area, Richard Brualdi (University of Wisconsin), Ljiljana Cvetković (University of Novi-Sad), Volker Mehrmann (Technische Universität-Berlin), François Robert (University of Grenoble), and anonymous referees, who have made substantial comments and suggestions on the entire manuscript. It is a pleasure for me to recognize their deep expertise in this area and their unselfish efforts to help the author with their comments.

Our warmest and sincere thanks are due to our secretaries (in the Institute for Computational Mathematics at Kent State University) Joyce Fuell, who began the typing of this manuscript, and to Misty Tackett, who completed, with great care, the typing of this manuscript, and to my wife, Esther, for her love, understanding and great forbearance in the writing of the book. Finally, we wish to thank Springer-Verlag for their strong and unfailing support over the years, of the author's efforts.

Richard S. Varga

March 3, 2004
Kent, Ohio

Contents

1. Basic Theory

1.1 Geršgorin's Theorem

It can be said that the original result of Geršgorin (1931), on obtaining an eigenvalue inclusion result for any $n \times n$ complex matrix in terms of n easily obtained disks in the complex plane, was a **sensation** at that time, which generated much enthusiasm. It is fitting that that the first result to be proved here is Geršgorin's original result. For notation, let $\mathbb{C}^n$ denote, for any positive integer n, the complex n-dimensional vector space of all column vectors $\mathbf{v} = [v_1, v_2, \cdots, v_n]^T$, where each entry v_i is a complex number, (i.e., $v_i \in \mathbb{C}, 1 \leq i \leq n$), and let $\mathbb{C}^{m \times n}$ denote the collection of all $m \times n$ rectangular matrices with complex entries. Specifically, for $A \in \mathbb{C}^{m \times n}$, we express A as $A = [a_{i,j}]$ and as

$$(1.1) \qquad A = \begin{bmatrix} a_{1,1} & a_{1,2} & \cdots & a_{1,n} \\ a_{2,1} & a_{2,2} & \cdots & a_{2,n} \\ \vdots & \vdots & & \vdots \\ a_{m,1} & a_{m,2} & \cdots & a_{m,n} \end{bmatrix},$$

where $a_{i,j} \in \mathbb{C}$ for all $1 \leq i \leq m$ and $1 \leq j \leq n$. (In a completely similar fashion, $\mathbb{R}^n$ and $\mathbb{R}^{m \times n}$ denote, respectively, column vectors and rectangular matrices with *real* entries. Also, I_n will denote the $n \times n$ *identity matrix*, whose diagonal entries are all unity and whose off-diagonal entries are all zero.)

For additional notation which is used throughout, the **spectrum** $\sigma(A)$ of a square matrix $A = [a_{i,j}] \in \mathbb{C}^{n \times n}$ is the collection of all eigenvalues of A, i.e.,

$$(1.2) \qquad \sigma(A) := \{\lambda \in \mathbb{C} : \det(\lambda I_n - A) = 0\},$$

and with

$$(1.3) \qquad N := \{1, 2, \cdots, n\},$$

we call

$$r_i(A) := \sum_{j \in N \setminus \{i\}} |a_{i,j}| \qquad (i \in N) \tag{1.4}$$

the i-th deleted absolute **row sum**[1] of A (with the convention that $r_1(A) := 0$ if $n = 1$). Further, we set

$$\begin{cases} \Gamma_i(A) := \{z \in \mathbb{C} : |z - a_{i,i}| \le r_i(A)\} & (i \in N), \\ \text{and} \\ \Gamma(A) := \bigcup_{i \in N} \Gamma_i(A). \end{cases} \tag{1.5}$$

Note that $\Gamma_i(A)$, called the $\boldsymbol{i^{\text{th}}}$**-Geršgorin disk** of A, is from (1.5) a closed disk in the complex plane $\mathbb{C}$, having center $a_{i,i}$ and radius $r_i(A)$. Then $\Gamma(A)$, defined in (1.5) as the union of the n Geršgorin disks, is called the **Geršgorin set**, and is evidently closed and bounded in $\mathbb{C}$. But $\Gamma(A)$ can have a comparatively interesting geometrical structure, as the next examples show.

For each matrix $\{A_i\}_{i=1}^3$ defined below, its spectrum is given[2] , and its associated Geršgorin set, $\Gamma(A_i)$, is shown in Figs. 1.1-1.3. Also, the eigenvalues of each matrix A_i are shown by "$\times$'s" in these figures.

$$A_1 = \begin{bmatrix} 1 & -1 \\ \\ 1 & -1 \end{bmatrix}, \ \sigma(A_1) = \{0; 0\}.$$

$$A_2 = \begin{bmatrix} 1 & i & 0 \\ 1/2 & 4 & i/2 \\ 1 & 0 & 7 \end{bmatrix},$$

$\sigma(A_2) = \{0.9641 - 0.1620i; 4.0641 + 0.1629i; 6.9718 - 0.0008i\}.$

$$A_3 = \begin{bmatrix} 0 & 4 & 0 & 0 & 0 & 0 & 0 \\ 1 & 2 & 0 & 0 & 0 & 0 & 0 \\ 0 & 1 & -2 & 0 & 0 & 0 & 0 \\ 0 & 0 & 1/8 & -i & 1/8 & 0 & 0 \\ 0 & 0 & 0 & 1/4 & -2i & 1/4 & 0 \\ 0 & 0 & 0 & 0 & 0 & 9/2 & 1/2 \\ 0 & 0 & 0 & 0 & 0 & 1/2 & -9/2 \end{bmatrix},$$

$\sigma(A_3) = \{3.2361; -1.2361; -2; -1.0323i; -1.9677i; 4.5277; -4.5277\}.$

[1] While this term is very precise, we shall use "row sum" throughout, for reasons of brevity. This notation also applies to "column sums", which are introduced in section 1.3.

[2] The numerical values for the non-integer real and imaginary parts of eigenvalues are always given, rounded to the first four decimal digits.

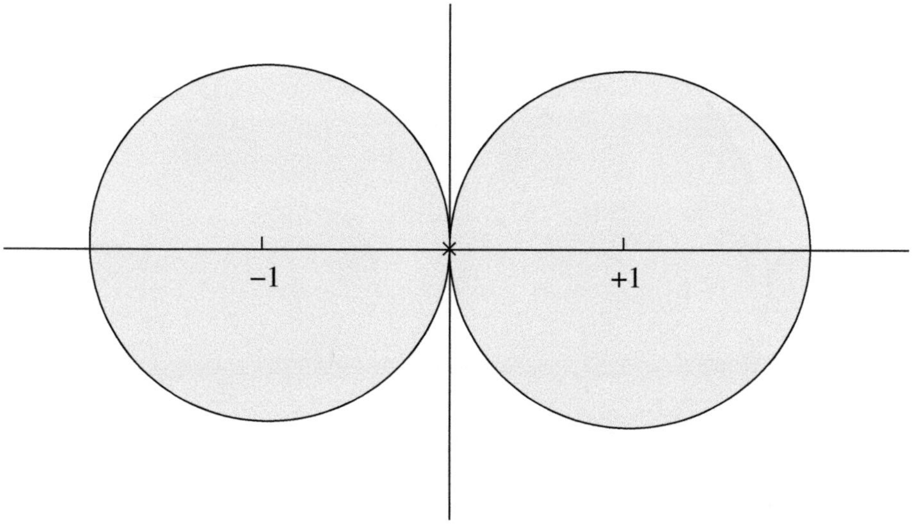

Fig. 1.1. $\Gamma(A_1)$ and $\sigma(A_1)$ for the matrix A_1

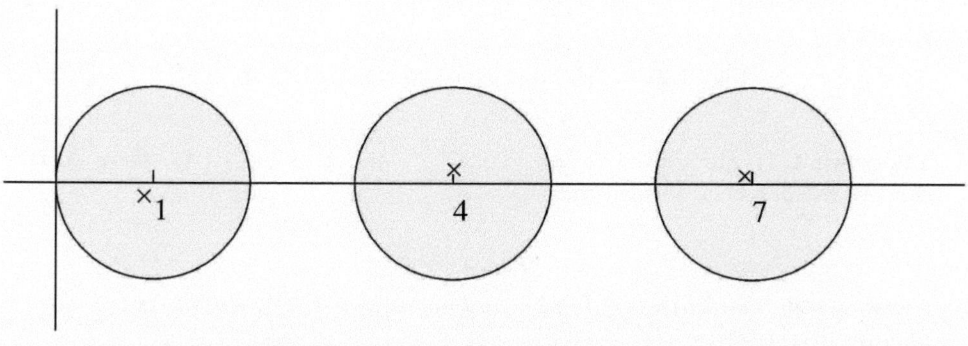

Fig. 1.2. $\Gamma(A_2)$ and $\sigma(A_2)$ for the matrix A_2

The figures for the Geršgorin sets $\Gamma(A_i)$ indicate that some eigenvalues can lie on the *boundary* of their associated Geršgorin set (cf. $\Gamma(A_1)$), that some Geršgorin disks each contain *exactly one* eigenvalue (cf. $\Gamma(A_2)$), and that some Geršgorin disks contain *no* eigenvalues (cf. $\Gamma_2(A_3)$). But, in all cases, all eigenvalues of A_i lie in the Geršgorin set $\Gamma(A_i)$. This latter statement holds for all $A = [a_{i,j}] \in \mathbb{C}^{n \times n}$ and is the original result of Geršgorin (1931), which we now prove.

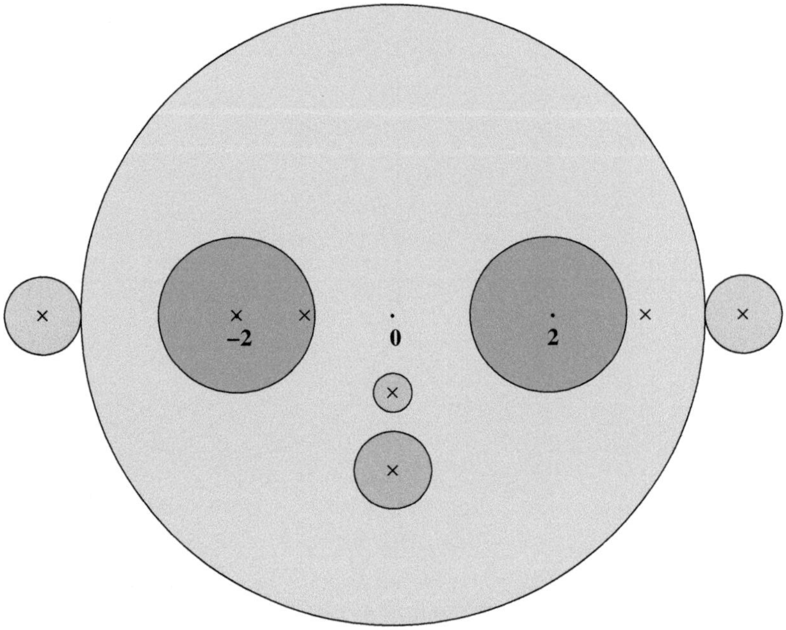

Fig. 1.3. $\Gamma(A_3)$ and $\sigma(A_3)$ for the matrix A_3

Theorem 1.1. *For any* $A = [a_{i,j}] \in \mathbb{C}^{n \times n}$ *and any* $\lambda \in \sigma(A)$*, there is a positive integer* k *in* N *such that*

$$|\lambda - a_{k,k}| \leq r_k(A). \tag{1.6}$$

Consequently (cf. (1.5)), $\lambda \in \Gamma_k(A)$*, and hence,* $\lambda \in \Gamma(A)$*. As this is true for each* $\lambda \in \sigma(A)$*, then*

$$\sigma(A) \subseteq \Gamma(A). \tag{1.7}$$

Proof. For any $\lambda \in \sigma(A)$, let $\mathbf{0} \neq \mathbf{x} = [x_1, x_2, \cdots, x_n]^T \in \mathbb{C}^n$ be an associated eigenvector, i.e., $A\mathbf{x} = \lambda\mathbf{x}$, so that $\sum_{j \in N} a_{i,j}x_j = \lambda x_i$ for each $i \in N$. Since $\mathbf{x} \neq \mathbf{0}$, there is a positive integer $k \in N$ for which $0 < |x_k| = \max\{|x_i| : i \in N\}$. For this k, we have $\sum_{i \in N} a_{k,i}x_i = \lambda x_k$, or equivalently,

$$(\lambda - a_{k,k})x_k = \sum_{i \in N \backslash \{k\}} a_{k,i}x_i.$$

Taking absolute values in the above equation and using the triangle inequality gives, with (1.4),

$$|\lambda - a_{k,k}| \cdot |x_k| \leq \sum_{i \in N \setminus \{k\}} |a_{k,i}| \cdot |x_i| \leq \sum_{i \in N \setminus \{k\}} |a_{k,i}| \cdot |x_k| = |x_k| \cdot r_k(A),$$

and dividing through by $|x_k| > 0$ gives (1.6). Then from (1.5), $\lambda \in \Gamma_k(A)$ and hence, $\lambda \in \Gamma(A)$. As this is valid for each $\lambda \in \sigma(A)$, then (1.7) follows. ■

The beauty and simplicity of Geršgorin's Theorem 1.1 is evident: for any matrix A in $\mathbb{C}^{n \times n}$, easy arithmetic operations on the entries of A give the row sums $\{r_i(A)\}_{i \in N}$, which in turn determine the n disks of (1.5), whose union must contain all eigenvalues of A! However, we note, from Figs. 1.1-1.3, that these easy operations may not give very precise information about these eigenvalues of A, but we also note that this union of disks applies equally well to all matrices $B = [b_{i,j}] \in \mathbb{C}^{n \times n}$ with $b_{i,i} = a_{i,i}$ and $r_i(B) = r_i(A)$ for all $i \in N$, so that $\sigma(B) \subseteq \Gamma(A)$ for *all* such B.

A useful consequence of Theorem 1.1 is the following. For notation, given any $A \in \mathbb{C}^{n \times n}$, the **spectral radius** $\rho(A)$ of A is defined by

$$\rho(A) := \max \{|\lambda| : \lambda \in \sigma(A)\}. \tag{1.8}$$

Corollary 1.2. *For any $A = [a_{i,j}] \in \mathbb{C}^{n \times n}$, then*

$$\rho(A) \leq \max_{i \in N} \sum_{j \in N} |a_{i,j}|. \tag{1.9}$$

Proof. Given any $\lambda \in \sigma(A)$, Theorem 1.1 gives the existence of a $k \in N$ such that $|\lambda - a_{k,k}| \leq r_k(A)$. But, by the reverse triangle inequality, $|\lambda| - |a_{k,k}| \leq |\lambda - a_{k,k}| \leq r_k(A)$; whence,

$$|\lambda| \leq |a_{k,k}| + r_k(A) = \sum_{j \in N} |a_{k,j}| \leq \max_{i \in N} \sum_{j \in N} |a_{i,j}|.$$

As this holds for each $\lambda \in \sigma(A)$, (1.9) follows. ■

We remark that (1.9) is just the statement (see (B.3) and (B.5) of Appendix B) that

$$\rho(A) \leq \|A\|_\infty, \tag{1.10}$$

where $\|A\|_\infty$ is the induced operator norm of A for the vector norm $\|\mathbf{x}\|_\infty := \max\{|x_j| : j \in N\}$, defined on $\mathbb{C}^n$.

Geršgorin's Theorem 1.1 will actually be shown below to be equivalent to the next result, the so-called *Strict Diagonal Dominance Theorem*, whose origin is even older than Geršgorin's Theorem 1.1 of 1931.

Definition 1.3. A matrix $A = [a_{i,j}] \in \mathbb{C}^{n\times n}$ is **strictly diagonally dominant** if

$$|a_{i,i}| > r_i(A) \qquad (\text{all } i \in N) \tag{1.11}$$

With this definition, we have

Theorem 1.4. *For any $A = [a_{i,j}] \in \mathbb{C}^{n\times n}$ which is strictly diagonally dominant, then A is nonsingular.*

Proof. Suppose, on the contrary, that $A = [a_{i,j}] \in \mathbb{C}^{n\times n}$ satisfies (1.11) and is singular, i.e., $0 \in \sigma(A)$. But from (1.6) with $\lambda = 0$, there exists a $k \in N$ such that $|\lambda - a_{k,k}| = |a_{k,k}| \leq r_k(A)$, which contradicts (1.11). ■

We have just shown that Theorem 1.4 follows from Theorem 1.1, and we now show that the **reverse** implication is also true. Assume that the result of Theorem 1.4 is true, and suppose that Theorem 1.1 is not valid, i.e., for some matrix $A = [a_{i,j}] \in \mathbb{C}^{n\times n}$, suppose that there exists a $\lambda \in \sigma(A)$ such that

$$|\lambda - a_{k,k}| > r_k(A) \ (\text{all } k \in N). \tag{1.12}$$

On setting $B := \lambda I_n - A := [b_{i,j}]$, where I_n is the identity matrix in $\mathbb{C}^{n\times n}$, then B is surely singular. On the other hand, the definition of B gives, with (1.4), that $r_k(B) = r_k(A)$ and that $|\lambda - a_{k,k}| = |b_{k,k}|$ for all $k \in N$, so that (1.12) becomes

$$|b_{k,k}| > r_k(B) \ (\text{all } k \in N).$$

But on applying Theorem 1.4, the above inequalities give that B is then nonsingular, a contradiction. Thus, we have shown that Theorems 1.1 and 1.4 are actually *equivalent*. This will be our **first recurring theme** in this book, namely, that a **nonsingularity result**, such as Theorem 1.4, induces an **eigenvalue inclusion result**, such as Theorem 1.1, and **conversely**.

That strictly diagonally dominant matrices are, from Theorem 1.4, nonsingular is an old and recurring result in matrix theory. This basic result can be traced back to at least Lévy (1881), Desplanques (1887), Minkowski (1900), and Hadamard (1903), whose contributions are not at all to be slighted. See the Bibliography and Discussion section, at the end of this chapter, for a more complete coverage of the history in this area.

It must be said, however, that Geršgorin's Theorem 1.1 gained immediate and wide recognition, as it could be easily applied to any square matrix. As a first extension of Theorem 1.1 (already considered in Geršgorin (1931)), let $\mathbf{x} = [x_1, x_2, \cdots, x_n]^T \in \mathbb{R}^n$ satisfy $x_i > 0$ for all $i \in N$, where we denote this by $\mathbf{x} > \mathbf{0}$. With this vector $\mathbf{x}$, define the matrix X, in $\mathbb{R}^{n\times n}$, by $X := \text{diag}[\mathbf{x}] := \text{diag}\,[x_1, x_2, \cdots, x_n]$, so that X is nonsingular. If $A = [a_{i,j}] \in \mathbb{C}^{n\times n}$, then $X^{-1}AX = [a_{i,j}x_j/x_i]$, and, as $X^{-1}AX$ is similar to A, then $\sigma(X^{-1}AX) = \sigma(A)$. In analogy with the definitions of (1.4) and (1.5), we set

(1.13) $$r_i^{\mathbf{x}}(A) := r_i(X^{-1}AX) = \sum_{j \in N \setminus \{i\}} \frac{|a_{i,j}|x_j}{x_i} \quad (i \in N,\ \mathbf{x} > \mathbf{0}),$$

and we call $r_i^{\mathbf{x}}(A)$ the i-th **weighted row sum** of A. In addition, we set[3]

(1.14) $$\begin{cases} \Gamma_i^{r^{\mathbf{x}}}(A) := \{z \in \mathbb{C} : |z - a_{i,i}| \leq r_i^{\mathbf{x}}(A)\}, \\ \text{and} \\ \Gamma^{r^{\mathbf{x}}}(A) := \bigcup_{i \in N} \Gamma_i^{r^{\mathbf{x}}}(A). \end{cases}$$

Now, $\Gamma_i^{r^{\mathbf{x}}}(A)$ of (1.14) is called the ***i*-th weighted Geršgorin disk** and $\Gamma^{r^x}(A)$ is the **weighted Geršgorin set**, where we note that the row sums $r_i^{\mathbf{x}}(A)$ of (1.13) are simply *weighted sums* of the absolute values of off-diagonal entries in the i-th row of A, for each $\mathbf{x} > \mathbf{0}$ in $\mathbb{R}^n$.

Applying Theorem 1.1 to $X^{-1}AX$ and using the fact that $\sigma(X^{-1}AX) = \sigma(A)$, we directly obtain

Corollary 1.5. *For any* $A = [a_{i,j}] \in \mathbb{C}^{n \times n}$ *and any* $\mathbf{x} > \mathbf{0}$ *in* $\mathbb{R}^n$, *then* (cf.(1.14))

(1.15) $$\sigma(A) \subseteq \Gamma^{r^{\mathbf{x}}}(A).$$

Of course, Geršgorin's Theorem 1.1 can be directly applied to $X^{-1}AX$, for *any* nonsingular X in $\mathbb{C}^{n \times n}$, to estimate $\sigma(A)$, and Corollary 1.5 is the first special case when $X = \text{diag}[x_1, x_2, \cdots, x_s]$, where $x_i > 0$ for all $i \in N$. This special case is important, in that determining the weighted row sums $\{r_i^{\mathbf{x}}(A)\}_{i \in N}$ requires now only slightly more computational effort than do the row sums $\{r_i(A)\}_{i \in N}$ of Geršgorin's Theorem 1.1.

At the other extreme, where X is any full nonsingular matrix in $\mathbb{C}^{n \times n}$, this substantially increases the work in finding the associated weighted row sums for the matrix $X^{-1}AX$, but it is important to know that there is a nonsingular matrix S in $\mathbb{C}^{n \times n}$ for which $S^{-1}AS =: J$ is in **Jordan normal form** (see Exercise 2 of this section), i.e., J is an upper bidiagonal matrix whose diagonal entries are eigenvalues of A, and whose upper bidiagonal entries are either 0 or 1. While possessing J would clearly give all the eigenvalues of A, and would do an optimal job of estimating the eigenvalues of A, it is in general much more computationally difficult to determine the Jordan normal form of A.

The next extension of Theorem 1.1 and Corollary 1.5 concerns the number of eigenvalues of A which lie in any **component** (i.e., a maximal connected subset) of $\Gamma^{r^{\mathbf{x}}}(A)$. Specifically, for $n \geq 2$, let S be a **proper subset** of $N = \{1, 2, \cdots, n\}$, i.e., $\emptyset \neq S \subsetneq N$, and denote the **cardinality** of S (i.e., the

[3] The superscript $r^{\mathbf{x}}$ in (1.14) may seem cumbersome, but it connects with the notation to be used in Chapter 5 on G-functions.

number of its elements) by $|S|$. For any $A = [a_{i,j}] \in \mathbb{C}^{n\times n}$ and for any $\mathbf{x} > \mathbf{0}$ in $\mathbb{R}^n$, write

$$(1.16) \qquad \Gamma_S^{r^{\mathbf{x}}}(A) := \bigcup_{i\in S} \Gamma_i^{r^{\mathbf{x}}}(A).$$

If $N\backslash S$ denotes the complement of S with respect to N and if $\emptyset$ denotes the empty set, the relation

$$(1.17) \qquad \Gamma_S^{r^{\mathbf{x}}}(A) \bigcap \Gamma_{N\backslash S}^{r^{\mathbf{x}}}(A) = \emptyset$$

states that the union of the Geršgorin disks, with indices belonging to S, is **disjoint** from the union of the remaining disks. We now establish another famous result of Geršgorin (1931).

Theorem 1.6. *For any $A = [a_{i,j}] \in \mathbb{C}^{n\times n}, n \geq 2$, and any $\mathbf{x} > \mathbf{0}$ in $\mathbb{R}^n$ for which the relation (1.17) is valid for some proper subset S of N, then $\Gamma_S^{r^{\mathbf{x}}}(A)$ contains exactly $|S|$ eigenvalues of A.*

Proof. Consider the set of matrices $A(t) := [a_{i,j}(t)]$ in $\mathbb{C}^{n\times n}$ where

$$a_{i,i}(t) := a_{i,i} \text{ and } a_{i,j}(t) := t\cdot a_{i,j} \text{ for } i \neq j \text{ (for all } i,j \in N),$$

with $0 \leq t \leq 1$. First, observe that

$$r_i^{\mathbf{x}}(A(t)) = \sum_{j\in N\backslash\{i\}} \frac{|a_{i,j}(t)|x_j}{x_i} = t\cdot \sum_{j\in N\backslash\{i\}} \frac{|a_{i,j}|x_j}{x_i} = t\cdot r_i^{\mathbf{x}}(A) \leq r_i^{\mathbf{x}}(A),$$

for all $t \in [0,1]$. Thus, with (1.14), $\Gamma_i^{r^{\mathbf{x}}}(A(t)) \subseteq \Gamma_i^{r^{\mathbf{x}}}(A)$ for all $t \in [0,1]$ and for all $i \in N$. These inclusions, coupled with the assumption (1.17), geometrically imply that

$$(1.18) \qquad \Gamma_S^{r^{\mathbf{x}}}(A(t)) \bigcap \Gamma_{N\backslash S}^{r^{\mathbf{x}}}(A(t)) = \emptyset \text{ for all } t \in [0,1],$$

while on the other hand, from (1.15) of Corollary 1.5, it follows that

$$\sigma(A(t)) \subseteq \Gamma^{r^{\mathbf{x}}}(A(t)) \text{ for all } t \in [0,1].$$

Now for $t = 0$, $A(0)$ is a diagonal matrix, so that the eigenvalues of the diagonal matrix $A(0)$ are just $\{a_{i,i}\}_{i=1}^n$. Hence, $\Gamma_S^{r^{\mathbf{x}}}(A(0)) = \{a_{i,i} : i \in S\}$ contains exactly $|S|$ eigenvalues of $A(0)$. Clearly, the entries of $A(t)$ vary continuously with t, and consequently (cf. Ostrowski (1960), Appendix K), the eigenvalues $\lambda_i(t)$ of $A(t)$ also vary continuously with $t \in [0,1]$. But because (1.18) holds for *all* $t \in [0,1]$, it is impossible, as t increases continuously from zero to unity, for $\Gamma_S^{r^{\mathbf{x}}}(A(t))$ to either gain or lose any eigenvalues of $A(t)$. Hence, $\Gamma_S^{r^{\mathbf{x}}}(A(t))$ contains exactly $|S|$ eigenvalues of $A(t)$ for every $t \in [0,1]$. Thus as $A(1) = A$, then $\Gamma_S^{r^{\mathbf{x}}}(A)$ contains exactly $|S|$ eigenvalues of A. ■

In particular, if one disk $\Gamma_i^{r^x}(A)$ for A is disjoint from the remaining disks $\Gamma_j^{r^x}(A)$, i.e., if $S = \{i\}$, then Theorem 1.6 asserts that this disk $\Gamma_i^{r^x}(A)$ contains exactly *one* eigenvalue of A. For example, for the particular matrix

$$A_2 = \begin{bmatrix} 1 & i & 0 \\ 1/2 & 4 & i/2 \\ 1 & 0 & 7 \end{bmatrix}$$

of Section 1.1, each of its Geršgorin disks is, by inspection, separated from the union of the remaining disks. Hence, we immediately have from Theorem 1.6 that

$$|\lambda_1 - 1| \leq 1, \ |\lambda_2 - 4| \leq 1, \ |\lambda_3 - 7| \leq 1,$$

where $\sigma(A_2) := \{\lambda_1, \lambda_2, \lambda_3\}$. Of course, since we have graphed the eigenvalues of A_2 in Fig. 1.2, we indeed see that each Geršgorin disk $\Gamma_i(A_2)$ contains exactly one eigenvalue of A_2. We remark that the above bounds for these three eigenvalues of A_2 can be *improved* by using weighted row sums from (1.14). See Exercise 3 of this section.

Exercises

1. Suppose (cf. (1.17)) that $\Gamma_S^{r^x}(A) \cap \Gamma_{N/S}^{r^x}(A)$ is not empty, but consists of $m \geq 1$ isolated points. Show that a more complicated form of Theorem 1.6 is still valid, where each isolated point may, or may not, be an eigenvalue (possibly multiple) of A.

2. Given any A in $\mathbb{C}^{n \times n}$, it is known that (cf. Horn and Johnson (1985), p.126) there is a nonsingular matrix S such that $S^{-1}AS =: J$ is in block diagonal **Jordan normal form**, i.e.,

$$J = \text{diag}[J_1(\lambda_1), J_2(\lambda_2), \cdots, J_k(\lambda_k)] \ ,$$

where, for all $1 \leq j \leq k$, $J_j(\lambda_j) = [\lambda_j]$ if $J_j(\lambda_j)$ is a 1×1 matrix, or otherwise, $J_j(\lambda_j)$ is the upper bidiagonal matrix

$$J_j(\lambda_j) = \begin{bmatrix} \lambda_j & 1 & & & \bigcirc \\ & \ddots & \ddots & & \\ & & \ddots & \ddots & \\ & & & \ddots & 1 \\ \bigcirc & & & & \lambda_j \end{bmatrix}.$$

Thus, $\{\lambda_j\}_{j=1}^k$ are the eigenvalues of A. If X is the $n \times n$ diagonal matrix $X := \text{diag}[1, \epsilon, \epsilon^2, \cdots, \epsilon^{n-1}]$, where $0 < \epsilon < 1$, show that the matrix

$$X^{-1}JX := \text{diag}\,[\widetilde{J}_1\,(\lambda_1), \widetilde{J}_2\,(\lambda_2), \cdots, \widetilde{J}_k\,(\lambda_k)],$$

where $\widetilde{J}_j\,(\lambda_j) = [\lambda_j]$ if $\widetilde{J}_j\,(\lambda_j)$ is a 1×1 matrix, and otherwise,

$$\widetilde{J}_j\,(\lambda_j) := \begin{bmatrix} \lambda_j & \epsilon & & & \bigcirc \\ & \ddots & \ddots & & \\ & & \ddots & \ddots & \\ & & & \ddots & \epsilon \\ \bigcirc & & & & \lambda_j \end{bmatrix}.$$

Further, deduce that any row sum of $X^{-1}JX$ is either 0 or ϵ.

3. For an $A = [a_{i,j}] \in \mathbb{C}^{n\times n}$, $n \geq 2$, suppose that there is an $\mathbf{x} > \mathbf{0}$ in $\mathbb{R}^n$ such that $r_1^{\mathbf{x}}(A) > 0$ and the first Geršgorin disk $\Gamma_1^{r^{\mathbf{x}}}(A)$ is disjoint from the remaining $(n-1)$ Geršgorin disks $\Gamma_j^{r^{\mathbf{x}}}(A)$, $2 \leq j \leq n$. From Theorem 1.6, there is a single eigenvalue λ of A in the disk $\Gamma_1^{r^{\mathbf{x}}}(A)$, i.e.,

$$|\lambda - a_{1,1}| \leq r_1^{\mathbf{x}}(A).$$

Show that it is always possible to find a $\mathbf{y} > \mathbf{0}$ in $\mathbb{R}^n$ such that $r_1^{\mathbf{y}}(A) < r_1^{\mathbf{x}}(A)$, while $\Gamma_1^{r^{\mathbf{y}}}(A)$ is still disjoint from the remaining $n-1$ disks $\Gamma_j^{r^{\mathbf{y}}}(A)$, $2 \leq j \leq n$, i.e., a tighter bound for λ can always be found! (Hint: Try increasing the first component of $\mathbf{x}$.)

4. If $A = [a_{i,j}] \in \mathbb{R}^{n\times n}$ is such that $|a_{i,i} - a_{j,j}| \geq r_i(A) + r_j(A)$, for all $i \neq j$ in N, then all the eigenvalues of A are real. (Geršgorin (1931)). (Hint: Use the fact that nonreal eigenvalues of A must occur in conjugate complex pairs.)

1.2 Extensions of Geršgorin's Theorem via Graph Theory

As we have seen in Theorem 1.4, the assumption of strict diagonal dominance of a matrix $A = [a_{i,j}] \in \mathbb{C}^{n\times n}$, i.e.,

$$|a_{i.i}| > r_i(A) = \sum_{j \in N \backslash \{i\}} |a_{i,j}| \quad (\text{all } i \in N),$$

implies that A is nonsingular. For $n \geq 2$, it is natural to ask if one of the above inequalities can be *weakened* to equality, as for example in

$$|a_{i,i}| > r_i(A) \text{ for } 1 \leq i \leq n-1, \text{ and } |a_{n,n}| = r_n(A),$$

while still ensuring that A is nonsingular. This may or may not be true: the following two matrices,

$$\begin{bmatrix} 2 & 1 \\ 2 & 2 \end{bmatrix} \text{ and } \begin{bmatrix} 1 & 0 \\ 0 & 0 \end{bmatrix},$$

satisfy the above weakened inequalities, where the first matrix is nonsingular, but the second is singular. This brings us to a graph-theoretic development which is connected with the notion of **irreducibility** for matrices, which we now summarize.

For notation, a matrix $P \in \mathbb{R}^{n \times n}$ is said to be a **permutation matrix** if there is a permutation ϕ, i.e., a 1-1 mapping of $N = \{1, 2, \cdots, n\}$ onto N, such that $P = [p_{i,j}] := [\delta_{i,\phi(j)}]$, where $\delta_{k,\ell} := \left\{ \begin{array}{l} 1 \text{ if } k = \ell \\ 0 \text{ if } k \neq \ell \end{array} \right\}$ is the familiar Kronecker delta function.

Definition 1.7. A matrix $A \in \mathbb{C}^{n \times n}, n \geq 2$, is **reducible** if there exist a permutation matrix $\tilde{P} \in \mathbb{R}^{n \times n}$ and a positive integer r, with $1 \leq r < n$, for which

$$\tilde{P} A \tilde{P}^T = \left[\begin{array}{c|c} A_{1,1} & A_{1,2} \\ \hline O & A_{2,2} \end{array} \right], \tag{1.19}$$

where $A_{1,1} \in \mathbb{C}^{r \times r}$, and where $A_{2,2} \in \mathbb{C}^{(n-r) \times (n-r)}$. If no such permutation exists, A is **irreducible**. If $A \in \mathbb{C}^{1 \times 1}$, then A is irreducible if its single entry is nonzero, and reducible otherwise.

It is convenient to continue in the reducible case of (1.19). On recursively applying this algorithm (based on reducibility) to the square matrices $A_{1,1}$ and $A_{2,2}$ and their progenies, we ultimately obtain a permutation matrix $P \in \mathbb{R}^{n \times n}$ and a positive integer m, with $2 \leq m \leq n$, such that

$$PAP^T = \begin{bmatrix} R_{1,1} & R_{1,2} & \cdots & R_{1,m} \\ O & R_{2,2} & \cdots & R_{2,m} \\ \vdots & & \ddots & \vdots \\ O & O & \cdots & R_{m,m} \end{bmatrix}, \tag{1.20}$$

The above matrix is called the **normal reduced form**[4] of A (cf. Berman and Plemmons (1994), p.43), where each matrix $R_{j,j}$, $1 \leq j \leq m$, in (1.20) is such that

$$(1.21) \qquad \begin{cases} i) \; R_{j,j} \text{ is a } p_j \times p_j \text{ irreducible matrix with } p_j \geq 2, \\ \text{or} \\ ii) \; R_{j,j} \text{ is a } 1 \times 1 \text{ matrix with } R_{j,j} = [a_{k,k}] \text{ for some } k \in N. \end{cases}$$

We note from the structure of the matrix in (1.19) that, for $n \geq 2$, the diagonal entries of A play no role in the irreducibility or reducibility of A. It can also be seen, for $n \geq 2$, that replacing an arbitrary nonzero nondiagonal entry of $A = [a_{i,j}] \in \mathbb{C}^{n \times n}$ by any nonzero complex number leaves the irreducible or reducible character of A *invariant.* This indicates that the concept of irreducibility is in fact a graph-theoretic one, which we now describe.

Given any $A = [a_{i,j}] \in \mathbb{C}^{n \times n}$, let $\{v_i, v_2, \cdots, v_n\}$ be any n distinct points, which are called **vertices**. For each nonzero entry $a_{i,j}$ of A, connect the vertex v_i to the vertex v_j by means of a **directed arc** $\overrightarrow{v_i v_j}$, directed from the initial vertex v_i to the terminal vertex v_j, as shown in Fig. 1.4. (If $a_{i,i} \neq 0$, $\overrightarrow{v_i v_i}$ is a **loop**, which is also shown in Fig. 1.4.) The collection of all

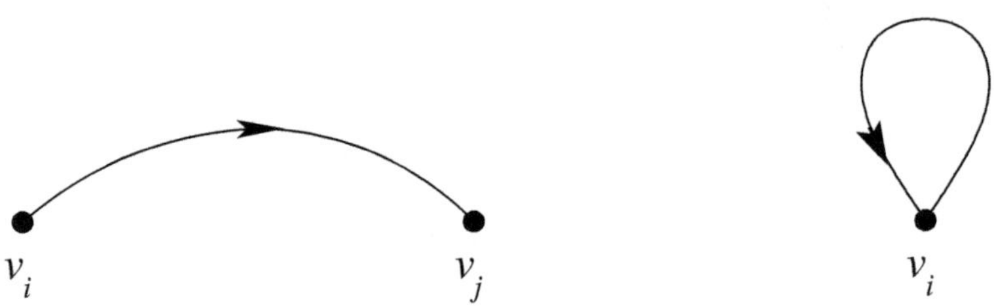

Fig. 1.4. Directed arcs

such directed arcs defines the **directed graph** $\mathbb{G}(A)$ of A. For additional notation, a **directed path** in $\mathbb{G}(A)$ is a collection of abutting directed arcs $\overrightarrow{v_{\ell_0} v_{\ell_1}}, \overrightarrow{v_{\ell_1} v_{\ell_2}}, \cdots, \overrightarrow{v_{\ell_{r-1}} v_{\ell_r}}$, connecting the initial vertex v_{ℓ_0} to the terminal vertex v_{ℓ_r}. Note that this directed path implies that

$$\prod_{k=0}^{r-1} a_{\ell_k, \ell_{k+1}} \neq 0.$$

[4] We remark that the normal reduced form of a reducible matrix is not necessarily unique. It is unique, up to permutations of the blocks in (1.20). We further note that (1.20) is also called in the literature the **Frobenius normal form** of a reducible matrix.

Definition 1.8. The directed graph $\mathbb{G}(A)$ of a matrix $A = [a_{i,j}] \in \mathbb{C}^{n \times n}$ is **strongly connected** if, for each ordered pair v_i and v_j of vertices, there is a directed path in $\mathbb{G}(A)$ with initial vertex v_i and terminal vertex v_j.

As examples, consider again the matrices A_1, A_2, and A_3 of Section 1.1. The directed graphs of these matrices are given below in Fig. 1.7. By inspection, $\mathbb{G}(A_1)$ and $\mathbb{G}(A_2)$ are strongly connected. On the other hand, $\mathbb{G}(A_3)$ is not strongly connected since there is no directed path, for example, from vertex v_2 to vertex v_3.

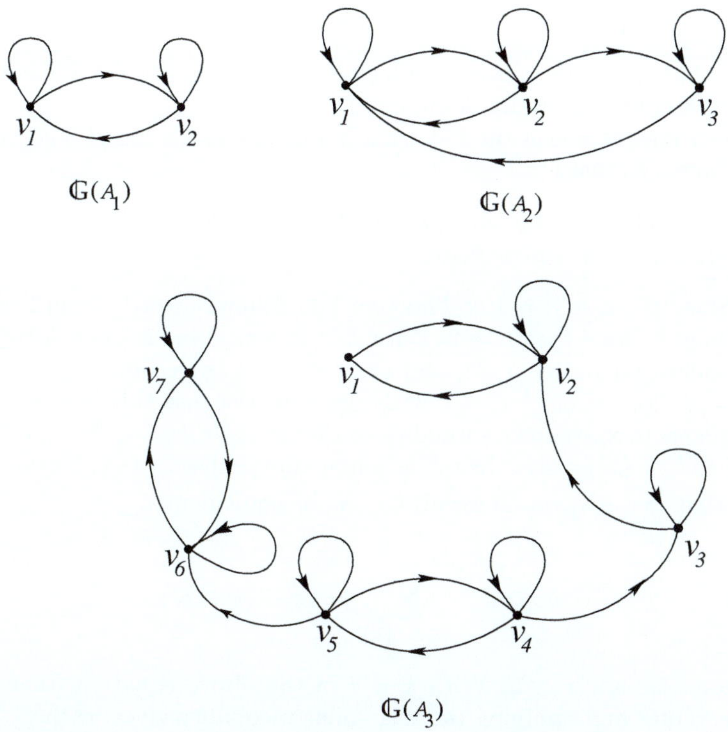

Fig. 1.5. Directed graphs $\mathbb{G}(A_i)$, $i = 1, 2, 3$, for the matrices A_i

It is easy to see that if $A = [a_{i,j}] \in \mathbb{C}^{n \times n}$ is such that its directed graph $\mathbb{G}(A)$ is strongly connected, then all of the off-diagonal entries in any row of A cannot simultaneously vanish. Also, if $\mathbb{G}(A)$ is strongly connected, then so is $\mathbb{G}(P^T AP)$ for any $n \times n$ permutation matrix P, since $\mathbb{G}(P^T AP)$ is precisely $\mathbb{G}(A)$ with a renumbered vertex set.

The following theorem shows the equivalence of the concepts introduced in Definitions 1.7 and 1.8. Its proof is left as Exercise 7 in this section.

Theorem 1.9. *For any* $A = [a_{i,j}] \in \mathbb{C}^{n\times n}$*, then* A *is irreducible if and only if its directed graph* $\mathbb{G}(A)$ *is strongly connected.*

We extend now the notion in (1.11) of a strictly diagonally dominant matrix in

Definition 1.10. A matrix $A = [a_{i,j}] \in \mathbb{C}^{n\times n}$ is an **irreducibly diagonally dominant matrix** if A is irreducible, if A is *diagonally dominant*, i.e.,

$$|a_{i,i}| \geq r_i(A) \text{ (all } i \in N), \tag{1.22}$$

and if strict inequality holds in (1.22) for at least one i.

Note that a matrix can be irreducible and diagonally dominant *without* being irreducibly diagonally dominant[5].

We can now establish the following important extension of Theorem 1.4, due to Taussky (1949).

Theorem 1.11. *For any* $A = [a_{i,j}] \in \mathbb{C}^{n\times n}$ *which is irreducibly diagonally dominant, then* A *is nonsingular.*

Proof. Since the case $n = 1$ of Theorem 1.11 follows directly from Definition 1.7, we may assume that $n \geq 2$. With $n \geq 2$, suppose, on the contrary, that A is singular. Then, $0 \in \sigma(A)$, and there exists a corresponding eigenvector $\mathbf{0} \neq \mathbf{x} = [x_1, x_2, \cdots, x_n]^T \in \mathbb{C}^n$ with $A\mathbf{x} = \mathbf{0}$. Because this last equation is homogeneous in $\mathbf{x}$, we may normalize so that $\max\{|x_i| : i \in N\} = 1$. Letting $S := \{j \in N : |x_j| = 1\}$, then S is a nonempty subset of N. Now, $A\mathbf{x} = \mathbf{0}$ implies that $\sum_{j\in N} a_{k,j}x_j = 0$ for all $k \in N$, or equivalently,

$$-a_{k,k}x_k = \sum_{j\in N\backslash\{k\}} a_{k,j}x_j \quad (k \in N).$$

Then, consider any $i \in S$. With $k = i$ in the above equation, then taking absolute values and applying the triangular inequality gives, with (1.4),

$$|a_{i,i}| \leq \sum_{j\in N\backslash\{i\}} |a_{i,j}| \cdot |x_j| \leq \sum_{j\in N\backslash\{i\}} |a_{i,j}| = r_i(A) \quad (i \in S).$$

Because the reverse inequality of the above must hold by hypothesis from (1.22), then

$$|a_{i,i}| = \sum_{j\in N\backslash\{i\}} |a_{i,j}| \cdot |x_j| = r_i(A) \qquad (i \in S). \tag{1.23}$$

[5] Unfortunately, this definition, while awkward, is in common use in the literature, and we bow to convention here.

But since strict inequality must hold, by hypothesis in (1.22), for at least one i, it follows that S must be a *proper* subset of N, i.e., $\emptyset \neq S \subsetneqq N$. On the other hand, for any $i \in S$, the terms in the sum of (1.23) cannot, by irreducibility, all vanish, so consider any $a_{i,j} \neq 0$ for some $j \neq i$ in N. It must follow from the second equality in (1.23) that $|x_j| = 1$; whence, j is also an element of S. Since A is irreducible, we can find a directed path in $\mathbb{G}(A)$ from the vertex v_i to *any* other vertex v_k, by means of directed arcs corresponding to the non-zero off-diagonal entries $a_{i,i_1}, a_{i_1,i_2}, \cdots, a_{i_{r-1},i_r}$, with $i_r = k$. Thus, $i_\ell \in S$ for each $1 \leq \ell \leq r$, and in particular, $i_r = k \in S$. But, as k was any integer with $k \in N$, then $S = N$, a contradiction. ■

We next give another important result of Taussky (1948) and Taussky (1949), which, by our first recurring theme, is exactly the equivalent eigenvalue inclusion result of Theorem 1.11. (We leave this claim to Exercise 5 of this section.) We also remark that only the second part of Theorem 1.12 appears in Taussky (1948) and Taussky (1949). For notation, let $\mathbb{C}_\infty := \mathbb{C} \cup \{\infty\}$ denote the extended complex plane. If T is any subset of $\mathbb{C}$, then $\overline{T}$ denotes the *closure* in $\mathbb{C}_\infty$ of T,

$$\partial T := \overline{T} \cap \overline{(\mathbb{C}_\infty \backslash T)}$$

denotes the *boundary* of T, and *int* $T := T \backslash \partial T$ denotes the *interior* of T.

Theorem 1.12. *Let $A = [a_{i,j}] \in \mathbb{C}^{n \times n}$ be irreducible. If $\lambda \in \sigma(A)$ is such that $\lambda \notin int\ \Gamma_i(A)$ for each $i \in N$, i.e., $|\lambda - a_{i,i}| \geq r_i(A)$ for each $i \in N$, then*

$$|\lambda - a_{i,i}| = r_i(A) \text{ for each } i \in N, \tag{1.24}$$

i.e., all the Geršgorin circles $\{z \in \mathbb{C} : |z - a_{i,i}| = r_i(A)\}$ pass through λ. In particular, if some eigenvalue λ of A lies on the boundary of $\Gamma(A)$, then (1.24) *holds.*

As an illustration of Theorem 1.12, consider the particular matrix

$$A_4 = \begin{bmatrix} 1 & -1 & 0 & 0 \\ 0 & i & -i & 0 \\ 0 & 0 & -1 & 1 \\ i & 0 & 0 & -i \end{bmatrix},$$

which can be verified to be irreducible and singular, so that $0 \in \sigma(A_4)$. As can be seen in Fig. 1.8, $0 \notin int\ \Gamma_i(A_4)$ for all $1 \leq i \leq 4$, and all Geršgorin circles $\{z \in \mathbb{C} : |z - a_{i,i}| = r_i(A_4)\}$ do indeed pass through 0, as required by Theorem 1.12, but 0 is *not* a boundary point of $\Gamma(A_4)$.

If $A = [a_{i,j}] \in \mathbb{C}^{n \times n}$ is irreducible, then Theorem 1.12 gives us that a **necessary condition**, for an eigenvalue λ of A to lie on the boundary of $\Gamma(A)$, is that all n Geršgorin circles pass through λ. But this necessary condition need not be **sufficient** to produce an eigenvalue of A, i.e., even if all n

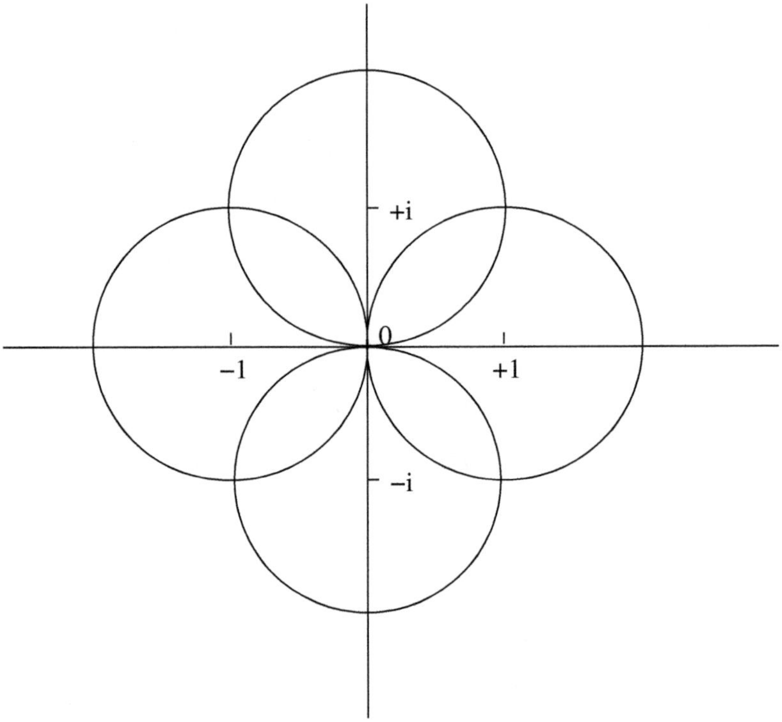

Fig. 1.6. $\Gamma(A_4)$ for the matrix A_4

Geršgorin circles pass through a single point z, then z need not be an eigenvalue of A. (See Exercise 6 of this section.) In any event, finding necessary **and** sufficient conditions for an eigenvalue of A to lie on the boundary of $\Gamma(A)$ is certainly interesting, and recent references on this are given in 1.3 of the Bibliography and Discussion of this chapter.

Next, a very familiar matrix in numerical analysis is the $n \times n$ tridiagonal matrix

$$A_5 = [a_{i,j}] = \begin{bmatrix} 2 & -1 & & & \bigcirc \\ -1 & 2 & -1 & & \\ & \ddots & \ddots & \ddots & \\ & & \ddots & 2 & -1 \\ \bigcirc & & & -1 & 2 \end{bmatrix}, \quad n \geq 1,$$

whose directed graph $\mathbb{G}(A_5)$ is shown in Fig. 1.7 for $n = 6$.

It is evident that $\mathbb{G}(A_5)$ is strongly connected, so that A_5 is irreducible from Theorem 1.9. To deduce that A_5 is nonsingular, simply note that $2 =$

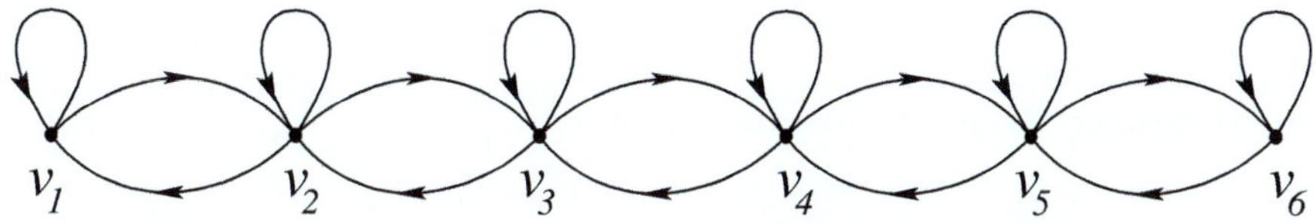

Fig. 1.7. $\mathbb{G}(A_5)$ for the matrix A_5

$|a_{i,i}| \geq r_i(A_5)$ for all $i \in N$, with strict inequality holding for $i = 1$ and $i = n$. Applying Theorem 1.11 then gives that A_5 is nonsingular, *independent* of the size of n.

Exercises

1. Given $A = [a_{i,j}] \in \mathbb{C}^{n \times n}$with $n \geq 2$, suppose that $|a_{i,i}| > r_i(A)$ $(1 \leq i < n)$ and $|a_{n,n}| = r_n(A)$. Show that A is nonsingular if $r_n(A) > 0$, and singular if $r_n(A) = 0$. (Cvetkovic (2001)) (Remark: This result is similar to that of Theorem 1.11, but irreducibility is not used here!)

2. Let $A = [a_{i,j}] \in \mathbb{C}^{n \times n}, n \geq 2$, and let $N := \{1, 2, \cdots, n\}$. Prove that A is irreducible if and only if, for every two disjoint nonempty subsets S and T of N with $S \cup T = N$, there is an element $a_{i,j} \neq 0$ of A with $i \in S$ and $j \in T$.

3. Let $A = [a_{i,j}] \in \mathbb{C}^{n \times n}$ with $n \geq 2$. If $\mathbb{G}(A)$ is the directed graph of A, show that A is strongly connected if, and only if, for *each* ordered pair (v_i, v_j) of vertices in $\mathbb{G}(A)$ with $i \neq j$, there is a directed path from v_i to v_j.

4. A matrix $A = [a_{i,j}] \in \mathbb{C}^{n \times n}$ is called **lower semistrictly diagonally dominant** if $|a_{i,i}| \geq r_i(A)$ (for all $i \in N$) and if $|a_{i,i}| > \sum_{j=1}^{i-1} |a_{i,j}|$ (for all $i \in N$). Similarly, $A = [a_{i,j}] \in \mathbb{C}^{n \times n}$ is called **semistrictly diagonally dominant** if there is a permutation matrix $P \in \mathbb{R}^{n \times n}$ such that PAP^T is lower semistrictly diagonally dominant. Show that a matrix A is irreducibly diagonally dominant if and only if A is semistrictly diagonally dominant. (Beauwens (1976)).

5. As an example of our first "recurring theme", show that a careful reformulation of the nonsingularity result of Theorem 1.11 gives the eigenvalue inclusion of Theorem 1.12, and, conversely, Theorem 1.12 implies the result of Theorem 1.11.

6. Consider the irreducible matrix $B = \begin{bmatrix} 1 & e^{i\theta_1} & 0 & 0 \\ 0 & i & e^{i\theta_2} & 0 \\ 0 & 0 & -1 & e^{i\theta_3} \\ e^{i\theta_4} & 0 & 0 & -i \end{bmatrix}$, where the θ_i's are all real numbers, so that the Geršgorin set $\Gamma(B)$ is the same as that shown in Fig. 1.6, and all Geršgorin circles pass through 0. Then, show that $z = 0$ is an eigenvalue of B only if $\sum_{i=1}^{4} \theta_i = \pi (\text{mod } 2\pi)$.

7. Give a proof of Theorem 1.9. (Hint: Prove that A is reducible (cf. Def. 1.7) if and only if $\mathbb{G}(A)$ is not strongly connected.)

1.3 Analysis Extensions of Geršgorin's Theorem and Fan's Theorem

Given $A = [a_{i,j}] \in \mathbb{C}^{n\times n}$, it is well known that $\sigma(A) = \sigma(A^T)$, where $A^T := [a_{j,i}]$. Thus, A^T is nonsingular if and only if A is nonsingular. Hence, applying Theorem 1.4 to A^T directly gives

Corollary 1.13. *For any $A = [a_{i,j}] \in \mathbb{C}^{n\times n}$ which satisfies*

$$(1.25) \qquad |a_{i,i}| > c_i(A) := r_i(A^T) = \sum_{j\in N\setminus\{i\}} |a_{j,i}| \ (\textit{all } i \in N),$$

then A is nonsingular.

Note that the quantities $c_i(A) = r_i(A^T)$ of (1.25) now depend on **column sums** of A, in contrast with the row sums $r_i(A)$ occurring in (1.11) of Theorem 1.4. Moreover, just as Theorem 1.1 and Theorem 1.4 are equivalent, so are Corollary 1.13 and

Corollary 1.14. *For any $A = [a_{i,j}] \in \mathbb{C}^{n\times n}$, then*

$$(1.26) \qquad \sigma(A) \subseteq \Gamma(A^T).$$

As obvious consequences of (1.7) of Theorem 1.1 and (1.26) of Corollary 1.14, we have

Theorem 1.15. *For any* $A = [a_{i,j}] \in \mathbb{C}^{n \times n}$, *let* $\Gamma(A^T) := \bigcup_{i \in N} \{z \in \mathbb{C} : |z - a_{i,i}| \leq c_i(A)\}$, *where* $c_i(A)$ *is defined in (1.25). Then,*

$$\sigma(A) \subseteq (\Gamma(A) \cap \Gamma(A^T)). \tag{1.27}$$

Note that the eigenvalue inclusion of (1.27) now depends upon the following $3n$ quantities, derived from the matrix $A = [a_{i,j}]$ in $\mathbb{C}^{n \times n}$:

$$\{a_{i,i}\}_{i=1}^{n}\ ,\ \{r_i(A)\}_{i=1}^{n}, \text{ and } \{c_i(A)\}_{i=1}^{n}. \tag{1.28}$$

As an example of the inclusion of (1.26) of Corollary 1.14, consider the particular matrix

$$A_2 = \begin{bmatrix} 1 & i & 0 \\ 1/2 & 4 & i/2 \\ 1 & 0 & 7 \end{bmatrix}, \tag{1.29}$$

previously considered in Section 1.1 of this chapter. The inclusion of (1.26), based on column sums of A_2, in this case is the set of three disks, shown below in Fig. 1.8 with solid boundaries. The corresponding inclusion of (1.7), based on row sums of A_2, is the set of three disks, shown in Fig. 1.8 with the dashed boundaries. Clearly, a better inclusion for the eigenvalues of A_2 is given by the corresponding intersection of (1.27) of Theorem 1.15, with one eigenvalue of A_2 in each shaded disk (cf. Theorem 1.6).

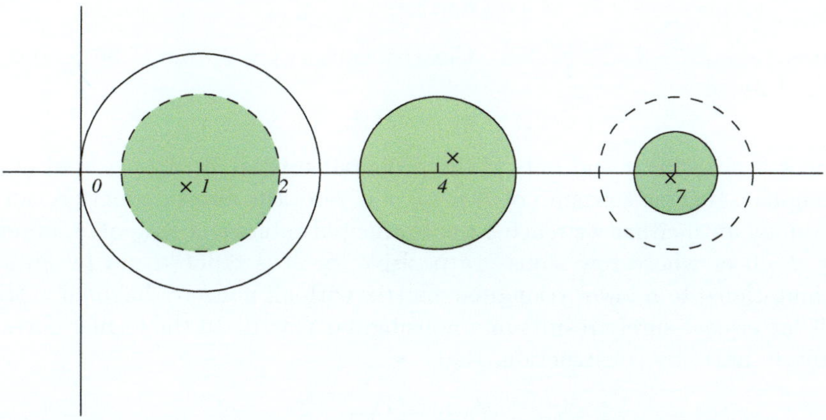

Fig. 1.8. $\Gamma(A_2), \Gamma(A_2^T)$, and $\Gamma(A_2) \cap \Gamma(A_2^T)$ (shaded) for the matrix A_2

Since both row and column sums of A enter respectively into separate sufficient conditions for the nonsingularity of A in Theorem 1.4 and Corollary 1.13, it is not unreasonable to expect that particular combinations of row *and* column sums do the same. Indeed, we have the following result of

Ostrowski (1951a), which we view as an **analysis extension** of Theorem 1.4 and Corollary 1.13. We give a complete proof of this extension, to show how typical analysis tools are utilized in this regard in linear algebra.

Theorem 1.16. *For any $A = [a_{i,j}] \in \mathbb{C}^{n \times n}$, and any α with $0 \leq \alpha \leq 1$, assume that*

$$|a_{i,i}| > (r_i(A))^{\alpha}(c_i(A))^{1-\alpha} \text{ (all } i \in N). \tag{1.30}$$

Then, A is nonsingular.

Proof. First, we remark that if ϕ is any permutation of the indices in N and if $P := [\delta_{i,\phi(j)}]$ is its associated permutation matrix in $\mathbb{R}^{n \times n}$, then the matrix $\tilde{A} := [\tilde{a}_{i,j}] := P^T AP$ has the *same* set of diagonal entries as does A and, moreover, the *same* set of row (and column) sums, as does A. Thus, the inequalities of (1.30) are *invariant* under such permutations of the set N.

Next, with the convention used in (1.4), the case $n = 1$ of (1.30) immediately gives that A is nonsingular. Then for $n \geq 2$, suppose that some $r_j(A) = 0$. By means of a suitable permutation of N, we may assume, without loss of generality, that $j = 1$. Then, $r_1(A) = 0$ implies that A has the form

$$A = \left[\begin{array}{c|c} a_{1,1} & 0 \cdots 0 \\ \hline a_{2,1} & \\ \vdots & A_{n-1,n-1} \\ a_{n,1} & \end{array}\right], \tag{1.31}$$

where $A_{n-1,n-1} \in \mathbb{C}^{(n-1)\times(n-1)}$. Clearly, since $a_{1,1} \neq 0$ from (1.30) and since from (1.31)

$$\det A = a_{1,1} \cdot \det A_{n-1,n-1},$$

then A is singular if and only if the principal submatrix $A_{n-1,n-1}$ of (1.31) is singular. If some row sum of $A_{n-1,n-1}$ is zero, the above reduction can be continued until either we reach a final principal submatrix $A_{p,p}$ of A, of order $p \geq 2$, all of whose row sums are positive, or A is reduced, up to suitable permutations, to a lower triangular matrix with all nonzero diagonal entries. This latter case surely results in a nonsingular matrix. In the former case, we similarly have, by construction, that

$$\det A = \left(\prod_{j=1}^{n-p} a_{j,j} \right) \cdot \det A_{p,p},$$

and again, A is singular if and only if $A_{p,p}$, with positive row sums, is singular. (It is important to note, by definition, that the i-th row (or column) sum of $A_{p,p}$ is *at most* the i-th row (or column) sum of the given matrix A). This reduction shows that we may assume, without loss of generality, that all row sums $r_i(A)$ of A are positive. Also, since the special case $\alpha = 1$ and $\alpha = 0$ of Theorem 1.16 reduce to the known results of Theorem 1.4 and Corollary 1.13, we may also assume that $0 < \alpha < 1$.

Suppose then, on the contrary, that $A = [a_{i,j}] \in \mathbb{C}^{n \times n}$ satisfies (1.30), has $r_i(A) > 0$ for all $i \in N$, and is singular, so that there is an $\mathbf{x} = [x_1, x_2, \cdots, x_n]^T \in \mathbb{C}^n$ with $\mathbf{x} \neq \mathbf{0}$ and with $A\mathbf{x} = \mathbf{0}$. This implies that $a_{i,i}x_i = -\sum_{j \in N \setminus \{i\}} a_{i,j}x_j$ for all $i \in N$, and, taking absolute values and applying the triangle inequality,

$$|a_{i,i}| \cdot |x_i| \leq \sum_{j \in N \setminus \{i\}} |a_{i,j}| \cdot |x_j| \qquad (i \in N).$$

Employing the inequality of (1.30) on the left side of the above inequality and writing $|a_{i,j}| = |a_{i,j}|^\alpha \cdot |a_{i,j}|^{1-\alpha}$ in the above sum, then

$$(1.32)\ (r_i(A))^\alpha (c_i(A))^{1-\alpha} |x_i| \leq \sum_{j \in N \setminus \{i\}} |a_{i,j}|^\alpha \cdot (|a_{i,j}|^{1-\alpha} \cdot |x_j|) \quad (i \in N),$$

where strict inequality holds above whenever $|x_i| > 0$, and thus for at least one $i \in N$. Applying Hölder's inequality to the last sum above with $p := 1/\alpha$ and $q := (1-\alpha)^{-1}$, we obtain

$$(1.33)\ (r_i(A))^\alpha (c_i(A))^{1-\alpha} |x_i| \leq \left(\sum_{j \neq i} |a_{i,j}|^{\alpha p} \right)^{1/p} \cdot \left(\sum_{j \neq i} |a_{i,j}| \cdot |x_j|^q \right)^{1/q},$$

for all $i \in N$. Note that $\left(\sum_{j \in N \setminus \{i\}} |a_{i,j}|^{\alpha p} \right)^{1/p} = (r_i(A))^\alpha$, since $p = 1/\alpha$. Hence, cancelling $(r_i(A))^\alpha > 0$ on both sides of (1.33) and raising both sides of (1.33) to the power q, we obtain, since $q = (1-\alpha)^{-1}$, that

$$c_i(A) \cdot |x_i|^q \leq \sum_{j \in N \setminus \{i\}} |a_{i,j}| \cdot |x_j|^q \qquad (i \in N),$$

where strict inequality holds above for at least one $i \in N$. Summing on all i in N in the above inequalities gives

$$(1.34) \qquad \sum_{i \in N} c_i(A) \cdot |x_i|^q < \sum_{i \in N} \left(\sum_{j \in N \setminus \{i\}} |a_{i,j}| \cdot |x_j|^q \right).$$

But interchanging the orders of summation in the (finite) double sum of (1.34) shows that this double sum reduces exactly to $\sum_{j \in N} c_j(A) \cdot |x_j|^q$, and substituting this in (1.34) gives an obvious contradiction to the strict inequality there. ■

Some easy consequences of Theorem 1.16 can be quickly drawn. If, for any $\mathbf{x} = [x_1, x_2, \cdots, x_n]^T > \mathbf{0}$ in $\mathbb{R}^n$, we set $X = \text{diag}\,[\mathbf{x}]$, then applying Theorem 1.16 to $X^{-1}AX = [a_{i,j}x_j/x_i]$ immediately gives

Corollary 1.17. *For any* $A = [a_{i,j}] \in \mathbb{C}^{n\times n}$, *any* $\mathbf{x} > \mathbf{0}$ *in* $\mathbb{R}^n$, *and any* α *with* $0 \leq \alpha \leq 1$, *assume that*

$$|a_{i,i}| > (r_i^{\mathbf{x}}(A))^{\alpha} \cdot (c_i^{\mathbf{x}}(A))^{1-\alpha} \quad (\text{all } i \in N), \tag{1.35}$$

where (cf. (1.25)) $c_i^{\mathbf{x}}(A) := r_i^{\mathbf{x}}(A^T)$. *Then,* A *is nonsingular.*

Just as Theorem 1.4 is equivalent to Theorem 1.1, we similarly obtain, from Corollary 1.17, the following corollary, which is actually equivalent to Corollary 1.17.

Corollary 1.18. *For any* $A = [a_{i,j}] \in \mathbb{C}^{n\times n}$, *any* $\mathbf{x} > \mathbf{0}$ *in* $\mathbb{R}^n$, *and any* α *with* $0 \leq \alpha \leq 1$, *then*

$$\sigma(A) \subseteq \bigcup_{i \in N} \left\{z \in \mathbb{C} : |z - a_{i,i}| \leq (r_i^{\mathbf{x}}(A))^{\alpha}(c_i^{\mathbf{x}}(A))^{1-\alpha}\right\}. \tag{1.36}$$

It is evident that the union of the disks in (1.36) gives an eigenvalue inclusion set for any matrix A, which is one of many such **Ostrowski sets**. But, there are two results in the literature which have very much the flavor of Corollary 1.18. We state these results below, leaving their proofs to exercises in this section. The first result is again due to Ostrowski (1951b).

Theorem 1.19. *For any* $A = [a_{i,j}] \in \mathbb{C}^{n\times n}$, *then*

$$\sigma(A) \subseteq \bigcup_{i \in N} \{z \in \mathbb{C} : |z - a_{i,i}| \leq \rho_i^{(1)}\}, \tag{1.37}$$

where

$$\rho_i^{(1)} := \alpha_i^{\frac{1}{q}} \Big(\sum_{j \in N\backslash\{i\}} |a_{i,j}|^p \Big)^{\frac{i}{p}} \quad (\text{all } i \in N), \tag{1.38}$$

where $p \geq 1$, $\frac{1}{p} + \frac{1}{q} = 1$ *and the positive numbers* $\{\alpha_i\}_{i=1}^n$ *satisfy* $\sum_{i \in N} \frac{1}{(1+\alpha_i)} \leq 1$.

The next similar result is due to Fan and Hoffman (1954).

Theorem 1.20. *For any* $A = [a_{i,j}] \in \mathbb{C}^{n\times n}$, *then*

$$\sigma(A) \subseteq \bigcup_{i\in N} \{z \in \mathbb{C} : |z - a_{i,i}| \le \rho_i^{(2)}\}, \tag{1.39}$$

where

$$\rho_i^{(2)} := \alpha \max_{j\in N\backslash\{i\}} |a_{i,j}| \quad (\text{all } i \in N), \text{ and} \tag{1.40}$$

where[6] α *is any positive number satisfying* $\displaystyle\sum_{i\in N} \frac{r_i(A)}{\max_{j\in N\backslash\{i\}} |a_{i,j}|} \le \alpha(1+\alpha)$.

We note that the eigenvalue inclusion sets of Corollary 1.18 and Theorems 1.19 and 1.20 all depend on the absolute values of off-diagonal entries of the matrix A, as well as on a variety of extra parameters. In (1.36) of Corollary 1.18, there are the parameters α and the n positive coefficients of $\mathbf{x} > \mathbf{0}$ in $\mathbb{R}^n$, in (1.38) of Theorem 1.19, there are the n parameters $\{\alpha_i\}_{i\in N}$ and p and q, while in (1.40) of Theorem 1.20, there is a single parameter α. It then becomes an interesting question as to how the eigenvalue inclusions of the above three results, with their extra parameters, compare with the eigenvalue inclusion of Corollary 1.5, which depends on the n positive coefficients of $\mathbf{x} > \mathbf{0}$ in $\mathbb{R}^n$. for a given matrix A in $\mathbb{C}^{n\times n}$. This question is succinctly and beautifully answered, in the irreducible case, in the following result of Fan (1958), where the Perron-Frobenius Theorem on nonnegative irreducible matrices, our second recurring theme, is applied.

Theorem 1.21. *For* $n \ge 2$, *let* $B = [b_{i,j}] \in \mathbb{R}^{n\times n}$ *be irreducible, with* $b_{i,j} \ge 0$ *for all* i, j *in* N *with* $i \ne j$, *and with* $b_{i,i} = 0$ *for all* $i \in N$. *Let* $\{\rho_i\}_{i=1}^n$ *be* n *positive numbers such that, for every matrix* $A = [a_{i,j}] \in \mathbb{C}^{n\times n}$ *with* $|a_{i,j}| = b_{i,j}$ (all $i, j \in N$ with $i \ne j$), *each eigenvalue of* A *must lie in at least one of the disks*

$$\{z \in \mathbb{C} : |z - a_{i,i}| \le \rho_i\} \quad (i \in N). \tag{1.41}$$

Then, there is an $\mathbf{x} = [x_1, x_2, \cdots, x_n]^T > \mathbf{0}$ *in* $\mathbb{R}^n$ (*where* $\mathbf{x}$ *is dependent on* A) *such that*

$$\rho_i \ge r_i^{\mathbf{x}}(A) \quad (\text{all } i \in N). \tag{1.42}$$

Remark 1. From (1.42), the disk $\Gamma_i^{r^{\mathbf{x}}}(A)$ of (1.14) is a *subset* of the disk $\{z \in \mathbb{C} : |z - a_{i,i}| \le \rho_i\}$ for *each* $i \in N$, which implies that

$$\Gamma^{r^{\mathbf{x}}}(A) \subseteq \bigcup_{i\in N} \{z \in \mathbb{C} : |z - a_{i,i}| \le \rho_i\}. \tag{1.43}$$

Remark 2. Fan's Theorem 1.21 directly applies to Corollary 1.18, and Theorems 1.19 and 1.20, in the irreducible case.

[6] If, for some i, $\max_{j\in N\backslash\{i\}} |a_{i,j}| = 0 = r_i(A)$, then $r_i(A)/\max_{j\in N\backslash\{i\}} |a_{i,j}|$ is defined to be zero.

Proof. By hypothesis, if $A = [a_{i,j}] \in \mathbb{C}^{n \times n}$ satisfies $|a_{i,j}| = b_{i,j}$, for all $i, j \in N$ with $i \neq j$, and if $|a_{i,i}| > \rho_i$ for all $i \in N$, then $z = 0$ cannot, from (1.41), be an eigenvalue of A; whence, A is nonsingular. With the nonsingular matrix $D := \text{diag}[\frac{1}{\rho_1}, \frac{1}{\rho_2}, \cdots, \frac{1}{\rho_n}]$, set $B' := DB = [b'_{i,j}]$, so that

$$b'_{i,j} = \frac{b_{i,j}}{\rho_i}, \quad \text{with } b'_{i,i} = 0 \quad (\text{all } i, j \in N). \tag{1.44}$$

Then, every matrix $A' = [a'_{i,j}] \in \mathbb{C}^{n \times n}$, with $|a'_{i,j}| = b'_{i,j}$ (all $i, j \in N$ with $i \neq j$) and with $|a'_{i,i}| > 1$ for all $i \in N$, is also nonsingular. As B is irreducible, so is B'. From the Perron-Frobenius Theoremon irreducible nonnegative $n \times n$ matrices (see Theorem C.1 of Appendix C), B' has a positive eigenvalue λ and a corresponding eigenvector $\mathbf{x} = [x_1, x_2, \cdots, x_n]^T > \mathbf{0}$ such that $\lambda x_i = (B'\mathbf{x})_i$ for all $i \in N$. With (1.44), this can be expressed (cf.(1.13)) as

$$\lambda \rho_i = \frac{1}{x_i} \sum_{j \in N \setminus \{i\}} |a_{i,j}| x_j = r_i^{\mathbf{x}}(A) \quad (\text{all } i \in N). \tag{1.45}$$

We next fix the diagonal entries of A' by setting $A' := B' - \lambda I_n = [a'_{i,j}] \in \mathbb{C}^{n \times n}$. Then, A' is necessarily singular, where $|a'_{i,j}| = b'_{i,j}$ for all $i \neq j$ in N, and where $|a'_{i,i}| = \lambda$ for all $i \in N$. On the other hand, we know that if $|a'_{i,i}| > 1$ for all $i \in N$, then A' is nonsingular. Hence, as A' is singular and as $|a'_{i,i}| = \lambda$ for all $i \in N$, it follows that $\lambda \leq 1$. Thus, (1.45) reduces to $\rho_i \geq r_i^{\mathbf{x}}(A)$ for all $i \in N$, the desired result of (1.42). ∎

It is important to say that the n inequalities of (1.42) hold for a *special* vector $\mathbf{x} > \mathbf{0}$ in $\mathbb{R}^n$, which depends on the matrix A. This means that the inclusion of (1.42) can **fail** for other choices of $\mathbf{x} > \mathbf{0}$.

The result of Fan's Theorem 1.21 in essence equates all matrices having their off-diagonal entries in absolute value given by a nonnegative matrix B. It is then natural to ask if there is a more **general structure** to such eigenvalue inclusion results (or, equivalently, nonsingularity results) where comparisons are made on the basis of the absolute values of only off-diagonal entries of a matrix. Such a structure, an offspring of Fan's Theorem 1.21, exists and is related to the concept of a **G-function**, due to P. Nowosad and A. Hoffman. This will be described fully in Chapter 5.

Exercises

1. Given any $A = [a_{i,j}] \in \mathbb{C}^{n \times n}$, $n \geq 2$, show (cf. (1.4) and (1.25)) that $\sum_{i \in N} r_i(A) = \sum_{j \in N} c_j(A)$ is always valid.

2. Consider the 2×2 singular matrix $A = \begin{bmatrix} 1 & 8 \\ 0.5 & 4 \end{bmatrix}$. With $\mathbf{x} = [1,1]^T$, consider the two associated disks from (1.36), where $0 \le \alpha \le 1$:

$$\{z \in \mathbb{C} : |z-1| \le 8^{\alpha}(0.5)^{1-\alpha}\} \text{ and } \{z \in \mathbb{C} : |z-4| \le (0.5)^{\alpha}8^{1-\alpha}\}.$$

These disks, having centers at $z = 1$ and $z = 4$, each give rise to an interval, on the real axis, which is dependent on α. Letting $\beta(\alpha)$ be defined as the *least real abscissa* of these two disks, i.e.,

$$\beta(\alpha) := \min\left[1 - 8^{\alpha} \cdot (0.5)^{1-\alpha}; 4 - (0.5)^{\alpha} \cdot 8^{1-\alpha}\right] \quad (\alpha \in [0,1]),$$

show that $\max\limits_{0 \le \alpha \le 1} \beta(\alpha) = 0 = \beta(1/4)$. Also, show that the choice, $\alpha = 1/4$, is, in terms of the eigenvalue inclusion of (1.36), *best possible.*

3. Give a proof of Theorem 1.19. (Ostrowski (1951b)). (Hint: Use Hölder's inequality!)

4. Give a proof of Theorem 1.20. (Fan and Hoffman (1954))

5. Assume that the matrix $A = [a_{i,j}] \in \mathbb{C}^{3\times 3}$ has row sums $r_i(A) = 1, 1 \le i \le 3$, and column sums $c_i(A) = 1, 1 \le i \le 3$, so that the sums in Exercise 1 above are equal. Show that there are uncountably many matrices $B = [b_{i,j}] \in \mathbb{C}^{3\times 3}$, with the same row and column sums. (Hint: Set $|a_{1,3}| = \epsilon$, where $0 < \epsilon < 1$. and determine the absolute values of the remaining nondiagonal entries of $A = [a_{i,j}] \in \mathbb{C}^{3\times 3}$ so that $r_i(A) = c_i(A) = 1$ for all $1 \le i \le 3$.) Generalize this to all $n \ge 3$.

6. Show that the double sum in (1.34) reduces, with the definition of $c_i(A)$ of (1.25), to $\sum\limits_{j\in N} c_j(A) \cdot |x_j|^q$.

7. If $A = [a_{i,j}] \in \mathbb{C}^{n\times n}$, if $0 \le \alpha \le 1$ and if

$$|a_{i,i}| > \alpha r_i(A) + (1-\alpha)c_i(A) \quad (\text{all } \ i \in N),$$

then A is nonsingular. (Hint: Use the generalized arithmetic-geometric mean inequality, i.e., for $a \ge 0, b \ge 0$ and $0 \le \alpha \le 1$, then $\alpha a + (1-\alpha)b \ge a^{\alpha}b^{1-\alpha}$, and apply Theorem 1.16.)

8. Derive the analog of Theorem 1.21 where $B = [b_{i,j}] \in \mathbb{R}^{n\times n}, n \geq 2$, is reducible. (Hint: Use the normal reduced form of B from (1.20).)

1.4 A Norm Derivation of Geršgorin's Theorem 1.1

The purpose of this section is to give a different way of deriving Geršgorin's Theorem 1.1. This approach will be further analyzed and extended in Chapter 6, but it is important now to note how this new derivation gives rise to different proofs of previous matrix eigenvalue inclusions.

Our first derivation of Geršgorin's Theorem 1.1, or its equivalent form in Theorem 1.4, was the so-called **classical derivation**. Another derivation, due to Householder (1956) and Householder (1964), can be made from the theory of norms; see Appendix B for background and notation. To describe this approach, let φ be any norm on $\mathbb{C}^n$, let $A \in \mathbb{C}^{n\times n}$, and assume that $\lambda \in \sigma(A)$. Then, there is an $\mathbf{x} \in \mathbb{C}^n$ with $\mathbf{x} \neq \mathbf{0}$ and with $A\mathbf{x} = \lambda\mathbf{x}$. Hence, for any fixed $B \in \mathbb{C}^{n\times n}$, this can be written equivalently as

$$(A - B)\mathbf{x} = (\lambda I_n - B)\mathbf{x}.$$

In particular, if $\lambda \notin \sigma(B)$, then $(\lambda I_n - B)$ is nonsingular, and hence

$$(\lambda I_n - B)^{-1}(A - B)\mathbf{x} = \mathbf{x}. \tag{1.46}$$

Now, for any matrix C in $\mathbb{C}^{n\times n}$, its **induced operator norm**, with respect to φ, is defined as usual (cf. (B.2) of Appendix B) by

$$\|C\|_\varphi := \sup_{\mathbf{x}\neq\mathbf{0}} \frac{\varphi(C\mathbf{x})}{\varphi(\mathbf{x})} = \sup_{\varphi(\mathbf{x})=1} \varphi(C\mathbf{x}),$$

and it directly follows from (1.46) that

$$\|(\lambda I_n - B)^{-1}(A - B)\|_\varphi \geq 1, \text{ for } \lambda \in \sigma(A)\backslash\sigma(B). \tag{1.47}$$

Thus, we define the following set in the complex plane:

$$G_\varphi(A;B) := \sigma(B) \cup \{z \in \mathbb{C} : z \notin \sigma(B) \text{ and } \|(zI - B)^{-1}(A - B)\|_\varphi \geq 1\}. \tag{1.48}$$

Next, consider any $\lambda \in \sigma(A)$. If $\lambda \in \sigma(B)$, then from (1.48), $\lambda \in G_\varphi(A;B)$. Similarly, if $\lambda \in \sigma(A)$ with $\lambda \notin \sigma(B)$, then, from (1.47), λ is contained in the second set defining $G_\varphi(A;B)$ in (1.48). This proves Householder's result[7] of

[7] Actually, Householder (1964), p.66, claimed that the last set in (1.48) *alone* contained all the eigenvalues of A. See Exercise 4 of this section for a counterexample.

Theorem 1.22. *For any $A \in \mathbb{C}^{n\times n}$, and any $B \in \mathbb{C}^{n\times n}$, let φ be any norm on $\mathbb{C}^n$. Then* (cf. (1.48)),

$$\sigma(A) \subseteq G_\varphi(A;B). \tag{1.49}$$

We call $G_\phi(A;B)$ the **Householder set** for A and B, which, from (1.49), is an eigenvalue inclusion set for any matrix A. We now deduce some set-theoretic properties of $G_\varphi(A;B)$.

Proposition 1.23. *For any $A \in \mathbb{C}^{n\times n}$, and any $B \in \mathbb{C}^{n\times n}$ let φ be any norm on $\mathbb{C}^n$. Then* (cf.(1.48)), *$G_\varphi(A;B)$ is a closed and bounded set in $\mathbb{C}$.*

Proof. We first show that $G_\varphi(A;B)$ is bounded. Suppose that $z \in G_\varphi(A;B)$ with $z \notin \sigma(B)$, so that $1 \leq \|(zI_n - B)^{-1}(A-B)\|_\varphi$. Consequently (cf. Appendix B, Proposition B.3),

$$1 \leq \|(zI_n - B)^{-1}(A-B)\|_\varphi \leq \|(zI_n - B)^{-1}\|_\varphi \cdot \|A-B\|_\varphi.$$

But this implies that $\|(zI_n - B)^{-1}\|_\varphi > 0$, so that

$$\|A-B\|_\varphi \geq \frac{1}{\|(zI_n - B)^{-1}\|_\varphi} = \inf_{\mathbf{y}\neq\mathbf{0}} \left\{ \frac{\varphi\left((zI_n - B)\,\mathbf{y}\right)}{\varphi(\mathbf{y})} \right\}.$$

Now, with the reverse triangle inequality, it follows that

$$\|A-B\|_\varphi \geq \inf_{\mathbf{y}\neq\mathbf{0}} \left\{ |z| - \frac{\varphi\left(B\mathbf{y}\right)}{\varphi(\mathbf{y})} \right\} = |z| - \|B\|_\varphi;$$

whence,

$$|z| \leq \|A-B\|_\varphi + \|B\|_\varphi, \text{ for any } z \text{ in } G_\varphi(A;B) \text{ with } z \notin \sigma(B). \tag{1.50}$$

But for the remaining case that $z \in G_\varphi(A;B)$ with $z \in \sigma(B)$, it is well known (see Appendix B, Proposition B.2) that $\rho(B) \leq \|B\|_\varphi$, so that $|z| \leq \|B\|_\varphi$. Thus, the inequality in (1.50) trivially holds also in this case. Hence,

$$|z| \leq \|A-B\|_\varphi + \|B\|_\varphi \text{ for any } z \in G_\varphi(A;B), \tag{1.51}$$

which shows that $G_\varphi(A;B)$ is a bounded set.

To show that $G_\varphi(A;B)$ is closed, suppose that $\{z_i\}_{i=1}^\infty$ is a sequence of points in $\mathbb{C}$ with $\lim_{i\to\infty} z_i = z$, where $z_i \in G_\varphi(A;B)$ for all $i \geq 1$. If $z \in \sigma(B)$, then, from (1.48), $z \in G_\varphi(A;B)$. If $z \notin \sigma(B)$, then $z_i \notin \sigma(B)$ for all i sufficiently large, so that $\|(z_iI-B)^{-1}(A-B)\|_\varphi \geq 1$ for all i sufficiently large. As the norm of a matrix is known to vary continuously with the entries of the matrix (cf. Ostrowski (1960), Appendix K), then also $\|(zI-B)^{-1}(A-B)\|_\varphi \geq 1$; whence, $z \in G_\varphi(A;B)$, i.e., $G_\varphi(A;B)$ is a closed set. ∎

To see how the norm approach of this section *directly* couples with Geršgorin's Theorem 1.1, consider the particular norm, the ℓ_∞-norm on $\mathbb{C}^n$, defined by

$$(1.52) \qquad \phi_\infty(\mathbf{u}) := \max_{i\in N} |u_i| \quad (\text{any } \ \mathbf{u} = [u_1, u_2, ..., u_n]^T \in \mathbb{C}^n).$$

In addition, if $A = [a_{i,j}] \in \mathbb{C}^{n\times n}$, set $\text{diag}[A] := \text{diag}[a_{1,1}, a_{2,2}, \cdots, a_{n,n}]$. Then, with $B := \text{diag}[A]$ and $\phi := \phi_\infty$, we establish

Corollary 1.24. *For any* $A = [a_{i,j}] \in \mathbb{C}^{n\times n}$, *then* (cf.(1.49) and (1.5))

$$(1.53) \qquad G_{\phi_\infty}(A; \text{diag}[A]) = \Gamma(A).$$

Remark. In other words, (1.49) of Theorem 1.22, for the particular norm of (1.52) and for B =diag$[A]$, exactly gives Geršgorin's Theorem 1.1!

Proof. Let z be any point of $G_{\phi_\infty}(A; \text{diag}[A])$ so that if $D := \text{diag}[A]$, then (cf.(1.50)) either $z \in \sigma(D)$, or $z \notin \sigma(D)$ with $\|(zI - D)^{-1}(A - D)\|_{\phi_\infty} \geq 1$. Assume that $z \notin \sigma(D)$, so that

$$(1.54) \qquad \|(zI - D)^{-1}(A - D)\|_{\phi_\infty} \geq 1,$$

where the matrix $(zI - D)^{-1}(A - D)$ can be expressed as

$$(1.55) \quad (zI - D)^{-1}(A - D) = [\alpha_{i,j}], \text{where} \begin{cases} \alpha_{i,i} := 0, i \in N, \\ \alpha_{i,j} := \dfrac{a_{i,j}}{z - a_{i,i}}, i \neq j \text{ in } N. \end{cases}$$

As mentioned before, it is known (cf. (B.6) of Appendix B) that, for any $C = [c_{i,j}] \in \mathbb{C}^{n\times n}$,

$$(1.56) \qquad \|C\|_{\phi_\infty} = \max_{i\in N} \left(\sum_{j\in N} |c_{i,j}| \right).$$

Applying (1.56) to (1.55) gives, with (1.55) and (1.4),

$$1 \leq \|(zI - D)^{-1}(A - D)\|_{\phi_\infty} = \max_{i\in N} \left(\frac{\sum\limits_{j\in N\backslash\{i\}} |a_{i,j}|}{|z - a_{i,i}|} \right) = \max_{i\in N} \left(\frac{r_i(A)}{|z - a_{i,i}|} \right).$$

Hence, there is a $j \in N$ with $|\, z - a_{j,j} \,| \leq r_j(A)$, so that (cf.(1.5)) $z \in \Gamma_j(A)$, and thus $z \in \Gamma(A)$. If $z \in \sigma(D)$, i.e., $z = a_{j,j}$ for some $j \in N$, then again $z \in \Gamma_j(A)$ and thus $z \in \Gamma(A)$. This shows that $G_{\phi_\infty}(A; \text{diag}[A]) \subseteq \Gamma(A)$. The reverse inclusion follows similarly, giving the desired result of (1.53). ■

In the same fashion, we show that applying Theorem 1.22, with a suitably defined norm φ on $\mathbb{C}^n$, also gives (1.15) of Corollary 1.5. To this end, for any $\mathbf{x} = [x_1, x_2, \cdots, x_n]^T > \mathbf{0}$ in $\mathbb{R}^n$, set $X := \text{diag}[x_1, x_2, \cdots, x_n]$, so that X is a nonsingular diagonal matrix. Then, define

$$\phi_\infty^{\mathbf{x}}(\mathbf{u}) := \phi_\infty(X\mathbf{u}) \quad (\text{for all } \mathbf{u} \in \mathbb{C}^n), \tag{1.57}$$

which is a norm on $\mathbb{C}^n$. (See Exercise 2 of this section.) Then the following result (see Exercise 3 of this section) similarly follows.

Corollary 1.25. *For any $A = [a_{i,j}] \in \mathbb{C}^{n\times n}$ and any $\mathbf{x} > \mathbf{0}$ in $\mathbb{R}^n$, then*

$$G_{\phi_\infty^{\mathbf{x}}}(A; \text{diag}[A]) = \Gamma^{r^{\mathbf{x}}}(A). \tag{1.58}$$

Exercises

1. For the matrix $A := \begin{bmatrix} 1 & i & 0 \\ 0 & 4 & 1 \\ -1 & 0 & 2 \end{bmatrix}$ and for $B = \text{diag}[1, 4, 2]$, determine $G_\varphi(A; B)$ of (1.48) for each of the three vector norms ℓ_1, ℓ_2, and ℓ_∞ on $\mathbb{C}^3$, defined for any $\mathbf{x} = [x_1, x_2, x_3]^T$ in $\mathbb{C}^3$ as

$$\ell_1(\mathbf{x}) := \sum_{j=1}^{3} |x_j|; \ell_2(\mathbf{x}) := \left(\sum_{j=1}^{3} |x_j|^2\right)^{\frac{1}{2}}; \ell_\infty(\mathbf{x}) = \max\{|x_j|, 1 \le j \le 3\}.$$

   Show that each of these norms give rise to the **same** set $G_\phi(A; B)$ in the complex plane, as shown in Fig. 1.9. (The eigenvalues of A are also shown in Fig. 1.9 as "$\times$'s.")

2. For any $\mathbf{x} > \mathbf{0}$ in $\mathbb{R}^n$, show that $\phi_\infty^{\mathbf{x}}(\mathbf{u})$ in (1.57) is a norm on $\mathbb{C}^n$.

3. With the norm of (1.57), establish the result of Corollary 1.25.

4. For the matrices

$$A = \begin{bmatrix} 1 & 0 & 0 \\ 1 & 5 & 1 \\ 1 & 1 & 5 \end{bmatrix} \quad \text{and} \quad B = \begin{bmatrix} 1 & 0 & 0 \\ 0 & 5 & 0 \\ 0 & 0 & 5 \end{bmatrix}$$

   verify that if ϕ is the ℓ_∞-norm on $\mathbb{C}^3$, then

$$\{z \in \mathbb{C} : z \notin \sigma(B) \text{ and } \|(zI - B)^{-1}(A - B)\|_\phi \geq 1\} = \{z \in \mathbb{C} : |z - 5| \leq 2\},$$

which does **not** contain the spectrum, $\sigma(A) = \{1, 4, 6\}$, of A. This shows that it is necessary to **add** $\sigma(B)$ in (1.48) to insure that $G_\varphi(A; B)$ covers the spectrum of A. (See also Householder, Varga and Wilkinson (1972).)

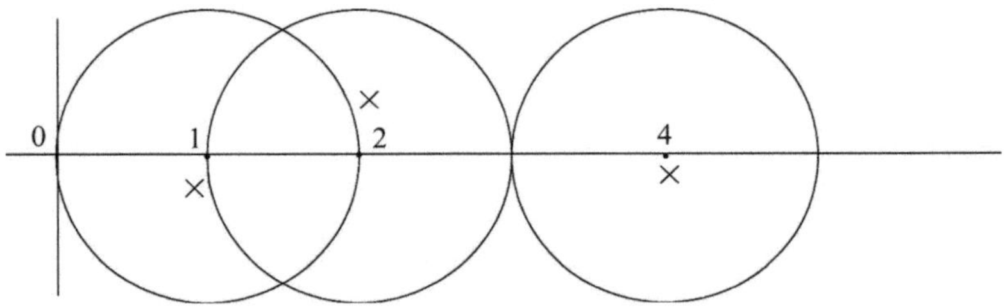

Fig. 1.9. $G_{\ell_j}(A, B), j{=}1, 2,$ and ∞, for the matrices A and B, given in Exercise 1

Bibliography and Discussion

1.1 The eigenvalue inclusion result of Theorem 1.1 of Geršgorin appeared in 1931, but, as was noted in the text, this result is actually equivalent to the nonsingularity result in Theorem 1.4 for strictly diagonally dominant matrices, a topic which can be traced back to the much earlier result of Lévy (1881), Desplanques (1887), Minkowski (1900), and Hadamard (1903), pp. 13-14. We remark that the results of Lévy (1881) were restricted to *real* matrices, as was that of Minkowski (1900), and Desplanques (1887) first obtained the general complex result. This general result was later independently obtained by Hadamard (1903) in his book, and was often referred in the literature as "Hadamard's theorem". We also note that Geršgorin (1931) paid homage to the earlier work of Lévy (1881), and also mentioned earlier work of R. Kusmin. See page 750 of Geršgorin (1931), which is reproduced in full in Appendix A, for the reader's convenience.

We remark that Brauer (1946) rediscovered Geršgorin's earlier Theorem 1.1 from 1931. In the same paper, Brauer deduces, from Geršgorin's Theorem 1.1, the strictly diagonally dominant Theorem 1.4, but does not obtain the equivalence of these two results. Brauer (1946) also obtained the case $|S| = 1$ of Geršgorin's Theorem 1.8.

On reading the literature, it is interesting to note that this general **equivalence** of eigenvalue inclusion results with nonsingularity results, which we now readily understand and use, seems not to have been in wide use until fairly recently. It has been kindly pointed out to me by Hans Schneider that this equivalence *does* appear in the isolated paper of Rohrbach (1931) but we see in Ostrowski (1951a) the nonsingularity result of Theorem 1.16, and a separate publication, by Ostrowski (1951b), for the associated eigenvalue inclusion result.

The idea of using diagonal similarity transformations $X^{-1}AX$, to improve the eigenvalue inclusion result of Theorem 1.1, is due again to Geršgorin (1931), as is the result of Theorem 1.6 on disjoint components of the union of Geršgorin disks.

This brings us to an important *consideration*, regarding our treatment of Geršgorin-type eigenvalue inclusions. Clearly, a goal of these Geršgorin-type eigenvalue inclusions might be to obtain estimates of the entire spectrum of a given matrix in $\mathbb{C}^{n\times n}$. Now, it is well known *theoretically* that, given *any* matrix $A = [a_{i,j}] \in \mathbb{C}^{n\times n}$, there exists a nonsingular matrix S in $\mathbb{C}^{n\times n}$ such that $S^{-1}AS$ is in **Jordan normal form**, so that $S^{-1}AS$ is an upper bidiagonal matrix, whose diagonal entries are the eigenvalues of A and whose entries, of the upper bidiagonal, are either 0 or 1. With this nonsingular matrix S, then one can directly read off the diagonal entries of $S^{-1}AS$, and obtain all the eigenvalues of A with, in theory, infinite precision! Practically, however, finding such a nonsingular matrix S is computationally nontrivial in general, and numerical methods for this, such as the QR iterative method, must confront the effects of rounding errors in actual computations and the termination of this iterative method which introduces errors. So on one hand, the Geršgorin-type eigenvalue inclusions, in their simplest forms, are easy to use, but give results which can be quite crude, while on the other hand, the QR method, for example, can give much more precise eigenvalue estimates, at the expense of more computations.

But, there is a middle ground! As will be shown in Chapter 6, where partitioned matrices come into play, one can go beyond the diagonal similarity matrices of Sec. 1.2 to block diagonal matrices, whose blocks are of relatively small orders, to improve Geršgorin-type eigenvalue inclusions, at the expense of more computational work.

It must be emphasized here that our approach in this book is a **theoretical one**, which does **not** attempt to obtain optimal methods for computing eigenvalues of general matrices!

1.2 The use of irreducibility as a tool in linear algebra was championed by Olga Taussky in her Theorem 1.12 of Taussky (1949). See also her famous paper Taussky (1988), "How I became a torch bearer for Matrix Theory." It is also clear that she knew that Geršgorin's Satz 1, from his 1931 paper (see Appendix A where this paper is reproduced) was *incorrect* because he was unaware of the notion of irreducibility!

As mentioned in this section, Taussky's Theorem 1.12 gives a **necessary** condition for an eigenvalue λ, of an irreducible matrix A, to lie in the boundary of the Geršgorin set $\Gamma(A)$, of (1.5), but it need not be **sufficient**. Finding **both** necessary and sufficient conditions for λ to be an eigenvalue of A which lies on the boundary of $\Gamma(A)$, where A is an irreducible matrix in $\mathbb{C}^{n \times n}$, has been recently given in Theorem 3.2 of Kolotolina (2003a), where many other interesting results are obtained. Related results of Kolotolina (2003b) on Brauer sets and Brualdi sets, are cited in the Bibliography and Discussion of Section 2.2.

1.3 The simultaneous use of both row and column sums to achieve better nonsingularity results, such as Theorem 1.16 and Corollary 1.17, or its associated eigenvalue inclusion result of Corollary 1.18, stems from the work of Ostrowski (1951a), while the eigenvalue inclusion results of Theorem 1.19 and 1.20 are due, respectively, to Ostrowski (1951b), and Fan and Hoffman (1954). The culminating result of Fan (1958) in Theorem 1.21, whose proof makes use of the Perron-Frobenius Theorem on nonnegative irreducible matrices, masterfully gives the characterization of the "best" Geršgorin disks, in a certain setting! Further generalizations of this appear in Chapter 5.

1.4 This norm-approach for obtaining matrix eigenvalue inclusion results first appeared in Householder (1956) and Householder (1964), where it was also showed that this norm approach can be directly used to obtain Geršgorin's Theorem 1.1. As we will see in Chapter 6, on partitioned matrices, this norm approach will prove to be very useful.

2. Geršgorin-Type Eigenvalue Inclusion Theorems

2.1 Brauer's Ovals of Cassini

We begin with the following nonsingularity result of Ostrowski (1937b), where $r_i(A)$ is defined in (1.4).

Theorem 2.1. *If $A = [a_{i,j}] \in \mathbb{C}^{n\times n}$, $n \geq 2$, and if*

$$(2.1) \qquad |a_{i,i}| \cdot |a_{j,j}| > r_i(A) \cdot r_j(A) \text{ (all } i \neq j \text{ in } N := \{1, 2, \cdots, n\}),$$

then A is nonsingular.

Remark. Note that if A is strictly diagonally dominant (cf.(1.11)), then (2.1) is valid. Conversely, if (2.1) is valid, then all but at most one of the inequalities of (1.11) must hold. Thus, Theorem 2.1 is a stronger result than Theorem 1.4.

Proof. Suppose, on the contrary, that $A = [a_{i,j}] \in \mathbb{C}^{n\times n}$ satisfies (2.1) and is singular, so that there is an $\mathbf{x} = [x_1, x_2, \cdots, x_n]^T$ in $\mathbb{C}^n$ with $\mathbf{x} \neq \mathbf{0}$, such that $A\mathbf{x} = \mathbf{0}$. On ordering the components of $\mathbf{x}$ by their absolute values, we can find s and t in N with $s \neq t$ such that $|x_t| > 0$ and

$$(2.2) \qquad |x_t| \geq |x_s| \geq \max\{|x_k| : k \in N \text{ with } k \neq s,\ k \neq t\},$$

(where the last term above is defined to be zero if $n = 2$). Then, $A\mathbf{x} = \mathbf{0}$ implies that $a_{i,i}x_i = -\sum_{j\in N\backslash\{i\}} a_{i,j}x_j$ for all $i \in N$. Taking absolute values and applying the triangle inequality,

$$(2.3) \qquad |a_{i,i}| \cdot |x_i| \leq \sum_{j\in N\backslash\{i\}} |a_{i,j}| \cdot |x_j| \quad \text{(all } i \in N).$$

On choosing $i = t$, the inequality of (2.3) becomes, with (2.2),

$$(2.4) \qquad |a_{t,t}| \cdot |x_t| \leq \sum_{j\in N\backslash\{t\}} |a_{t,j}| \cdot |x_j| \leq r_t(A) \cdot |x_s|.$$

If $|x_s| = 0$, then (2.4) reduces to $|a_{t,t}| \cdot |x_t| = 0$, implying that $|a_{t,t}| = 0$. But this contradicts the fact, from (2.1), that $|a_{i,i}| > 0$ for all $i \in N$.

Next, assume that $|x_s| > 0$. On choosing $i = s$ in (2.3), we similarly obtain

$$|a_{s,s}| \cdot |x_s| \leq r_s(A) \cdot |x_t|.$$

On multiplying the above inequality with that of (2.4), then

$$|a_{t,t}| \cdot |a_{s,s}| \cdot |x_t| \cdot |x_s| \leq r_t(A) \cdot r_s(A) \cdot |x_t| \cdot |x_s|,$$

and as $|x_t| \cdot |x_s| > 0$, this gives $|a_{t,t}| \cdot |a_{s,s}| \leq r_t(A) \cdot r_s(A)$, which contradicts (2.1). ■

The result of Theorem 2.1 was later rediscovered by Brauer (1947), who used this to deduce the following Geršgorin-type eigenvalue inclusion theorem, which is, by our first recurring theme, equivalent to the result of Theorem 2.1, and hence, needs no proof!

Theorem 2.2. *For any $A = [a_{i,j}] \in \mathbb{C}^{n \times n}, n \geq 2$, and any $\lambda \in \sigma(A)$, there is a pair of distinct integers i and j in N such that*

$$\lambda \in K_{i,j}(A) := \{z \in \mathbb{C} : |z - a_{i,i}| \cdot |z - a_{j,j}| \leq r_i(A) \cdot r_j(A)\}. \tag{2.5}$$

As this is true for each λ in $\sigma(A)$, then

$$\sigma(A) \subseteq \mathcal{K}(A) := \bigcup_{\substack{i,j \in N \\ i \neq j}} K_{i,j}(A). \tag{2.6}$$

The quantity[1] $K_{i,j}(A)$, defined in (2.5), is called the (i,j)-th **Brauer Cassini oval** for the matrix A, while $\mathcal{K}(A)$ of (2.6) is called the **Brauer set.** There are now $\binom{n}{2} = \frac{n(n-1)}{2}$ such Cassini ovals for the eigenvalue inclusion of (2.6), as compared with n Geršgorin disks (cf. (1.5)) of the eigenvalue inclusion of (1.7) of Theorem 1.1. Moreover, the compact set $K_{i,j}(A)$ of (2.5) is more complicated than the Geršgorin set, as $K_{i,j}(A)$ can consist of two **disjoint components** if $|a_{i,i} - a_{j,j}| > 2(r_i(A) \cdot r_j(A))^{1/2}$. (See Exercise 1 of this section.)

To make this clearer, the boundaries of particular Cassini ovals, in a different notation, determined from

$$K(-1,+1,r) := \left\{z \in \mathbb{C} : |z-1| \cdot |z+1| \leq r^2\right\}, \tag{2.7}$$

are shown in Fig. 2.1 for the particular values $r = 0$, $r = 0.5$, $r = 0.9$, $r = 1$, $r = 1.2$, $r = \sqrt{2}$, and $r = 2$. For $r = 0$, this set consists of the two distinct points $z = -1$ and $z = +1$, while for $r = 0.5$, this set consists of two disjoint nearly circular disks which are centered about -1 and $+1$. (See Exercise 2 of this section.) For all $r \geq 1$, this set is a closed and connected set in the complex plane $\mathbb{C}$, which is convex for all $r \geq \sqrt{2}$. (See Exercise 4 of this section.)

[1] We use "K" here for "Cassini"!

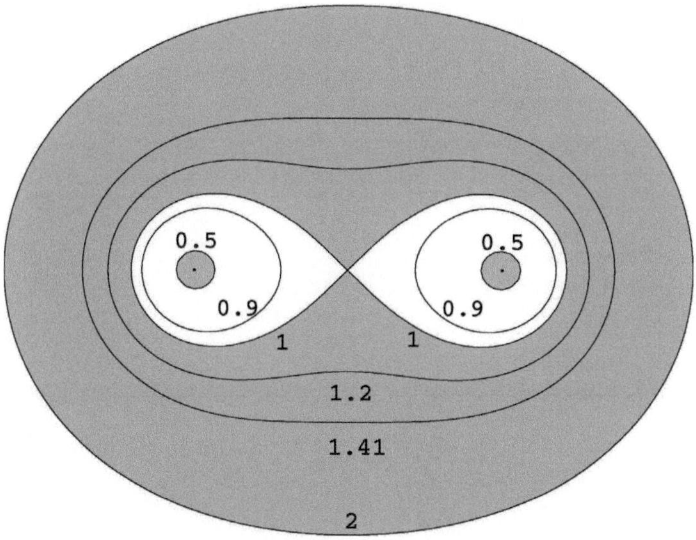

Fig. 2.1. E.T. or Cassini Ovals for $K(-1, +1, r)$ of (2.7)

It is interesting to note that $\Gamma(A)$, the union of the n Geršgorin disks in (1.5), and $\mathcal{K}(A)$, the union of the $\binom{n}{2}$ Brauer Cassini ovals in (2.6), both depend solely on the **same** $2n$ numbers, namely,

$$\{a_{i,i}\}_{i=1}^{n} \text{ and } \{r_i(A)\}_{i=1}^{n}, \tag{2.8}$$

derived from the matrix A, to obtain the eigenvalue inclusion Theorems 1.1 and 2.2. It is of theoretical interest to ask which of the sets $\Gamma(A)$ and $\mathcal{K}(A)$ is smaller, as the smaller set would give a "tighter" estimate for the spectrum $\sigma(A)$. That $\mathcal{K}(A) \subseteq \Gamma(A)$ holds in **all** cases is a result, not well known, which was stated by Brauer (1947). As its proof is simple and as the idea of the proof will be used later in Theorem 2.9 of Section 2.3, its proof is given here.

Theorem 2.3. *For any* $A = [a_{i,j}] \in \mathbb{C}^{n\times n}$, $n \geq 2$, *then* (cf. (2.6) and (1.5))

$$\mathcal{K}(A) \subseteq \Gamma(A). \tag{2.9}$$

Remark. This establishes that the Brauer set $\mathcal{K}(A)$, for any matrix A, is always a subset of its associated Geršgorin set $\Gamma(A)$, but for $n > 3$, there are more, $\binom{n}{2}$, Brauer Cassini ovals to determine, as opposed to the n associated Geršgorin disks.

Proof. Let i and j be any distinct integers in N of (2.1), and let z be any point of $K_{i,j}(A)$. Then from (2.5),

$$|z - a_{i,i}| \cdot |z - a_{j,j}| \leq r_i(A) \cdot r_j(A). \tag{2.10}$$

If $r_i(A) \cdot r_j(A) = 0$, then $z = a_{i,i}$ or $z = a_{j,j}$. But, as $a_{i,i} \in \Gamma_i(A)$ and $a_{j,j} \in \Gamma_j(A)$ from (1.5), then $z \in \Gamma_i(A) \cup \Gamma_j(A)$. If $r_i(A) \cdot r_j(A) > 0$, we have from (2.10) that

$$\left(\frac{|z - a_{i,i}|}{r_i(A)}\right) \cdot \left(\frac{|z - a_{j,j}|}{r_j(A)}\right) \leq 1. \tag{2.11}$$

As the factors on the left in (2.11) cannot both exceed unity, then at least one of these factors is at most unity, i.e., $z \in \Gamma_i(A)$ or $z \in \Gamma_j(z)$. Hence, it follows in either case that $z \in \Gamma_i(A) \cup \Gamma_j(A)$, so that

$$K_{i,j}(A) \subseteq \Gamma_i(A) \cup \Gamma_j(A). \tag{2.12}$$

As (2.12) holds for any i and j $(i \neq j)$ in N, we see from (1.5) and (2.6) that

$$\mathcal{K}(A) := \bigcup_{\substack{i,j \in N \\ i \neq j}} K_{i,j}(A) \subseteq \bigcup_{\substack{i,j \in N \\ i \neq j}} \{\Gamma_i(A) \cup \Gamma_j(A)\} = \bigcup_{\ell \in N} \Gamma_\ell(A) =: \Gamma(A),$$

the desired result of (2.9). ■

We remark that the case of equality in the inclusion of (2.12) is covered (cf. Varga and Krautstengl (1999)) in

$$\begin{cases} K_{i,j}(A) = \Gamma_i(A) \cup \Gamma_j(A) \text{ if and only if} \\ r_i(A) = r_j(A) = 0, \text{ or if } r_i(A) = r_j(A) > 0 \text{ and } a_{i,i} = a_{j,j}. \end{cases} \tag{2.13}$$

It is important to remark that while the Brauer set $\mathcal{K}(A)$ for any $A = [a_{i,j}] \in \mathbb{C}^{n \times n}$ with $n \geq 2$, is always a *subset* (cf.(2.9)) of the Geršgorin set $\Gamma(A)$, there is considerably more work, when n is large, in determining the $\binom{n}{2}$ Brauer Cassini ovals, than is the case in determining the associated n Geršgorin disks, i.e., getting the sharper inclusion of (2.9) *may* come at the price of more computations!

To illustrate the result of (2.9) of Theorem 2.3, consider the 4×4 irreducible matrix

$$B = \begin{bmatrix} 1 & 1 & 0 & 0 \\ 1/2 & i & 1/2 & 0 \\ 0 & 0 & -1 & 1 \\ 1 & 0 & 0 & -i \end{bmatrix}, \tag{2.14}$$

whose row sums $r_i(B)$ are all unity. The boundary of the Geršgorin set $\Gamma(B)$ is the outer closed curve in Fig. 2.2, composed of four circular arcs, while the Brauer set $\mathcal{K}(B)$, from (2.6), is the shaded inner set in Figure 2.2. (The curves in the shaded portion of this figure correspond to internal boundaries of the six Brauer Cassini ovals.) That $K(B) \subseteq \Gamma(B)$, from (2.9) of Theorem 2.3, is geometrically evident from Fig. 2.2!

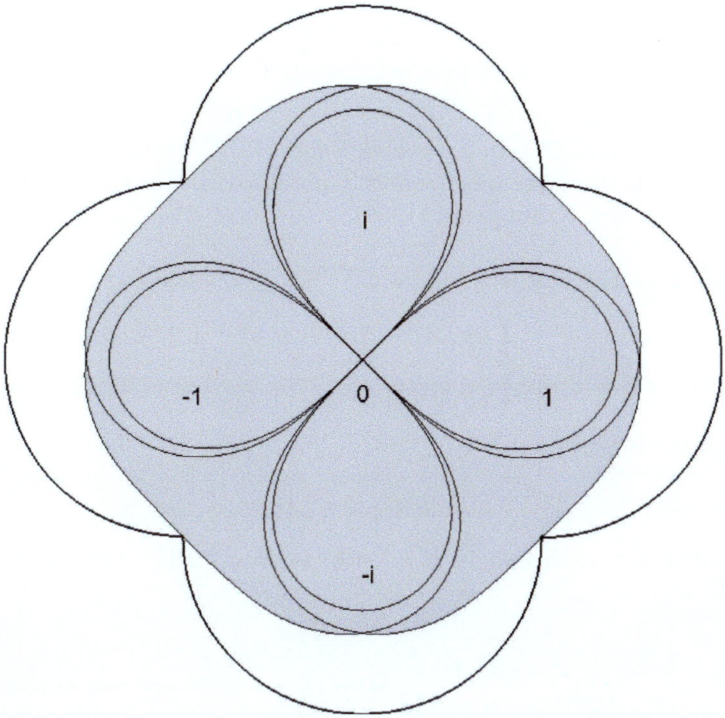

Fig. 2.2. The Brauer set $\mathcal{K}(B)$ (shaded) and the Geršgorin set $\Gamma(B)$ for the matrix B of (2.14)

As mentioned above, $\Gamma(A)$ and $\mathcal{K}(A)$ depend solely on the same $2n$ numbers of (2.8) which are derived from the matrix A, but there is a continuum of matrices (for $n \geq 2$ and some $r_i(A) > 0$) which give rise to the same numbers in (2.8). More precisely, we define the **equiradial set** for A as

(2.15) $$\omega(A) := \left\{B = [b_{i,j}] \in \mathbb{C}^{n\times n} : b_{i,i} = a_{i,i} \text{ and } r_i(B) = r_i(A),\ (i \in N)\right\},$$

and we also define

(2.16) $$\hat{\omega}(A) := \left\{B = [b_{i,j}] \in \mathbb{C}^{n\times n} : b_{i,i} = a_{i,i} \text{ and } r_i(B) \leq r_i(A),\ (i \in N)\right\}$$

as the **extended equiradial set** for A, so that $\omega(A) \subseteq \hat{\omega}(A)$. We note, from the final inequality in (2.5), that the eigenvalue inclusion of (2.6) is then valid for all matrices in $\omega(A)$ or $\hat{\omega}(A)$, i.e., with (2.9) and with the definitions of

(2.17) $$\sigma(\omega(A)) := \bigcup_{B\in\omega(A)} \sigma(B), \quad \text{and } \sigma(\hat{\omega}(A)) := \bigcup_{B\in\hat{\omega}(A)} \sigma(B),$$

it follows that

$$\sigma(\omega(A)) \subseteq \sigma(\hat{\omega}(A)) \subseteq \mathcal{K}(A). \tag{2.18}$$

Again, we are interested in the *sharpness* of the set inclusions in (2.18), which is covered in Theorem 2.4 below. Engel (1973) obtained (2.19) of Theorem 2.4, which was later independently obtained with essentially the same proof, in Varga and Krautstengl (1999) in the form given below.

Theorem 2.4. *For any* $A = [a_{i,j}] \in \mathbb{C}^{n \times n}$, $n \geq 2$, *then*

$$\sigma(\omega(A)) = \begin{cases} \partial\mathcal{K}(A) = \partial K_{1,2}(A) \text{ if } n = 2, \text{ and} \\ \mathcal{K}(A) \text{ if } n \geq 3, \end{cases} \tag{2.19}$$

and, in general, for any $n \geq 2$,

$$\sigma(\hat{\omega}(A)) = \mathcal{K}(A). \tag{2.20}$$

Proof. For $n = 2$, each matrix B in $\omega(A)$ is, from (2.15), necessarily of the form

$$B = \begin{bmatrix} a_{1,1} & r_1(A)e^{i\psi_1} \\ r_2(A)e^{i\psi_2} & a_{2,2} \end{bmatrix}, \text{ with } \psi_1, \psi_2 \text{ arbitrary real numbers.} \tag{2.21}$$

If λ is any eigenvalue of B, then $\det(B - \lambda I) = 0$, so that from (2.21),

$$(a_{1,1} - \lambda)(a_{2,2} - \lambda) = r_1(A) \cdot r_2(A)e^{i(\psi_1+\psi_2)}.$$

Hence,

$$|a_{i,i} - \lambda| \cdot |a_{2,2} - \lambda| = r_1(A) \cdot r_2(A). \tag{2.22}$$

As (2.22) corresponds to the case of equality in (2.5), we see that $\lambda \in \partial K_{1,2}(A)$. Since this is true for any eigenvalue λ of any B in $\omega(A)$ and since, from (2.6), $K_{1,2}(A) = \mathcal{K}(A)$ in this case $n = 2$, then $\sigma(\omega(A)) \subseteq \partial K_{1,2}(A) = \partial\mathcal{K}(A)$. Moreover, it is easily seen that *each* point of $\partial K_{1,2}(A)$ is, for suitable choices of real ψ_1 and ψ_2, an eigenvalue of *some* B in (2.21), so that $\sigma(\omega(A)) = \partial K_{1,2}(A) = \partial\mathcal{K}(A)$, the desired result of the first part of (2.19).

To establish the second part of (2.19), first assume that $n \geq 4$, and consider a matrix $B = [b_{i,j}]$ in $\mathbb{C}^{n \times n}$, which has the partitioned form

$$B = \left[\begin{array}{c|c} B_{1,1} & B_{1,2} \\ \hline O & B_{2,2} \end{array}\right], \tag{2.23}$$

where

$$B_{1,1} := \begin{bmatrix} a_{1,1} & se^{i\psi_1} \\ te^{i\psi_2} & a_{2,2} \end{bmatrix}, \text{ with } 0 \leq s \leq r_1(A),\ 0 \leq t \leq r_2(A), \tag{2.24}$$

with ψ_1 and ψ_2 arbitrary real numbers, and with $b_{j,j} = a_{j,j}$ for all $1 \leq j \leq n$. Now, for any choices of s and t with $s \in [0, r_1(A)]$ and $t \in [0, r_2(A)]$, the entries of the block $B_{1,2}$ can be chosen so that the row sums, $r_1(B)$ and $r_2(B)$, in the first two rows of B, equal those of A. Similarly, because $n \geq 4$, the row sums of the matrix $B_{2,2}$ of (2.23) can be chosen to be the same as those in the remaining row sums of A. Thus, by our construction, the matrix B of (2.23) is an element of $\omega(A)$. (We remark that this construction fails to work in the case $n = 3$, unless $r_3(A) = 0$). But from the partitioned form in (2.23), it is evident that

$$\sigma(B) = \sigma(B_{1,1}) \cup \sigma(B_{2,2}). \tag{2.25}$$

Then from the parameters s, t, ψ_1, and ψ_2 in $B_{1,1}$ in (2.24), it can be seen from the definition in (2.5) that for *each* $z \in K_{1,2}(A)$, there are choices for these parameters such that z is an eigenvalue of $B_{1,1}$. In other words, the eigenvalues of $B_{1,1}$, on varying the real numbers ψ_1 and ψ_2, and s and t with $0 \leq s \leq r_1(A)$ and $0 \leq t \leq r_2(A)$, **fill out** $K_{1,2}(A)$, where we note that the remaining eigenvalues of B (namely, those of $B_{2,2}$) must still lie, from (2.18), in $\mathcal{K}(A)$. As this applies to *any* Cassini oval $K_{i,j}(A)$ for $i \neq j$, upon a suitable permutation of the rows and columns of B of (2.23) which moves row i into row 1 and row j into row 2, then $\sigma(\omega(A)) = \mathcal{K}(A)$, for all $n \geq 4$.

For the remaining case $n = 3$ of (2.19), any matrix B in $\omega(A)$ can be expressed as

$$B = \begin{bmatrix} a_{1,1} & se^{i\psi_1} & (r_1(A) - s)e^{i\psi_2} \\ te^{i\psi_3} & a_{2,2} & (r_2(A) - t)e^{i\psi_4} \\ ue^{i\psi_5} & (r_3(A) - u)e^{i\psi_6} & a_{3,3} \end{bmatrix}, \tag{2.26}$$

where

$$\begin{cases} 0 \leq s \leq r_1(A), \; 0 \leq t \leq r_2(A), \text{ and } 0 \leq u \leq r_3(A), \text{ and} \\ \{\psi_i\}_{i=1}^{6} \text{ are arbitrary real numbers.} \end{cases} \tag{2.27}$$

Now, fix any complex number z in the Brauer Cassini oval $K_{1,2}(A)$, i.e., (cf. (2.5)), let z satisfy

$$|z - a_{1,1}| \cdot |z - a_{2,2}| \leq r_1(A) \cdot r_2(A). \tag{2.28}$$

If $r_1(A) = 0$, then from (2.28), $z = a_{1,1}$ or $z = a_{2,2}$. On the other hand, $r_1(A) = 0$ implies that the first row of B is $[a_{1,1}, 0, 0]$, so that $a_{1,1} = z$ is an eigenvalue of B. As the same argument also applies to the case $r_2(A) = 0$, we assume that $r_1(A) \cdot r_2(A) > 0$ in (2.28). Then, let s, with $0 \leq s \leq r_1(A)$, be such that $|z - a_{1,1}| \cdot |z - a_{2,2}| = sr_2(A)$, and select a real number ψ such that

$$(a_{1,1} - z) \cdot (a_{2,2} - z) = sr_2(A)\epsilon^{i\psi}.$$

For $\alpha := r_2(A) + |a_{2,2} - z|$ (so that $\alpha > 0$), the matrix $\tilde{B}$, defined by

$$\tilde{B} := \begin{bmatrix} a_{1,1} & se^{i\psi} & (r_1(A)-s) \\ r_2(A) & a_{2,2} & 0 \\ \frac{r_2(A)\cdot r_3(A)}{\alpha} & \frac{(a_{2,2}-z)r_3(A)}{\alpha} & a_{3,3} \end{bmatrix}, \tag{2.29}$$

can be verified to be in the set $\omega(A)$ of (2.15). But a calculation directly shows that $\det(\tilde{B}-zI) = 0$, so that z is an eigenvalue of $\tilde{B}$. Hence, each z in $K_{1,2}(A)$ is an eigenvalue of some B in $\omega(A)$. Consequently, as this construction for $n = 3$ can be applied to *any* point of *any* Cassini oval $K_{i,j}(A)$ with $i \neq j$, then $\sigma(\omega(A)) = \mathcal{K}(A)$, which completes the proof of (2.19).

The proof of the remaining equality in (2.20) is similar to the above proof, and is left as an exercise. (See Exercise 5 of this section.) ∎

From Theorem 2.4, we remark, for any $A = [a_{i,j}] \in \mathbb{C}^{n\times n}$ for $n \geq 3$, that the Brauer set $\mathcal{K}(A)$ does a **perfect job** of estimating the spectra of all matrices in the equiradial set $\omega(A)$ or in the extended equiradial set $\hat{\omega}(A)$, which is, in general, **not** the case for the Geršgorin set $\Gamma(A)$. In fact, for matrix B of (2.14), one sees in Fig. 2.2 that there are four unshaded curved domains of $\Gamma(B)$ which contain **no** eigenvalue of the set $\omega(B)$ or $\hat{\omega}(B)$.

Exercises

1. Show that the Brauer Cassini oval $K_{i,j}(A), i \neq j$, of (2.5) consists of two disjoint compact sets if $|a_{i,i} - a_{j,j}| > 2(r_i(A)\cdot r_j(A))^{1/2}$.

2. Consider the level curve $\{z \in \mathbb{C} : |z-\alpha|\cdot|z-\beta| = \rho\}$ where $\beta \neq \alpha$. From Exercise 1, this level curve consists of two disjoint curves, for all ρ with $0 < \rho < |\beta-\alpha|^2/4$. Show, more precisely, that each of these level curves is nearly a circle, where the radius of each circle is asymptotically $\frac{\rho}{|\beta-\alpha|}$, as $\rho \to 0$.

3. For the Brauer Cassini oval determined from

$$K(-1,+1,r) := \{z \in \mathbb{C} : |z-1|\cdot|z+1| \leq r^2\},$$

show the following statements are valid:
a. $K(-1,+1,r)$ has two closed components for all $0 \leq r < 1$;
b. $K(-1,+1,r)$ is a closed and connected set for all $r \geq 1$;
c. $K(-1,+1,r)$ is a **convex** set for all $r \geq \sqrt{2}$.

4. Given $K_{i,j}(A), i \neq j$, of (2.5), show that the result of (2.13) is valid.

5. With the definition of the extended equiradial set $\hat{\omega}(A)$ of (2.16), establish the result of (2.20) of Theorem 2.4.

6. State and prove an analog of Geršgorin's Theorem 1.6 (on disjoint subsets of $\Gamma(A)$), for Brauer's $n(n-1)/2$ Cassini ovals of an $n \times n$ matrix $A = [a_{i,j}] \in \mathbb{C}^{n \times n}$, $n \geq 2$. (Hint: From (2.12), if $\Gamma_i(A)$ and $\Gamma_j(A)$ are disjoint, then $K_{i,j}(A)$ must consist of two disjoint components.)

7. Given any $n \times n$ matrix A with $n \geq 2$, show, as is suggested from Fig. 2.2, that a common point z of the boundaries, $\partial\Gamma_i(A)$ and $\partial\Gamma_j(A)$ $(i \neq j)$, of two Geršgorin disks, is a point of $\partial K_{i,j}(A)$.

2.2 Higher-Order Lemniscates

Given a matrix $A = [a_{i,j}] \in \mathbb{C}^{n \times n}$, let $\{i_j\}_{j=1}^m$ be any m distinct positive integers from $N := \{1, 2, \cdots, n\}$, so that $n \geq m$. Then, the **lemniscate**[2] **of order m**, derived from $\{i_j\}_{j=1}^m$ and the $2n$ numbers $\{a_{i,i}\}_{i=1}^n$ and $\{r_i(A)\}_{i=1}^n$, is the compact set in $\mathbb{C}$ defined by

$$(2.30) \qquad \ell_{i_1,\cdots,i_m}(A) := \left\{ z \in \mathbb{C} : \prod_{j=1}^{m} |z - a_{i_j,i_j}| \leq \prod_{j=1}^{m} r_{i_j}(A) \right\},$$

and their union, called the **lemniscate set** for A, denoted by

$$(2.31) \quad \mathcal{L}_{(m)}(A) := \bigcup_{1 \leq i_1, i_2, \cdots, i_m \leq n} \ell_{i_1,i_2,\cdots,i_m}(A) \quad (\{i_j\}_{j=1}^m \text{ are distinct in } N),$$

is over all $\binom{n}{m}$ such choices of $\{i_j\}_{j=1}^m$ from N. As special cases, the Geršgorin disks $\Gamma_i(A)$ of (1.5) are lemniscates of order 1, while the Brauer Cassini ovals $K_{i,j}(A)$ of (2.5) are lemniscates of order 2, so that with (1.5) and (2.6)), we have

$$\mathcal{L}_{(1)}(A) = \Gamma(A) \text{ and } \mathcal{L}_{(2)}(A) = \mathcal{K}(A).$$

As examples of lemniscates, let

$$\begin{cases} a_1 := 1 + i, \ a_2 := -1 + i, \ a_3 := 0, \\ a_4 := -\frac{1}{4} - 1.4i, \text{ and } a_5 := +\frac{1}{4} - 1.4i, \end{cases}$$

and consider the lemniscate boundary of order 5 of

$$(2.32) \qquad \ell_5(\{a_i\}_{i=1}^5; \rho) := \{z \in \mathbb{C} : \prod_{i=1}^{5} |z - a_i| = \rho\} \quad (\rho \geq 0).$$

[2] The classical definition of a lemniscate (cf. Walsh (1969), p. 54), is the curve, corresponding to the case of equality in (2.30). The above definition of a lemniscate then is the union of this curve and its interior.

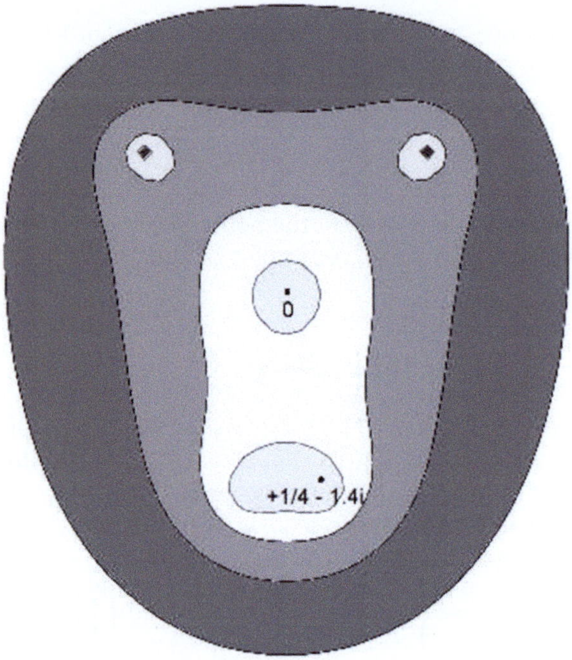

Fig. 2.3. Bear or lemniscate boundaries for $\ell_5(\{a_i\}_{i=1}^5; \rho)$ of (2.32) for $\rho = 3, 10, 20, 50$

These lemniscate boundaries appear in Fig. 2.3 for the particular values $\rho = 3, 10, 20, 50$.

When one considers the proof of Geršgorin's result (1.7) or the proof of Brauer's result (2.6), the difference is that the former focuses on **one** row of the matrix A, while the latter focuses on **two** distinct rows of the matrix A. But from the result of (2.9) of Theorem 2.3, this would seem to suggest that "using more rows in A gives better eigenvalue inclusion results for the spectrum of A". Alas, it turns out that $\mathcal{L}_{(m)}(A)$, as defined in (2.31), **fails**, in general for $m > 2$, to give a set in the complex plane which contains the spectrum of each A in $\mathbb{C}^{n \times n}, n \geq m$, as the following example (attributed to Morris Newman in Marcus and Minc (1964), p.149) shows. (See also Horn and Johnson (1985), p.382, for a nice treatment of this example.) It suffices to consider the 4×4 matrix

$$(2.33) \qquad A := \begin{bmatrix} 1 & 1 & 0 & 0 \\ 1 & 1 & 0 & 0 \\ 0 & 0 & 1 & 0 \\ 0 & 0 & 0 & 1 \end{bmatrix}, \text{ where } \sigma(A) = \{0, 1, 1, 2\},$$

where $a_{i,i} = 1$ for $1 \leq i \leq 4$ and where $r_1(A) = r_2(A) = 1; r_3(A) = r_4(A) = 0$. On choosing $m = 3$ in (2.30), then, for any choice of three distinct integers $\{i_1, i_2, i_3\}$ from $\{1, 2, 3, 4\}$, the product $r_{i_1}(A) \cdot r_{i_2}(A) \cdot r_{i_3}(A)$ is necessarily zero. Thus, the associated lemniscate in (2.30), for the matrix A of (2.33), always reduces to the set of points z for which $|z - 1|^3 = 0$, so that $z = 1$ is its sole point. Hence, with (2.31), $\mathcal{L}_{(3)}(A) = \{1\}$, which **fails** to contain $\sigma(A)$ in (2.33). (The same argument also gives $\mathcal{L}_{(4)}(A) = \{1\}$, and this failure can be extended to all matrices of order $n \geq 3$.)

To obtain a suitable compact set in the complex plane $\mathbb{C}$, based on higher-order lemniscates, which will include all eigenvalues of **any** given matrix A, such as in (2.33), we describe below a modest extension of an important work of Brualdi (1982), which introduced the notion of a cycle[3], from the directed graph of A, to obtain an eigenvalue inclusion region for any A. This extension is also derived from properties of the directed graph of the matrix A.

Given $A = [a_{i,j}] \in \mathbb{C}^{n \times n}$, $n \geq 1$, let $\mathbb{G}(A)$ be its directed graph, as described in Section 1.2, on n distinct vertices $\{v_i\}_{i=1}^n$, which consists of a (directed) arc $\overrightarrow{v_i v_j}$, from vertex v_i to vertex v_j, only if $a_{i,j} \neq 0$. (This directed graph $\mathbb{G}(A)$ allows loops, as in Section 1.2.) A **strong cycle** γ in $\mathbb{G}(A)$ is defined as a sequence $\{i_j\}_{j=1}^{p+1}$ of integers in N such that $p \geq 2$, the elements of $\{i_j\}_{j=1}^p$ are all *distinct* with $i_{p+1} = i_1$, and $\overrightarrow{v_{i_1} v_{i_2}}, \cdots, \overrightarrow{v_{i_p} v_{i_{p+1}}}$ are arcs of $\mathbb{G}(A)$. This implies that the associated entries of A, namely

$$a_{i_1,i_2}, a_{i_2,i_3}, \cdots, a_{i_p,i_{p+1}}, \text{ are all nonzero (where } i_{p+1} = i_1).$$

It is convenient to express this strong cycle in standard cyclic permutation notation (cf. Birkhoff and MacLane (1960), p.133)

$$\gamma := (i_1 \;\; i_2 \;\; \cdots \;\; i_p) \qquad \text{for } p \geq 2, \tag{2.34}$$

where $i_1, i_2, \cdots, i_p$ are distinct integers in N, and where γ is regarded as the permutation mapping defined by $\gamma(i_1) := i_2, \gamma(i_2) := i_3, \cdots$, and $\gamma(i_p) := i_1$. We also say that this strong cycle γ *passes through* the vertices $\{v_{i_j}\}_{j=1}^p$, and that γ has **length** p, with $p \geq 2$. (Note that a loop in $\mathbb{G}(A)$ cannot be a strong cycle since its length would be unity.) If there is a vertex v_i of $\mathbb{G}(A)$ for which there is no strong cycle passing though v_i, then we define its associated **weak cycle** γ simply as $\gamma = (i)$, independent of whether or not $a_{i,i} = 0$, and we say that γ passes through the vertex v_i. Next, on defining the **cycle set** $C(A)$ to be the set of all strong and weak cycles of $\mathbb{G}(A)$, then for each vertex v_i of $\mathbb{G}(A)$, there is always a cycle of $C(A)$ which passes through v_i. For example, if $n = 1$, then $C(A) = (1)$, and there is a unique (weak) cycle through vertex v_1.

Continuing, assume that $A = [a_{i,j}] \in \mathbb{C}^{n \times n}$, $n \geq 2$, is reducible. From the discussion in Section 1.2, there is a permutation matrix $P \in \mathbb{R}^{n \times n}$ and

[3] More precisely, the word "circuit" is used in Brualdi (1982) for what is called a "cycle" above. Our usage here agrees with that of Horn and Johnson (1985), p.383.

a positive integer m, with $2 \leq m \leq n$, such that PAP^T is in the **normal reduced form** of

$$(2.35) \qquad PAP^T = \begin{bmatrix} R_{1,1} & R_{1,2} & \cdots & R_{1,m} \\ O & R_{2,2} & \cdots & R_{2,m} \\ \vdots & & \ddots & \vdots \\ O & O & \cdots & R_{m,m} \end{bmatrix},$$

where each matrix $R_{j,j}$, $1 \leq j \leq m$, in (2.35) is such that

$$(2.36) \qquad \begin{cases} i) \ R_{j,j} \text{ is a } p_j \times p_j \text{ irreducible matrix with } p_j \geq 2, \\ \text{or} \\ ii) \ R_{j,j} \text{ is a } 1 \times 1 \text{ matrix with } R_{j,j} = [a_{k,k}] \text{ for some } k \in N. \end{cases}$$

Of course, if the given matrix $A \in \mathbb{C}^{n \times n}$, $n \geq 2$, is irreducible, we can view this as the case $m = 1$ of (2.35) and (2.36i).

Some easy observations follow. The existence of an $R_{j,j} = [a_{k,k}]$ in (2.36ii) is, by definition, equivalent to the statement that vertex v_k of $\mathbb{G}(A)$ has no strong cycle through it. Similarly, the existence of an $R_{j,j}$ satisfying (2.36i) implies that there is at least one strong cycle through *each* vertex v_k of $\mathbb{G}(A)$, associated with the irreducible submatrix $R_{j,j}$. (See Exercise 2 of this section.) We also note from (2.35) that

$$\sigma(A) = \bigcup_{k=1}^{m} \sigma(R_{k,k}),$$

so that the upper triangular blocks $R_{j,k} (j < k \leq m)$ in (2.35), if they exist, have no effect on the eigenvalues of A. Because of this last observation, we define the new rows sums $\tilde{r}_i(A)$ of A as

$$(2.37) \qquad \tilde{r}_i(A) := r_\ell(R_{j,j}),$$

if the ith row of A corresponds to the ℓth row of $R_{j,j}$ in (2.35). These new row sums are the old row sums if A is irreducible. Also, we see that (2.37) implies that $\tilde{r}_i(A) = 0$ for any vertex v_i corresponding to a weak cycle in $\mathbb{G}(A)$, (which is consistent with the convention used in (1.4)). Similarly, $\tilde{r}_i(A) > 0$ for each row, corresponding to an irreducible matrix $R_{j,j}$ of (2.36i).

To summarize, these new definitions, of strong and weak cycles and of modified row sums, are all results from a more detailed study of the directed graph $\mathbb{G}(A)$ of A.

With the above notations, given any $A = [a_{i,j}] \in \mathbb{C}^{n \times n}$, $n \geq 1$, if (cf.(2.34)) $\gamma = (i_1 \ i_2 \ \cdots \ i_p)$, with distinct elements $\{i_j\}_{j=1}^p$ and with $p \geq 2$, is a **strong cycle** in $\mathbb{G}(A)$, its associated **Brualdi lemniscate**, $\mathcal{B}_\gamma(A)$, of order p, is defined by

$$(2.38) \qquad \mathcal{B}_\gamma(A) := \{z \in \mathbb{C} : \prod_{i \in \gamma} |z - a_{i,i}| \leq \prod_{i \in \gamma} \tilde{r}_i(A)\}.$$

If $\gamma = (i)$ is a **weak cycle** in $\mathbb{G}(A)$, its associated **Brualdi lemniscate** $\mathcal{B}_\gamma(A)$ is defined by

$$\mathcal{B}_\gamma(A) := \{z \in \mathbb{C} : |z - a_{i,i}| = \tilde{r}_i(A) = 0\} = \{a_{i,i}\}. \tag{2.39}$$

The **Brualdi set** for A is then defined as

$$\mathcal{B}(A) := \bigcup_{\gamma \in C(A)} B_\gamma(A). \tag{2.40}$$

We now establish our new extension of a result of Brualdi (1982).

Theorem 2.5. *For any $A = [a_{i,j}] \in \mathbb{C}^{n \times n}$ and any eigenvalue λ of A, there is a (strong or weak) cycle γ in the cycle set $C(A)$ such that* (cf.(2.38) or (2.39))

$$\lambda \in \mathcal{B}_\gamma(A). \tag{2.41}$$

Consequently (cf.(2.40)),

$$\sigma(A) \subseteq \mathcal{B}(A). \tag{2.42}$$

Proof. If $n = 1$, then $A = [a_{1,1}] \in \mathbb{C}^{1\times 1}$, and $C(A)$ consists of the sole weak cycle $\gamma = (1)$. From (2.39), $\mathcal{B}_\gamma(A) = \{a_{1,1}\}$, and the sole eigenvalue of A, i.e., $a_{1,1}$, is exactly given by $\mathcal{B}_\gamma$. Thus, (2.41) and (2.42) trivially follow. Next, assume that $n \geq 2$. Let λ be an eigenvalue of A, and assume that $\lambda = a_{k,k}$ for some $k \in N$. Then, there is a (strong or weak) cycle γ in $C(A)$ such that $k \in \gamma$. If γ is a weak cycle through vertex v_k, then $\mathcal{B}_\gamma(A) = \{a_{k,k}\} = \lambda$, so again, (2.41) is satisfied. If γ is a strong cycle in $C(A)$ which passes through v_k, then with the choice of $z := a_{k,k}$, we see from (2.38) that $a_{k,k} \in \mathcal{B}_\gamma(A)$, and thus $\lambda = a_{k,k} \in \mathcal{B}(A)$. If each eigenvalue of A is a diagonal entry of A, the preceding argument gives $\sigma(A) \subseteq \mathcal{B}(A)$, the desired result of (2.42).

Next, assume that λ is an eigenvalue of A with $\lambda \neq a_{j,j}$ for *any* $j \in N$. If A is reducible, it follows from the dichotomy in (2.36) that λ must be an eigenvalue of some irreducible matrix $R_{j,j}$, of order p_j, with $2 \leq p_j \leq n$. Similarly, if A is irreducible (i.e., the case $m = 1$ of (2.35)), then λ is an eigenvalue of the irreducible matrix $R_{1,1} = A$, of order n. To simplify notations, we assume that $A = R_{j,j}$ is irreducible. (This means that we will use below the old row sums $r_i(A)$, rather than the new row sums $\tilde{r}_i(A)$ of $R_{j,j}$.) Writing $A\mathbf{x} = \lambda\mathbf{x}$, where $\mathbf{x} \in \mathbb{C}^n$ with $\mathbf{x} = [x_1, x_2, \cdots, x_n]^T \neq \mathbf{0}$, assume that $x_i \neq 0$. Then, $(A\mathbf{x})_i = \lambda x_i$ gives

$$(\lambda - a_{i,i})\, x_i = \sum_{j \in N\backslash\{i\}} a_{i,j} x_j.$$

As $(\lambda - a_{i,i})x_i \neq 0$, all the products $a_{i,j}x_j$ in the above sum cannot be zero. Hence, there is a $k \in N$, with $k \neq i$, such that

$$|x_k| = \max\left\{|x_j| : j \in N \text{ with } j \neq i \text{ and } a_{i,j}x_j \neq 0\right\}.$$

Thus, $|x_k| > 0$ and $a_{i,k} \neq 0$, so that

$$|\lambda - a_{i,i}| \cdot |x_i| \leq \sum_{j \in N \backslash \{i\}} |a_{i,j}| \cdot |x_j| \leq r_i(A) \cdot |x_k|, \text{ with } k \neq i.$$

Calling $i := i_1$ and $k := i_2$, we can repeat the above process, starting with

$$(\lambda - a_{i_2,i_2})x_{i_2} = \sum_{j \in N \backslash \{i_2\}} a_{i_2,j} x_j,$$

and there is similarly an i_3, with $|x_{i_3}| \neq 0$, and $a_{i_2,i_3} \neq 0$ such that

$$|\lambda - a_{i_2,i_2}| \cdot |x_{i_2}| \leq r_{i_2}(A) \cdot |x_{i_3}|, \text{ where } i_3 \neq i_2,$$

and where

$$|x_{i_3}| = \max\{|x_j| : j \in N \text{ with } j \neq i_2 \text{ and } a_{i_2,j} x_j \neq 0\}.$$

If $i_3 = i_1$, the process terminates, having produced the distinct integers i_1, and i_2, with $i_3 = i_1$, and with a_{i_1,i_2} and a_{i_2,i_1} nonzero. Thus, with the notation of (2.34), the strong cycle $\gamma := (i_1 \ \ i_2)$ has been produced. If $i_3 \neq i_1$, this process continues, but eventually terminates (since N is a finite set) when an $i_{p+1} \in N$ is found which is equal to some previous i_ℓ. In either case, there is a sequence $\{i_j\}_{j=\ell}^{p}$, with $p \geq 2$, of distinct integers in N, with $i_{p+1} = i_\ell$. But, this sequence also produces the following nonzero entries of A:

$$a_{i_\ell,i_{\ell+1}}, a_{i_{\ell+1},i_{\ell+2}}, \cdots, a_{i_p,i_{p+1}}, \text{ with } i_{p+1} = i_\ell,$$

so that, with our notation of (2.34), $\gamma := (i_\ell \ \ i_{\ell+1} \ \ \cdots \ \ i_p)$ is then a strong cycle of A. Thus, we have that

$$|\lambda - a_{i_j,i_j}| \cdot |x_{i_j}| \leq r_{i_j}(A) \cdot |x_{i_{j+1}}| \ , j = \ell, \ell+1, \cdots, p, \text{ where } i_{p+1} = i_\ell,$$

where the x_{i_j}'s are all nonzero. Taking the products of the above gives

$$\left(\prod_{j=\ell}^{p} |\lambda - a_{i_j,i_j}|\right) \cdot \left(\prod_{j=\ell}^{p} |x_{i_j}|\right) \leq \left(\prod_{j=\ell}^{p} r_{i_j}(A)\right) \cdot \left(\prod_{j=\ell}^{p} |x_{i_{j+1}}|\right).$$

But, as $x_{i_{p+1}} = x_{i_\ell}$, then $\prod_{j=\ell}^{p} |x_{i_j}| = \prod_{j=\ell}^{p} |x_{i_{j+1}}| > 0$, so that, on cancelling these products in the above display, we have

$$\prod_{j=\ell}^{p} |\lambda - a_{i_j,i_j}| \leq \prod_{j=\ell}^{p} r_{i_j}(A).$$

Thus (cf. (2.38)), $\lambda \in \mathcal{B}_\gamma(A)$, giving (2.41) . ■

As an application of Theorem 2.5, consider the matrix B of (2.33), which is already in normal reduced form of (2.35). Its cycle set $C(A)$ consists of the strong cycle $\gamma_1 = (1\ \ 2)$, and the weak cycles $\gamma_2 = (3)$ and $\gamma_3 = (4)$. It follows from (2.40) that

$$\mathcal{B}(B) = \{z \in \mathbb{C} : |z-1|^2 \leq 1\} \cup \{1\} \cup \{1\},$$

which now nicely contains $\sigma(B) = \{0, 1, 1, 2\}$.

We next state the equivalent nonsingularity result associated with Theorem 2.5, which slightly extends the corresponding result in Brualdi (1982).

Theorem 2.6. *If $A = [a_{i,j}] \in \mathbb{C}^{n \times n}$, if $C(A)$ is the set of all strong and weak cycles in $\mathbb{G}(A)$, and if* (cf.(2.38))

$$\prod_{i \in \gamma} |a_{i,i}| > \prod_{i \in \gamma} \tilde{r}_i(A) \qquad (\text{all } \gamma \in C(A)), \tag{2.43}$$

then A is nonsingular.

Now, we come to an analog, Theorem 2.7, of Taussky's nonsingularity result of Theorem 1.11, which also appears in Brualdi (1982). (Its proof, which uses the irreducibility of A, is left as Exercise 3 of this section, as it follows along the lines of the proofs of Theorems 1.11 and 2.5.)

Theorem 2.7. *If $A = [a_{i,j}] \in \mathbb{C}^{n \times n}$ is irreducible, and if*

$$\prod_{i \in \gamma} |a_{i,i}| \geq \prod_{i \in \gamma} r_i(A) \qquad (\text{all } \gamma \in C(A)), \tag{2.44}$$

with strict inequality holding for some $\gamma \in C(A)$, then A is nonsingular.

Then for completeness, the equivalent eigenvalue inclusion result associated with Theorem 2.7 is the following result, which is a generalization of Taussky's Theorem 1.12.

Theorem 2.8. *If $A = [a_{i,j}] \in \mathbb{C}^{n \times n}$ is irreducible, and if λ, an eigenvalue of A, is such that $\lambda \notin \text{int } \mathcal{B}_\gamma(A)$ for any $\gamma \in C(A)$, i.e.,*

$$\prod_{i \in \gamma} |\lambda - a_{i,i}| \geq \prod_{i \in \gamma} r_i(A) \qquad (\text{all } \gamma \in C(A)),$$

then

$$\prod_{i \in \gamma} |\lambda - a_{i,i}| = \prod_{i \in \gamma} r_i(A) \qquad (\text{all } \gamma \in C(A)), \tag{2.45}$$

i.e., λ is on the boundary of the Brualdi lemniscate $\mathcal{B}_\gamma(A)$, for each γ in $C(A)$. In particular, if some eigenvalue λ of A lies on the boundary of the Brualdi set $\mathcal{B}(A)$ of (2.40)*, then* (2.45) *holds.*

It is worthwhile to examine again the matrix B of (2.33), which is reducible. Theorem 2.5 gives us (cf. (2.42)) that the spectrum of B can be enclosed in the Brualdi set $\mathcal{B}(A)$, which depends on cycles of different lengths, determined from the directed graph of B. This raises the following obvious question: If higher-order lemniscates $\mathcal{L}_m(B)$ of (2.31), for a fixed m, fail to work for the reducible matrix B of (2.33), can they be successfully applied to arbitrary irreducible matrices? To answer this, consider the matrix

$$(2.46) \qquad D = \begin{bmatrix} 1 & 1 & \epsilon & \epsilon \\ 1 & 1 & 0 & 0 \\ \epsilon & 0 & 1 & 0 \\ \epsilon & 0 & 0 & 1 \end{bmatrix}, \text{ with } \epsilon > 0,$$

which is obtained by adding some nonzero entries to the matrix of (2.33). The directed graph $\mathbb{G}(D)$ of D is (omitting diagonal loops) given in Fig. 2.4, so that D is irreducible, where its cycle set $C(A)$ consists of the strong cycles

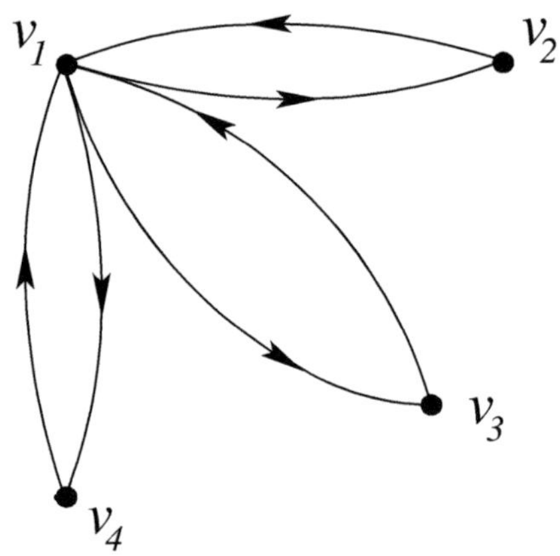

Fig. 2.4. The directed graph, $\mathbb{G}(D)$, without loops, for the matrix D of (2.46)

$\gamma_1 = (1\ \ 2)$, $\gamma_2 = (1\ \ 3)$, and $\gamma_3 = (1\ \ 4)$, where $r_1(D) = 1 + 2\epsilon, r_2(D) = 1$, and $r_3(D) = r_4(D) = \epsilon$. If we consider the lemniscate of order 4 for this matrix D, we obtain, from (2.31) that

$$\mathcal{L}_{(4)}(D) = \{z \in \mathbb{C} : |z - 1|^4 \leq \epsilon^2(1 + 2\epsilon)\},$$

which is the disk $\{z \in \mathbb{C} : |z - 1| \leq \sqrt{\epsilon}(1 + 2\epsilon)^{1/4}\}$. But, as

$$\sigma(D) = \{1 - (1 + 2\epsilon^2)^{1/2}, 1, 1, 1 + (1 + 2\epsilon^2)^{1/2}\},$$

it can be verified (see Exercise 5 of this section) that

$$\sigma(D) \not\subset \mathcal{L}_{(4)}(D), \text{ for } any\ \epsilon > 0.$$

Similarly, it can be verified (see Exercise 5 of this section) that (cf. (2.31))

$$\sigma(D) \not\subset \mathcal{L}_{(3)}(D), \text{ for } any\ \epsilon > 0.$$

In other words, even with an irreducible matrix, lemniscates of a *fixed* order $m \geq 3$, applied to this irreducible matrix, can **fail** to capture the spectrum of this matrix, while the Brualdi sets always work!

We note that the eigenvalue inclusion of Theorem 2.5, applied to the matrix $A = [a_{i,j}] \in \mathbb{C}^{n \times n}, n \geq 1$, now depends on all the quantities of

$$\{a_{i,i}\}_{i=1}^{n}, \ \{\tilde{r}_i(A)\}_{i=1}^{n}, \text{ and the cycle set } \mathcal{C}(A), \tag{2.47}$$

which are derived from the matrix A, its directed graph $\mathbb{G}(A)$, and its normal reduced form (cf.(2.35)). It is of course of interest to see how the Brualdi set $\mathcal{B}(A)$ compares with the Brauer set $\mathcal{K}(A)$. This will be done in the next section. We also ask, in the spirit of Theorem 2.4, if the union of the spectra of all matrices which match the data of (2.47), fills out the Brualdi set $\mathcal{B}(A)$ of (2.47). This will be precisely answered in Section 2.4.

Exercises

1. A matrix $A = [a_{i,j}] \in \mathbb{C}^{n \times n}$, $n \geq 2$, is said to be **weakly irreducible** (see Brualdi (1982)) if there is a strong cycle through each vertex v_i of $\mathbb{G}(A)$, the directed graph of A. Show that irreducibility and weak irreducibility are equivalent for all matrices in $\mathbb{C}^{n \times n}$ if $n = 2$ or $n = 3$. For $n \geq 4$, show that irreducibility implies weak irreducibility, but not conversely in general.

2. Given $A = [a_{i,j}] \in \mathbb{C}^{n \times n}$, $n \geq 2$, show that the existence of an irreducible submatrix $R_{j,j}$ in (2.36i) and (2.35) implies that there is a strong cycle through **each** vertex of $\mathbb{G}(A)$, associated with the submatrix $R_{j,j}$.

3. Give a complete proof of Theorem 2.7. (Hint: Follow the general outline of the proofs of Theorem 1.11 and 2.5.)

4. Consider the matrix $B(\epsilon) := \begin{bmatrix} 1\,1\,0\,0 \\ 1\,1\,0\,0 \\ 0\,0\,1\,\epsilon \\ 0\,0\,\epsilon\,1 \end{bmatrix}$, where $\epsilon > 0$.
   a. Show that the eigenvalues of $B(\epsilon)$ are $\{0, 1-\epsilon, 1+\epsilon, 2\}$.
   b. Show that $B(\epsilon)$ is weakly irreducible, in the sense of Exercise 1, for any $\epsilon > 0$.
   c. Show that the (strong) cycles of $B(\epsilon)$ are $\gamma_1 = (1\ \ 2)$ and

$\gamma_2 = (3 \;\; 4)$, and that $\mathcal{B}_{\gamma_1}(B(\epsilon)) \cup \mathcal{B}_{\gamma_2}(B(\epsilon))$ does include all the eigenvalues of $B(\epsilon)$.

5. Consider the matrix D of (2.46). With the definition of $\mathcal{L}_{(m)}(A)$ in (2.31),
   a. Show that $\sigma(D) \not\subset \mathcal{L}_{(4)}(D)$ for *any* $\epsilon > 0$;
   b. Show that $\sigma(D) \not\subset \mathcal{L}_{(3)}(D)$ for *any* $\epsilon > 0$;
   c. Determine $\mathcal{B}(D)$ from (2.40), and verify that $\sigma(D) \subseteq \mathcal{B}(D)$, for *any* $\epsilon > 0$.

6. Let $\gamma = (1 \;\; 2 \;\; 3 \;\; 4)$ be a strong cycle from $\mathcal{C}(A)$ for the matrix A. For any of the following five permutations of γ, i.e., $(1 \;\; 2 \;\; 4 \;\; 3)$, $(1 \;\; 3 \;\; 2 \;\; 4)$, $(1 \;\; 3 \;\; 4 \;\; 2)$, $(1 \;\; 4 \;\; 2 \;\; 3)$ and $(1 \;\; 4 \;\; 3 \;\; 2)$, show that the Brualdi lemniscate set $\mathcal{B}_\gamma(A)$ of (2.38) is *unchanged*. What does this say about the quantities of (2.40)?

7. Consider the irreducible matrix $E = \begin{bmatrix} 1 & 1 & 1 \\ 2 & 4 & 0 \\ 1 & 0 & 2 \end{bmatrix}$, which has an eigenvalue $\lambda = 0$, and for which its cycle set $C(E)$ consists of the two strong cycles $\gamma_1 = (1 \;\; 2)$ and $\gamma_2 = (1 \;\; 3)$. Show that **not** all three Brauer Cassini ovals pass through $z = 0$ (Zhang and Gu (1994)). However, show that $\lambda = 0$ is a boundary point of its Brualdi set, $\mathcal{B}(E) = \mathcal{B}_{\gamma_1}(E) \cup \mathcal{B}_{\gamma_2}(E)$, and that $\mathcal{B}_{\gamma_1}(E)$ and $\mathcal{B}_{\gamma_2}(E)$ both pass through $z = 0$, as dictated by Theorem 2.8.

8. Let $A = [a_{i,j}] \in \mathbb{C}^{n \times n}$ be irreducible, with the property that A has two distinct rows for which each row has two nonzero non-diagonal entries. (This implies that $n \geq 3$.) If $\lambda \in \sigma(A)$ is such that (cf.(2.6)) $\lambda \in \partial\mathcal{K}(A)$, show that $\lambda \in \partial K_{i,j}(A)$ for all $i \neq j$, with $1 \leq i, j \leq n$. (Rein (1967)). Note that the 3×3 matrix of Exercise 7 does not satisfy the above hypotheses.

9. The definitions of the Brualdi lemniscates in (2.38) and (2.39) make use of the new row sums $\{\tilde{r}_i(A)\}_{i=1}^n$ of (2.37). Show that if these Brualdi lemniscates are defined with the *old* row sums $\{r_i(A)\}_{i=1}^n$, the result of Theorem 2.5 is still valid. (Hint: use (2.49).)

2.3 Comparison of the Brauer Sets and the Brualdi Sets

Our new result here is very much in the spirit of Theorem 2.3.

Theorem 2.9. *For any $A = [a_{i,j}] \in \mathbb{C}^{n \times n}, n \geq 2$, then, with the definitions of* (2.6) *and* (2.40),

$$\mathcal{B}(A) \subseteq \mathcal{K}(A). \tag{2.48}$$

Remark. This establishes that the Brualdi set, for any matrix A, is always a subset of its associated Brauer set $\mathcal{K}(A)$. Note also that the restriction $n \geq 2$ is necessary for the definition of the Brauer set $\mathcal{K}(A)$, as in Theorem 2.2.

Proof. Consider any cycle γ of the cycle set $C(A)$. If γ is a **weak cycle**, i.e., $\gamma = (i)$ for some $i \in N$, then (cf.(2.39)) $\mathcal{B}_\gamma(A) = \{a_{i,i}\}$. Now, the Brauer Cassini oval $K_{i,j}(A)$, for any $j \neq i$, is, from (2.5),

$$K_{i,j}(A) := \{z \in \mathbb{C} : |z - a_{i,i}| \cdot |z - a_{j,j}| \leq r_i(A) \cdot r_j(A)\},$$

so that $a_{i,i} \in K_{i,j}(A)$. Thus, $\mathcal{B}_\gamma(A) \subseteq K_{i,j}(A)$ for any $j \neq i$; whence, $\mathcal{B}_\gamma(A) \subseteq \mathcal{K}(A)$.

Next, assume that γ is a **strong cycle** from $C(A)$, where we point out that the new row sums $\{\tilde{r}_i(A)\}_{i=1}^n$ for A, from (2.37) in the reducible case, and the old row sums $\{r_i(A)\}_{i=1}^n$ for A, necessarily satisfy

$$0 < \tilde{r}_i(A) \leq r_i(A) \qquad (\text{all } i \in \gamma), \tag{2.49}$$

with $\tilde{r}_i(A) = r_i(A) > 0$ for all $i \in N$ if A is irreducible. If the strong cycle γ has length 2, i.e., $\gamma = (i_1 \;\; i_2)$ where $i_3 = i_1$, then it follows from (2.38) that its associated Brualdi lemniscate is

$$\mathcal{B}_\gamma(A) = \{z \in \mathbb{C} : |z - a_{i_1,i_1}| \cdot |z - a_{i_2,i_2}| \leq \tilde{r}_{i_1}(A) \cdot \tilde{r}_{i_2}(A)\}.$$

Hence, with (2.5) and the inequalities of (2.49), it follows that

$$\mathcal{B}_\gamma(A) \subseteq K_{i_1,i_2}(A).$$

Next, assume that this strong cycle γ has length $p > 2$, i.e., $\gamma = (i_1 \;\; i_2 \;\; \cdots \;\; i_p)$, with $i_{p+1} = i_1$, where the associated new row sums $\{\tilde{r}_{i_j}(A)\}_{j=1}^p$ are all positive. From (2.38), the associated Brualdi lemniscate is

$$\mathcal{B}_\gamma(A) := \left\{ z \in \mathbb{C} : \prod_{j=1}^p |z - a_{i_j,i_j}| \leq \prod_{j=1}^p \tilde{r}_{i_j}(A) \right\}. \tag{2.50}$$

Let z be any point of $\mathcal{B}_\gamma(A)$. On squaring the inequality in (2.50), we have

$$|z-a_{i_1,i_1}|^2 \cdot |z-a_{i_2,i_2}|^2 \cdots |z-a_{i_p,i_p}|^2 \leq \tilde{r}_{i_1}^2(A) \cdot \tilde{r}_{i_2}^2(A) \cdots \tilde{r}_{i_p}^2(A).$$

As these $\tilde{r}_{i_j}(A)$'s are all positive, we can equivalently express the above inequality as

$$(2.51)\qquad \left(\frac{|z-a_{i_1,i_1}| \cdot |z-a_{i_2,i_2}|}{\tilde{r}_{i_1}(A) \cdot \tilde{r}_{i_2}(A)}\right)\left(\frac{|z-a_{i_2,i_2}| \cdot |z-a_{i_3,i_3}|}{\tilde{r}_{i_2}(A) \cdot \tilde{r}_{i_3}(A)}\right)\cdots \cdot \left(\frac{|z-a_{i_p,i_p}| \cdot |z-a_{i_1,i_1}|}{\tilde{r}_{i_p}(A) \cdot \tilde{r}_{i_1}(A)}\right) \leq 1.$$

As the factors on the left of (2.51) cannot all exceed unity, then at least one of the factors is at most unity. Hence, there is an ℓ with $1 \leq \ell \leq p$ such that

$$|z-a_{i_\ell,i_\ell}| \cdot |z-a_{i_{\ell+1},i_{\ell+1}}| \leq \tilde{r}_{i_\ell}(A) \cdot \tilde{r}_{i_{\ell+1}}(A),$$

(where if $\ell = p$, then $i_{\ell+1} = i_1$). But from the definition in (2.5) and from (2.49), we see that $z \in K_{i_\ell,i_{\ell+1}}(A)$. Hence, as z is any point of $\mathcal{B}_\gamma(A)$, it follows that

$$\mathcal{B}_\gamma(A) \subseteq \bigcup_{j=1}^{p} K_{i_j,i_{j+1}}(A) \quad (\text{where } i_{p+1} = i_1).$$

Thus, from (2.40) and the above display,

$$\mathcal{B}(A) := \bigcup_{\gamma \in \mathcal{C}(A)} \mathcal{B}_\gamma(A) \subseteq \bigcup_{\substack{i,j \in N \\ i \neq j}} K_{i,j}(A) := \mathcal{K}(A),$$

the desired result of (2.48). ■

It is important to remark that while the Brualdi set $\mathcal{B}(A)$ is always a *subset* (cf.(2.48)) of the Brauer set $\mathcal{K}(A)$, for any matrix $A = [a_{i,j}] \in \mathbb{C}^{n \times n}$ with $n \geq 2$, there may be substantially *more* computational work in determining all the Brualdi lemniscate sets $\mathcal{B}_\gamma(A)$, than there is in determining the $\binom{n}{2}$ Brauer Cassini ovals. That is, getting the possibly sharper inclusion in (2.48) *may* come at the price of more computations. This is illustrated in the next paragraph.

We next show that there are many cases where **equality** holds in (2.48) of Theorem 2.9. Consider any matrix $A = [a_{i,j}] \in \mathbb{C}^{n \times n}$, with $n \geq 2$, for which *every* nondiagonal entry $a_{i,j}$ of A is nonzero. The matrix A is then clearly irreducible. Next, for any $n \geq 2$, on defining

$$(2.52)\qquad \mathcal{P}_n := \{ \text{ the set of all cycles of length at least two, from the integers } (1, 2, \cdots, n) \},$$

it can be verified (see Exercise 2 of this section) that the **cardinality** of $\mathcal{P}_n$ (i.e., the number of elements in $\mathcal{P}_n$), denoted by $|\mathcal{P}_n|$, is given by

(2.53)
$$|\mathcal{P}_n| = \sum_{k=2}^{n} \binom{n}{k} (k-1)! \,.$$

Then, each strong cycle γ of $\mathbb{G}(A)$, given by $\gamma = (i_1 \;\; i_2 \;\; \cdots \;\; i_p)$ with $p \geq 2$, can be associated with an element in $\mathcal{P}_n$, i.e.,

$$\gamma = (i_1 \;\; i_2 \;\; \cdots \;\; i_p) \in \mathcal{P}_n, \text{ where } 2 \leq p \leq n.$$

In this case, as *each* Brauer Cassini oval $K_{i,j}(A), i \neq j$, corresponds to a cycle (of length 2) in $\mathcal{B}(A)$, it follows from (2.40) that $\mathcal{K}(A) \subseteq \mathcal{B}(A)$, but as the reverse inclusion holds in (2.48), then

(2.54)
$$\mathcal{B}(A) = \mathcal{K}(A).$$

In other words, the Brualdi set $\mathcal{B}(A)$ need not, in general, be a proper subset of the Brauer set $\mathcal{K}(A)$. But what is most striking here is that, for a 10×10 complex matrix A, all of whose off-diagonal entries are nonzero, there are, from (2.53),

$$1,112,073 \text{ distinct cycles in } \mathcal{C}(A).$$

Thus, determining the Brualdi set $\mathcal{B}(A)$ would require finding 1,112,073 Brualdi lemniscates, a daunting task for a 10×10 matrix!. Fortunately, as $\mathcal{B}(A) = \mathcal{K}(A)$, only $\binom{10}{2} = 45$ of these cycles, corresponding to the Brauer Cassini ovals, are needed to determine $\mathcal{B}(A)$, while only 10 Geršgorin disks are needed for $\Gamma(A)$.

The example, in the previous paragraph, where the given irreducible matrix $A = [a_{i,j}] \in \mathbb{C}^{n \times n}$, $n > 2$, has all its off-diagonal entries nonzero, shows that it suffices to consider only the union of the Brauer Cassini ovals for A, in order to obtain a valid eigenvalue inclusion set for A. This **reduction** is noteworthy, and leads to the following considerations.

Let $A = [a_{i,j}] \in \mathbb{C}^{4 \times 4}$ be an irreducible matrix with zero diagonal entries, for which $\gamma_1 = (1 \;\; 2 \;\; 3 \;\; 4)$ and $\gamma_2 = (1 \;\; 2 \;\; 4 \;\; 3)$ are elements of the cycle set $C(A)$. (Note that $\mathcal{B}_{\gamma_1}(A) = \mathcal{B}_{\gamma_2}(A)$ from (2.38) and Exercise 6 of Section 2.2.) Then, A is irreducible and its directed graph, based on γ_1 and γ_2, is shown in Fig. 2.5. But, it is evident from this directed graph that $\gamma_3 = (1 \;\; 2 \;\; 3)$, $\gamma_4 = (1 \;\; 2 \;\; 4)$, and $\gamma_5 = (3 \;\; 4)$ are the remaining elements of $C(A)$. Moreover, using the technique of the proof of Theorem 2.9, it can be verified (see Exercise 3 of this section) that

(2.55)
$$\mathcal{B}(A) := \bigcup_{j=1}^{5} \mathcal{B}_{\gamma_j}(A) = \mathcal{B}_{\gamma_3}(A) \cup \mathcal{B}_{\gamma_4}(A) \cup \mathcal{B}_{\gamma_5}(A),$$

i.e., the higher-order lemniscates $B_{\gamma_1}(A)$ and $B_{\gamma_2}(A)$, of order 4, are *not needed*, from (2.55), in determining $\mathcal{B}(A)$. From this, we can speak of a **reduced cycle set** $\hat{C}(A)$ of an irreducible matrix A, in which particular higher-order Brualdi lemniscates are omitted from $C(A)$, but with $\hat{C}(A)$ having the property that

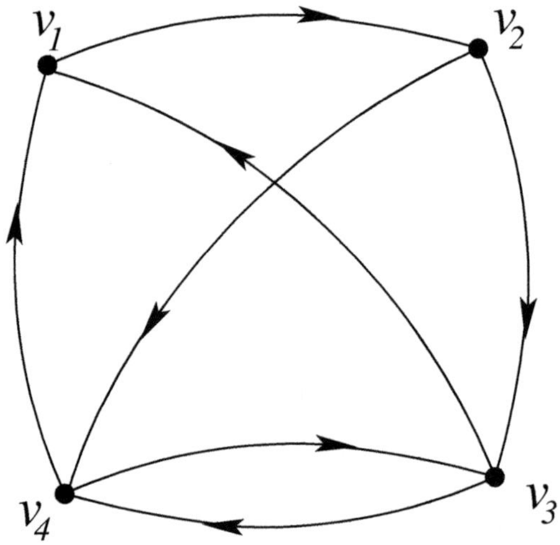

Fig. 2.5. The directed graph, $\mathbb{G}(A)$, for the cycles $\gamma_1 = (1 \;\; 2 \;\; 3 \;\; 4)$ and $\gamma_2 = (1 \;\; 2 \;\; 4 \;\; 3)$

$$\bigcup_{\gamma \in \hat{C}(A)} \mathcal{B}_\gamma(A) = \mathcal{B}(A).$$

We now give the following new result of Brualdi and Varga, Theorem 2.10, which provides a theoretical setting for such cycle reductions. (We note that the reduction of (2.55) is in fact the special case, $m = 2$, of this theorem.) For notation, if $\gamma = (i_1 \;\; i_2 \;\; \cdots \;\; i_p)$ is a cycle of $C(A)$, then

$$V(\gamma) := \bigcup_{j=1}^{p} \{i_j\} \text{ is its } \textbf{vertex set} \; ,$$

where we note, in this irreducible case, that all cycles are necessarily *strong* cycles.

Theorem 2.10. *Given an irreducible matrix* $A = [a_{i,j}] \in \mathbb{C}^{n \times n}, n \geq 2$, *let* $C(A)$ *be its cycle set, and let* $\{\gamma_j\}_{j=1}^{s}$, *with* $s \geq 2$, *be distinct cycles of* $C(A)$ *such that*

i) $V(\gamma_1) = \bigcup_{j=2}^{s} V(\gamma_j)$, *and*

ii) there is a positive integer m *such that each vertex from* γ_1 *appears exactly* m *times in* $\bigcup_{j=2}^{s} V(\gamma_j)$.

Then (cf.(2.38)),

$$\mathcal{B}_{\gamma_1}(A) \subseteq \bigcup_{j=2}^{s} \mathcal{B}_{\gamma_j}(A). \tag{2.56}$$

Remark 1. This means that the Brualdi lemniscate $\mathcal{B}_{\gamma_1}(A)$ can be *deleted* from the rualdi set (cf.(2.40)) $\mathcal{B}(A)$, since $\bigcup_{\gamma \in C(A)/\gamma_1} \mathcal{B}_\gamma(A) = \mathcal{B}(A)$.

Remark 2. Noting that the *length* of a cycle γ in $C(A)$ is the same as the *cardinality* of its vertex set $V(\gamma)$, then the hypotheses of Theorem 2.10 imply that

$$\text{length}(\gamma_1) \geq \text{length}(\gamma_j), \text{ for each } j \text{ with } 2 \leq j \leq s.$$

Thus, removing γ_1 from $C(A)$ removes a generally *higher-order* lemniscate from $C(A)$.

Proof. Since A is irreducible, then $r_i(A) > 0$ for all $i \in N$. Thus for any $\gamma \in C(A)$, we can equivalently express its associated Brualdi lemniscate (cf.(2.38)) as

$$\mathcal{B}_\gamma(A) = \{z \in \mathbb{C} : \prod_{i \in \gamma} \left(\frac{|z - a_{i,i}|}{r_i(A)} \right) \leq 1\}.$$

Now for any $z \in \mathbb{C}$, the hypotheses i) and ii) of Theorem 2.10 above directly give us that

$$\prod_{k \in \gamma_1} \left(\frac{|z - a_{k,k}|}{r_k(A)} \right)^m = \prod_{j=2}^{s} \left(\prod_{k \in \gamma_j} \left(\frac{|z - a_{k,k}|}{r_k(A)} \right) \right). \tag{2.57}$$

Hence, for any $z \in \mathcal{B}_{\gamma_1}(A)$, the product for γ_1, on the left in (2.57), is at most unity. Thus, not all products $\prod_{k \in \gamma_j} \left(\frac{|z - a_{k,k}|}{r_k(A)} \right)$, for $2 \leq j \leq s$ on the right in (2.57), can exceed unity. Therefore, there is an i, with $2 \leq i \leq s$, such that $\prod_{k \in \gamma_i} \left(\frac{|z - a_{k,k}|}{r_k(A)} \right) \leq 1$, which implies that $z \in \mathcal{B}_{\gamma_i}(A)$, and this gives the desired inclusion of (2.56). ■

To conclude this section, the use of the Brualdi set $\mathcal{B}(A)$, rather than the Brauer set $\mathcal{K}(A)$ or the Geršgorin set $\Gamma(A)$ to estimate the spectrum of A, seems to be more suitable in practical applications in cases where the cycle set $\mathcal{C}(A)$ has few elements, or, more precisely, when its *reduced cycle set* has few elements.

Exercises

1 For the tridiagonal $n \times n$ matrix A, associated with the directed graph of Fig. 1.7, show that its Brualdi set (cf. (2.40)) is, for any $n \geq 4$, just

$$\mathcal{B}(A) = \{z \in \mathbb{C} : |z - 2| \leq 2\}.$$

Also, show for $n \geq 4$ that $\mathcal{B}(A) = \mathcal{K}(A) = \Gamma(A)$.

2. Prove that the formula in (2.53) is valid. (Hint: Use the result of Exercise 6 of Section 2.2.)

3. For the matrix $A = [a_{i,i}] \in \mathbb{C}^{4 \times 4}$, whose directed graph is shown in Fig. 2.5, verify that (2.55) is valid. (Hint: Apply Theorem 2.10 for the case $m = 2$.)

4. For any complex numbers $\{z_j\}_{j=1}^p$ with $p \geq 2$, and for any p nonnegative real numbers $\{\rho_j\}_{j=1}^p$, define the Brualdi-like sets

$$S_p := \left\{ z \in \mathbb{C} : \prod_{j=1}^{p} |z - z_j| \leq \prod_{j=1}^{p} \rho_j \right\}, \text{ and}$$

$$S_{p \backslash k} := \left\{ z \in \mathbb{C} : \prod_{\substack{j=1 \\ j \neq k}}^{p} |z - z_j| \leq \prod_{\substack{j=1 \\ j \neq k}}^{p} \rho_j \right\},$$

for any k with $1 \leq k \leq p$. Then, it is known (Karow (2003)) that

$$S_p \subseteq \bigcup_{k=1}^{p} S_{p \backslash k}.$$

Show that the above inclusion can be deduced as a special case of $m = p - 1$ of Theorem 2.10.

2.4 The Sharpness of Brualdi Lemniscate Sets

Given any matrix $A = [a_{i,j}] \in \mathbb{C}^{n \times n}, n \geq 1$, we have from (2.42) of Theorem 2.5 that

$$\sigma(A) \subseteq \mathcal{B}(A), \tag{2.58}$$

where the associated Brualdi set $\mathcal{B}(A)$ is determined, in (2.40), from the quantities

$$\{a_{i,i}\}_{i=1}^n, \quad \{\tilde{r}_i(A)\}_{i=1}^n, \text{ and the cycle set } \mathcal{C}(A) \text{ of A.} \tag{2.59}$$

It is again evident that any matrix $B = [b_{i,j}] \in \mathbb{C}^{n \times n}$, having the identical quantities of (2.59), has its eigenvalues also in $\mathcal{B}(A)$, i.e., with notations similar to the equiradial set and the extended equiradial set of (2.15) and (2.16), if

$$\omega_{\mathcal{B}}(A) := \{B = [b_{i,j}] \in \mathbb{C}^{n\times n} : b_{i,i} = a_{i,i}, \tilde{r}_i(B) = \tilde{r}_i(A), \text{ for all } i \in N, \text{ and } \mathcal{C}(B) = \mathcal{C}(A)\} \tag{2.60}$$

denotes the **Brualdi radial set** for A, where $\sigma(\omega_{\mathcal{B}}(A)) := \bigcup_{B\in\omega_{\mathcal{B}}(A)} \sigma(B)$, and if the **extended Brualdi radial set** is given by

$$\hat{\omega}_{\mathcal{B}}(A) := \{B = [b_{i,j}] \in \mathbb{C}^{n\times n} : b_{i,i} = a_{i,i}, 0 \leq \tilde{r}_i(B) \leq \tilde{r}_i(A), \text{ for all } i \in N, \text{ and } \mathcal{C}(B) = \mathcal{C}(A)\} \tag{2.61}$$

for A, where $\sigma(\hat{\omega}_{\mathcal{B}}(A)) := \bigcup_{B\in\hat{\omega}_{\mathcal{B}}(A)} \sigma(B)$, it follows, in analogy with (2.18), that

$$\sigma(\omega_{\mathcal{B}}(A)) \subseteq \sigma(\hat{\omega}_{\mathcal{B}}(A)) \subseteq \mathcal{B}(A). \tag{2.62}$$

(We note that $\tilde{r}_i(A) > 0$ for any row i associated with a strong cycle of $\mathbb{G}(A)$.)

It is of theoretical interest to ask if equality can hold throughout in (2.62). The answer, in general, is **no**, as the following simple example shows. Consider the matrix

$$D = \begin{bmatrix} 1 & 1 & 0 \\ \frac{1}{2} & -1 & \frac{1}{2} \\ \frac{1}{2} & 0 & 1 \end{bmatrix}, \tag{2.63}$$

so that

$$r_1(D) = r_2(D) = 1, \text{ and } r_3(D) = \frac{1}{2}.$$

The directed graph $\mathbb{G}(D)$, without loops, is then given in Fig. 2.6, so that D is irreducible, and the cycle set of D is $C(D) = (1\ \ 2) \cup (1\ \ 2\ \ 3)$. Now, any matrix E in $\omega_{\mathcal{B}}(D)$ can be expressed from (2.60) as

$$E = \begin{bmatrix} 1 & e^{i\theta_1} & 0 \\ (1-s)e^{i\theta_2} & -1 & se^{i\theta_3} \\ \frac{1}{2}e^{i\theta_4} & 0 & 1 \end{bmatrix}, \tag{2.64}$$

where s satisfies $0 < s < 1$ and where $\{\theta_i\}_{i=1}^4$ are any real numbers. (Note that letting $s = 0$ or $s = 1$ in (2.64) does *not* preserve the cycles of $\mathcal{C}(D)$.) With $\gamma_1 := (1\ \ 2)$ and $\gamma_2 := (1\ \ 2\ \ 3)$, we see from (2.38) that

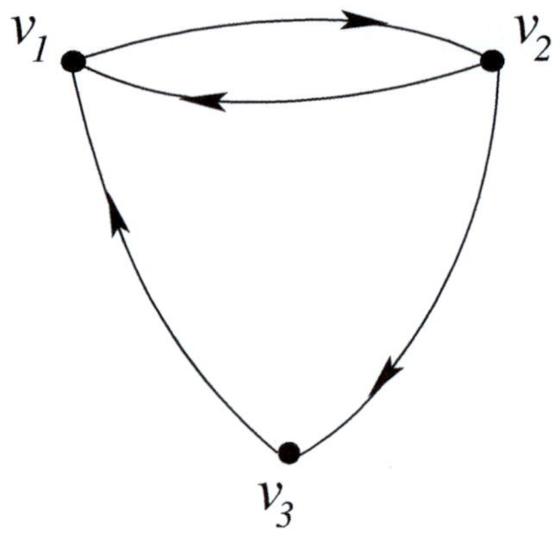

Fig. 2.6. The directed graph, $\mathbb{G}(D)$, without loops, for the matrix D of (2.63)

$$(2.65)\qquad \begin{cases} \mathcal{B}_{\gamma_1}(D) = \{z \in \mathbb{C} : |z^2 - 1| \leq 1\}, \text{ and} \\ \mathcal{B}_{\gamma_2}(D) = \left\{z \in \mathbb{C} : |z-1|^2 \cdot |z+1| \leq 1/2\right\}, \end{cases}$$

where the lemniscate $\mathcal{B}_{\gamma_2}(D)$ consists of two disjoint components. These lemniscates are shown in Fig. 2.7.

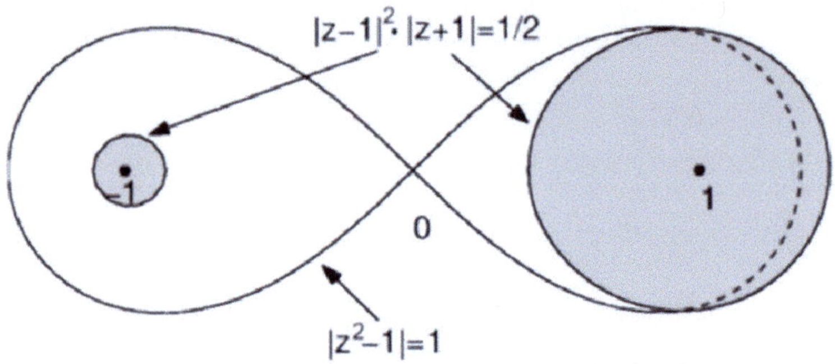

Fig. 2.7. The lemniscates $\mathcal{B}_{\gamma_1}(D)$ and $\mathcal{B}_{\gamma_2}(D)$ (shaded) for the matrix D of (2.63)

It can be seen, from (2.65) and from Fig. 2.7, that $z = 0$ is a boundary point of the compact sets $\mathcal{B}_{\gamma_1}(D)$ and $\mathcal{B}(D) := \mathcal{B}_{\gamma_1}(D) \cup \mathcal{B}_{\gamma_2}(D)$. Suppose that we can find an s with $0 < s < 1$ and real values of $\{\theta_1\}_{i=1}^{4}$ for which an associated matrix E of (2.64) has eigenvalue 0. This implies that $\det E = 0$, which, by direct calculations with (2.64), gives

$$0 = \det E = -1 - (1-s)^{i(\theta_1+\theta_2)} + \frac{1}{2} s e^{i(\theta_1+\theta_3+\theta_4)}, \text{ or}$$

$$1 = \left\{ -(1-s)e^{i(\theta_1+\theta_2)} + \frac{1}{2}se^{i(\theta_1+\theta_3+\theta_4)} \right\}. \tag{2.66}$$

But as $0 < s < 1$, the right side of (2.66) is in modulus at most

$$(1-s) + \frac{1}{2}s = \frac{2-s}{2} < 1,$$

so that $\det E \neq 0$ for any E in $\omega_{\mathcal{B}}(D)$, i.e., $0 \notin \sigma(\omega_{\mathcal{B}}(D))$. A similar argument shows (cf. (2.61)) that $0 \notin \sigma(\hat{\omega}_{\mathcal{B}}(D))$. But as $0 \in \mathcal{B}(D)$, we have

$$\sigma(\omega_{\mathcal{B}}(D)) \subseteq \sigma(\hat{\omega}_{\mathcal{B}}(D)) \subsetneqq \mathcal{B}(D). \tag{2.67}$$

But, in order to achieve equality in the last inequality in (2.67), suppose that we *allow* s to be zero in (2.64), noting from (2.65), that the parameter s plays *no role* in $\mathcal{B}(D) = \mathcal{B}_{\gamma_1}(D) \cup \mathcal{B}_{\gamma_2}(D)$. Then, on setting $s = 0$ in (2.64), the matrix E of (2.64) becomes

$$\hat{E} = \begin{bmatrix} 1 & e^{i\theta_1} & 0 \\ e^{i\theta_2} & -1 & 0 \\ \frac{1}{2}e^{i\theta_4} & 0 & 1 \end{bmatrix}, \tag{2.68}$$

and on choosing $\theta_1 = 0$ and $\theta_2 = \pi$, then $z = 0$ is an eigenvalue of $\hat{E}$, where $\hat{E}$ is the **limit** of matrices E in (2.64) when $s \downarrow 0$. We note that the directed graph of $\mathbb{G}(\hat{E})$ is shown in Fig. 2.8 , so that $\mathcal{C}(\hat{E}) \neq \mathcal{C}(D)$, but $\hat{E}$ remains an element of $\omega(D)$ of (2.15).

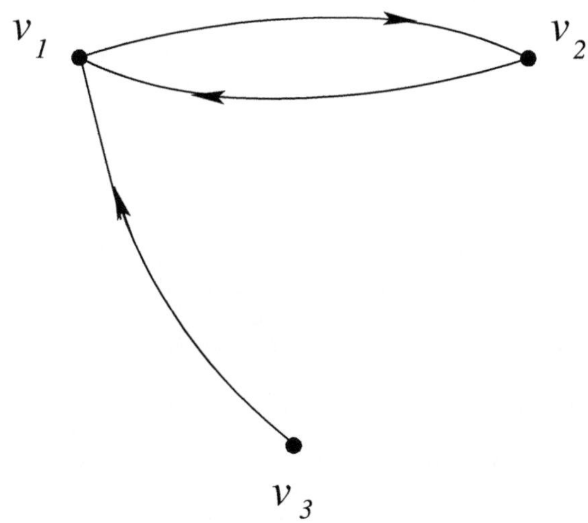

Fig. 2.8. The directed graph, $\mathbb{G}(\hat{E})$, with no loops, for the matrix of (2.68)

This example suggests that we consider the **closures** of the sets $\omega_{\mathcal{B}}(A)$ and $\hat{\omega}_{\mathcal{B}}(A)$ of (2.60) and (2.61), where $A = [a_{i,j}] \in \mathbb{C}^{n\times n}, n \geq 1$, is any matrix:

$$\overline{\omega}_{\mathcal{B}}(A) := \{B = [b_{i,j}] \in \mathbb{C}^{n\times n} : \text{there is a sequence of matrices } \{E_j\}_{j=1}^{\infty} \text{ in } \omega_{\mathcal{B}}(A), \text{for which } B = \lim_{j\to\infty} E_j\}. \tag{2.69}$$

and

$$\overline{\hat{\omega}}_{\mathcal{B}}(A) := \{B = [b_{i,j}] \in \mathbb{C}^{n\times n} : \text{there is a sequence of matrices } \{E_j\}_{j=1}^{\infty} \text{in } \hat{\omega}_{\mathcal{B}}(A), \text{ for which } B = \lim_{j\to\infty} E_j\}. \tag{2.70}$$

This brings us to a recent result (cf. Varga (2001a)) of

Theorem 2.11. *For any $A = [a_{i,j}] \in \mathbb{C}^{n\times n}$, then*

$$\partial\mathcal{B}(A) \subseteq \sigma(\overline{\omega}_{\mathcal{B}}(A)) \subseteq \sigma(\overline{\hat{\omega}}_{\mathcal{B}}(A)) = \mathcal{B}(A), \tag{2.71}$$

i.e., each boundary point of the Brualdi set $\mathcal{B}(A)$ is an eigenvalue of some matrix in $\overline{\omega}_{\mathcal{B}}(A)$, and each point of $\mathcal{B}(A)$ is an eigenvalue of some matrix in $\overline{\hat{\omega}}_{\mathcal{B}}(A)$.

Remark: This establishes the sharpness of the Brualdi set $\mathcal{B}(A)$ for the given matrix A, as the final equality in (2.71) gives that the spectra of matrices in $\hat{\omega}_{\mathcal{B}}(A)$ are **dense** in $\mathcal{B}(A)$.

Proof. Since $\sigma(\omega_{\mathcal{B}}(A)) \subseteq \sigma(\hat{\omega}_{\mathcal{B}}(A))$ from (2.62), it follows that their closures, of (2.69) and (2.70), necessarily satisfy $\sigma(\overline{\omega}_{\mathcal{B}}(A)) \subseteq \sigma(\overline{\hat{\omega}}_{\mathcal{B}}(A))$, giving the middle inclusion of (2.71). It suffices to establish the first inclusion and the final equality in (2.71).

First, suppose that γ in $C(A)$ is a weak cycle. Then, $\gamma = (i)$ for some $i \in N$, where its associated Brualdi lemniscate is, from (2.39), $\mathcal{B}_\gamma(A) = \{a_{i,i}\}$. Moreover, from (2.36*ii*) and (2.35), we see that either $n = 1$, or $n \geq 2$ with A reducible, and that $a_{i,i}$ is an eigenvalue of A. If all cycles γ in $C(A)$ are weak cycles, then the Brualdi set $\mathcal{B}(A)$ for A satisfies $\mathcal{B}(A) = \bigcup_{i=1}^{n} a_{i,i}$. Thus, as this case gives us that $\tilde{r}_i(A) = 0$ for all $i \in N$, then with (2.36*ii*), it follows from (2.34) that each matrix in $\omega_{\mathcal{B}}(PAP^T)$ or $\hat{\omega_{\mathcal{B}}}(PAP^T)$ is upper triangular with diagonal entries $\{a_{i,i}\}_{i=i}^{n}$. Hence,

$$\partial\mathcal{B}(A) = \sigma(\omega_{\mathcal{B}}(A)) = \sigma(\hat{\omega}_{\mathcal{B}}(A)) = \mathcal{B}(A) = \bigcup_{i=1}^{n} a_{i,i},$$

the case of equality in (2.71).

Consider any strong cycle γ in $\mathcal{C}(A)$. From our discussion in Section 2.3, we can express γ as an element of $\mathcal{P}_n$ of (2.52), i.e.,

$$\gamma = (i_1 \ \ i_2 \ \ \cdots \ \ i_p), \text{ where } 2 \leq p \leq n. \tag{2.72}$$

Without loss of generality, we can assume, after a suitable permutation of the rows and columns of A, that

$$\gamma = (1 \ \ 2 \ \ \cdots \ \ p), \tag{2.73}$$

noting that this permutation leaves unchanged the collection of diagonal entries, row sums, and cycles of A. This permutated matrix, also called A, then has the partitioned form

$$A = \left[\begin{array}{ccc|ccc} a_{1,1} & \cdots & a_{1,p} & a_{1,p+1} & \cdots & a_{1,n} \\ \vdots & & & & & \vdots \\ a_{p,1} & & a_{p,p} & a_{p,p+1} & & a_{p,n} \\ \hline a_{p+1,1} & & a_{p+1,p} & a_{p+1,p+1} & & a_{p+1,n} \\ \vdots & & & & & \vdots \\ a_{n,1} & \cdots & a_{n,p} & a_{n,p+1} & \cdots & a_{n,n} \end{array}\right] = \left[\begin{array}{c|c} A_{1,1} & A_{1,2} \\ \hline A_{2,1} & A_{2,2} \end{array}\right], \tag{2.74}$$

where the matrices $A_{1,2}, A_{2,1}$, and $A_{2,2}$ are not present in (2.74) if $p = n$. We also assume, for notational convenience, that $r_i(A) = \tilde{r}_i(A)$ for $1 \leq i \leq p$. This means that any entry of $A_{1,2}$, which arises from an upper triangular block of the normal reduced form of A in (2.35), is simply set to zero.

Our aim below is to construct a special matrix $B(t) = [b_{i,j}(t)] \in \mathbb{C}^{n \times n}$, whose entries depend continuously on the parameter t in $[0, 1]$, such that

$$\begin{cases} b_{i,i}(t) = a_{i,i}, \ r_i(B(t)) = r_i(A), \ \text{for all } i \in N, \text{ and all } t \in [0,1], \\ \text{and} \\ \mathcal{C}(B(t)) = \mathcal{C}(A) \text{ for all } 0 < t \leq 1. \end{cases} \tag{2.75}$$

To this end, write

$$B(t) := \left[\begin{array}{c|c} B_{1,1}(t) & B_{1,2}(t) \\ \hline A_{2,1} & A_{2,2} \end{array}\right], \tag{2.76}$$

i.e., the rows $p+1 \leq \ell \leq n$ of $B(t)$ are exactly those of A, and are independent of t. We note from (2.73) that

$$a_{1,2} \cdot a_{2,3} \cdots a_{p-1,p} \cdot a_{p,1} \neq 0, \tag{2.77}$$

and the entries of the first p rows of $B(t)$ are defined, for all $t \in [0, 1]$, to satisfy

$$
\begin{cases}
b_{i,i}(t) := a_{i,i} \text{ for all } 1 \le i \le p; \\
|b_{i,i+1}(t)| := (1-t)r_i(A) + t|a_{i,i+1}|, \text{ and } |b_{i,j}(t)| := t|a_{i,j}| \\
\qquad\qquad\qquad\qquad\qquad\qquad (j \ne i, i+1), \text{for all } 1 \le i < p; \\
|b_{p,1}(t)| := (1-t)r_p(A) + t|a_{p,1}|, \text{ and } |b_{p,j}(t)| := t|a_{p,j}| \text{ (all } j \ne 1, p).
\end{cases}
\tag{2.78}
$$

By definition, the entries of $B(t)$ are all continuous in the variable t of $[0,1]$, and $B(t)$ and A have the same diagonal entries. Moreover, it can be verified (see Exercise 1 of this section) that $B(t)$ and A have the same row sums for all $0 \le t \le 1$, and, as $a_{i,j} \ne 0$ implies $b_{i,j}(t) \ne 0$ for all $0 < t \le 1$, then $B(t)$ and A have the same cycles in their directed graphs for all $0 < t \le 1$. Also, from (2.60), $B(t) \in \omega_{\mathcal{B}}(A)$ for all $0 < t \le 1$, and from (2.69), $B(0) \in \overline{\omega}_{\mathcal{B}}(A)$. Hence, from (2.62),

$$\sigma(B(t)) \subseteq \mathcal{B}(A) \text{ for all } 0 < t \le 1. \tag{2.79}$$

But as $\mathcal{B}(A)$ is a closed set from (2.38) - (2.40), and as the eigenvalues of $B(t)$ are continuous functions of t, for $0 \le t \le 1$, we further have, for the limiting case $t = 0$, that

$$\sigma(B(0)) \subseteq \mathcal{B}(A),$$

where, from the definitions in (2.78),

$$B(0) = \left[\begin{array}{c|c} B_{1,1}(0) & O \\ \hline A_{2,1} & A_{2,2} \end{array}\right], \tag{2.80}$$

with

$$B_{1,1}(0) = \begin{bmatrix} a_{1,1} & r_1(A)e^{i\theta_1} & & & \\ & a_{2,2} & r_2(A)e^{i\theta_2} & & \\ & & \ddots & \ddots & \\ & & & a_{p-1,p-1} & r_{p-1}(A)e^{i\theta_{p-1}} \\ r_p(A)e^{i\theta_p} & & & & a_{p,p} \end{bmatrix}. \tag{2.81}$$

We note from (2.78) that the nondiagonal entries in the first p rows of $B(t)$ are defined only in terms of their moduli, which allows us to *fix* the arguments of certain nondiagonal nonzero entries of $B_{1,1}(0)$ through the factors $\{e^{i\theta_j}\}_{j=1}^p$ where the $\{\theta_j\}_{j=1}^p$ are contained in $[0, 2\pi]$. (These factors appear in $B_{1,1}(0)$ of (2.81).) The partitioned form of $B(0)$ gives us that

$$\sigma(B(0)) = \sigma(B_{1,1}(0)) \cup \sigma(A_{2,2}), \tag{2.82}$$

and, from the special cyclic-like form of $B_{1,1}(0)$ in (2.81), it is easily seen that each eigenvalue λ of $B_{1,1}(0)$ satisfies

$$\prod_{i=1}^{p} |\lambda - a_{i,i}| = \prod_{i=1}^{p} r_i(A), \tag{2.83}$$

for all real choices of $\{\theta_j\}_{j=1}^{p}$ in $[0, 2\pi]$ in (2.81). But (2.83), when coupled with the definition of $\mathcal{B}_\gamma(A)$ in (2.38), immediately gives us that $\lambda \in \partial\mathcal{B}_\gamma(A)$, and, as all different choices of the real numbers $\{\theta_j\}_{j=1}^{p}$, in $B_{1,1}(0)$ of (2.81), give eigenvalues of $B_{1,1}(0)$ which cover the *entire boundary* of $\mathcal{B}_\gamma(A)$, we have

$$\bigcup_{\theta_1,\cdots,\theta_p real} \sigma(B_{1,1}(0)) = \partial\mathcal{B}_\gamma(A). \tag{2.84}$$

This can be used as follows. Let z be any boundary point of $\mathcal{B}(A)$ of (2.40). As $\mathcal{B}(A)$ is the union of a finite number of closed sets $\mathcal{B}_\gamma(A)$, this implies that there is a cycle γ of $\mathcal{C}(A)$ with $z \in \partial\mathcal{B}_\gamma(A)$, where $\mathcal{B}_\gamma(A)$ is defined in (2.38). As the result of (2.84) is valid for any γ of $\mathcal{C}(A)$, then each boundary point z of $\mathcal{B}(A)$ is an eigenvalue of some matrix in $\overline{\omega}_{\mathcal{B}}(A)$ of (2.69), i.e.,

$$\partial\mathcal{B}(A) \subseteq \sigma(\overline{\omega}_{\mathcal{B}}(A)), \tag{2.85}$$

which is the desired first inclusion of (2.71).

To investigate how the eigenvalues of $\overline{\hat{\omega}}_{\mathcal{B}}(A)$ of (2.70) fill out $\mathcal{B}(A)$, we modify the definition of the matrix $B(t)$ of (2.75) and (2.78). Let $\{\tau_i\}_{i=1}^{p}$ be any positive numbers such that

$$0 < \tau_i \leq r_i(A) \quad (1 \leq i \leq p), \tag{2.86}$$

and let $\tilde{B}(t) = [\tilde{b}_{i,j}(t)] \in \mathbb{C}^{n\times n}$ have the same partitioned form as $B(t)$ of (2.76), but with (2.78) replaced by

$$\begin{cases} \tilde{b}_{i,i}(t) := a_{i,i} \text{ for all } 1 \leq i \leq p; \\ |\tilde{b}_{i,j}(t)| := \dfrac{\tau_i}{r_i(A)}|b_{i,j}(t)| \quad (j \neq i), \text{ for } 1 \leq i \leq p, \text{ and } t \in [0,1]. \end{cases} \tag{2.87}$$

Then, $\tilde{B}(t)$ and A have the same diagonal entries, the row sums of $\tilde{B}(t)$ now satisfy $r_j(\tilde{B}(t)) = \tau_j$ for all $1 \leq j \leq p$, all $0 \leq t \leq 1$, and $\tilde{B}(t)$ and A have the same cycles for all $0 < t \leq 1$. From (2.61), $\tilde{B}(t) \in \hat{\omega}_{\mathcal{B}}(A)$ for all $0 < t \leq 1$, and from (2.70), $\tilde{B}(0) \in \overline{\hat{\omega}}_{\mathcal{B}}(A)$. In analogy with (2.80), we have

$$\tilde{B}(0) = \left[\begin{array}{c|c} \tilde{B}_{1,1}(0) & O \\ \hline A_{2,1} & A_{2,2} \end{array}\right],$$

with

(2.88) $$\tilde{B}_{1,1}(0) = \begin{bmatrix} a_{1,1} & \tau_1 e^{i\theta_1} & & & \\ & a_{2,2} & \tau_2 e^{i\theta_2} & & \\ & & \ddots & \ddots & \\ & & & a_{p-1,p-1} & \tau_{p-1} e^{i\theta_{p-1}} \\ \tau_p e^{i\theta_p} & & & & a_{p,p} \end{bmatrix},$$

where

(2.89) $$\sigma(\tilde{B}(0)) = \sigma(\tilde{B}_{1,1}(0)) \cup \sigma(A_{2,2}).$$

It similarly follows that any eigenvalue λ of $\tilde{B}_{1,1}(0)$ in (2.88) now satisfies

(2.90) $$\prod_{j=1}^{p} |\lambda - a_{i,i}| = \prod_{i=1}^{p} \tau_i,$$

for any choice of the real numbers $\{\theta_j\}_{j=1}^p$ in $\tilde{B}_{1,1}(0)$ of (2.88). Writing

$$\tilde{B}_{1,1}(0) = \tilde{B}_{1,1}(0; \tau_1, \cdots, \tau_p; \theta_1, \cdots, \theta_p)$$

to show this matrix's dependence on these parameters τ_i and θ_i, we use the fact that $\{\tau_i\}_{i=1}^p$ are any numbers satisfying (2.86) and that $\{\theta_i\}_{i=1}^p$ are any real numbers in $[0, 2\pi]$. Hence, it follows, from the definition of $\mathcal{B}_\gamma(A)$ in (2.50) and closure considerations, that all the eigenvalues of $\tilde{B}_{1,1}(0; \tau_1, \cdots, \tau_p; \theta_1, \cdots, \theta_p)$ **fill out** $\mathcal{B}_\gamma(A)$, i.e.,

(2.91) $$\overline{\left\{ \bigcup_{\substack{\{0<\tau_i \le r_i(A)\}_{i=1}^p \\ \{\theta_i\}_{i=1}^p \in [2\pi]}} \sigma(\tilde{B}_{1,1}(0; \tau_1, \cdots, \tau_p; \theta_1, \cdots, \theta_p)) \right\}} = \mathcal{B}_\gamma(A).$$

As this holds for any $\gamma \in \mathcal{C}(A)$, where $\tilde{B}(0) \in \overline{\tilde{\omega}}_{\mathcal{B}}(A)$, then

(2.92) $$\sigma(\overline{\tilde{\omega}}_{\mathcal{B}}(A)) = \mathcal{B}(A),$$

the desired final result of (2.71). ■

Exercises

1. Verify that $\mathcal{B}(t)$, of (2.78), and A of (2.76) have the same row sums for all $0 \le t \le 1$, and the same cycles in their directed graphs for all $0 < t \le 1$.

2. Show, for all real choices of $\{\theta_j\}_{j=1}^p$ in $[0, 2\pi]$ in (2.81), that the eigenvalues of $B_{1,1}(0)$ in (2.81) fill out $\partial\mathcal{B}_\gamma(A)$ in (2.84).

3. Show, for all real choices of $\{\theta_j\}_{j=1}^{\rho}$ in $[0, 2\pi]$ and all choices of $\{\tau_j\}_{j=1}^{\rho}$ satisfying (2.86), that the eigenvalues of $\tilde{B}_{1,1}(0)$ of (2.88) are dense in $\mathcal{B}_\gamma(A)$ of (2.38).

4. Consider the familiar $n \times n$ irreducible tridiagonal matrix

$$A = \begin{bmatrix} 2 & -1 & & & \\ -1 & 2 & -1 & & \\ & \ddots & \ddots & \ddots & \\ & & -1 & 2 & -1 \\ & & & -1 & 2 \end{bmatrix},$$

whose Brualdi set, from Exercise 1 of Section 2.3, is $\mathcal{B}(A) = \{z \in \mathbb{C} : |z-2| \leq 2\}$ for any $n \geq 4$. While $z = 0$ is a boundary point of $\mathcal{B}(A)$, show, using Theorem 2.8, that $z = 0$ is *not* an eigenvalue of A. However, show that $z = 0$ *is* an eigenvalue of a *specific* matrix in $\overline{\omega}_{\mathcal{B}}(A)$ of (2.69), for any $n \geq 4$.

5. Assume that $A = [a_{i,j}] \in \mathbb{C}^{n \times n}$, $n \geq 2$, is irreducible and that its cycle set $C(A)$ consists of only one cycle, $\gamma = (1 \;\; 2 \;\; \cdots \;\; n)$. Show in this case (cf. (2.71)) that

$$\partial\mathcal{B}(A) = \sigma(\omega_{\mathcal{B}}(A)) \subsetneqq \mathcal{B}(A).$$

2.5 An Example

To illustrate the result of Theorem 2.11, consider the matrix

$$E = \begin{bmatrix} 1 & e^{i\theta_1} & 0 & 0 \\ \frac{e^{i\theta_2}}{2} & i & \frac{e^{i\theta_3}}{2} & 0 \\ 0 & 0 & -1 & e^{i\theta_4} \\ e^{i\theta_5} & 0 & 0 & -i \end{bmatrix}, \qquad (2.93)$$

where $\{\theta_i\}_{i=1}^{5}$ are any real numbers in $[0, 2\pi]$. Then, E is irreducible, with cycle set $\mathcal{C}(E) = (1 \;\; 2) \cup (1 \;\; 2 \;\; 3 \;\; 4)$, and with row sums $r_i(E) = 1$ for all $1 \leq i \leq 4$. In this case, we have from (2.40) that the Brualdi set $\mathcal{B}(E)$ is the union of the two closed lemniscates

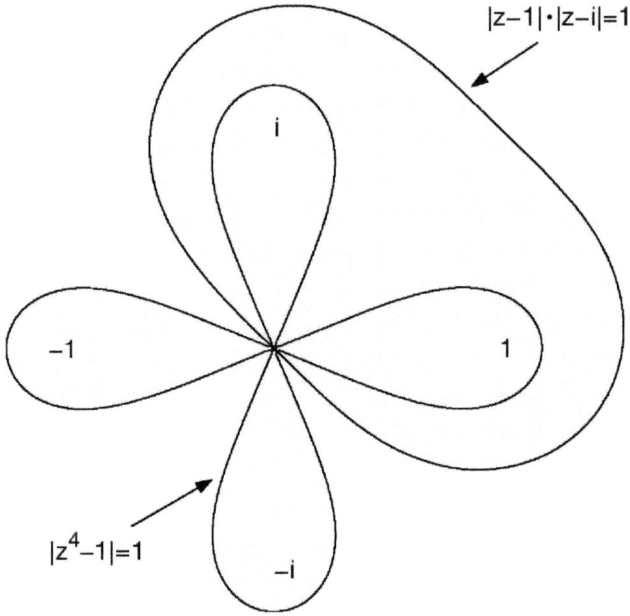

Fig. 2.9. The lemniscates, $\mathcal{B}_{\gamma_1}(E)$ and $\mathcal{B}_{\gamma_2}(E)$, for the matrix E of (2.93)

$$(2.94)\qquad \begin{cases} \mathcal{B}_{\gamma_1}(E) := \{z \in \mathbb{C} : |z-1| \cdot |z-i| \leq 1\} = K_{1,2}(E), \\ \quad \text{and} \\ \mathcal{B}_{\gamma_2}(E) := \{z \in \mathbb{C} : |z^4 - 1| \leq 1\}. \end{cases}$$

These sets are shown in Fig. 2.9, where $\mathcal{B}_{\gamma_2}(E)$ has the shape of a four-leaf clover.

To show how the eigenvalues of E fill out $\mathcal{B}(E)$, we take random numbers s from the open interval $(0,1)$ and random values of $\{\theta_i\}_{i=1}^5$ from $[0, 2\pi]$, and, using Matlab 6, the eigenvalues of these matrices are plotted in Fig. 2.10. Fig. 2.10 shows indeed that these eigenvalues of A tend to fill-out $\mathcal{B}(E)$.

The following **near paradox** arose from Theorem 2.11. As an example, the matrix E of (2.93) is irreducible, and it is known from Theorem 2.8 that a **necessary** condition for a boundary point z of $\mathcal{B}(E)$ to be an eigenvalue of E is that z is a boundary point of *each* of the lemniscates $\mathcal{B}_{\gamma_1}(E)$ and $\mathcal{B}_{\gamma_2}(E)$ of (2.94). (This is a generalization of Taussky's Theorem 1.12 on Geršgorin disks, to lemniscates.) But from Fig. 2.9, it is apparent that $z = 0$ is the *only* point for which $\partial\mathcal{B}_{\gamma_1}(E)$ and $\partial\mathcal{B}_{\gamma_2}(E)$ have a common point. Yet, (2.71) of Theorem 2.11 gives the nearly **contradictory result** that **each** point of $\partial\mathcal{B}(E)$ is an eigenvalue of **some** matrix in $\overline{\omega}_{\mathcal{B}}(E)$. The difference, of course, lies in the fact that the result of Theorem 2.8 applies to the *fixed* matrix

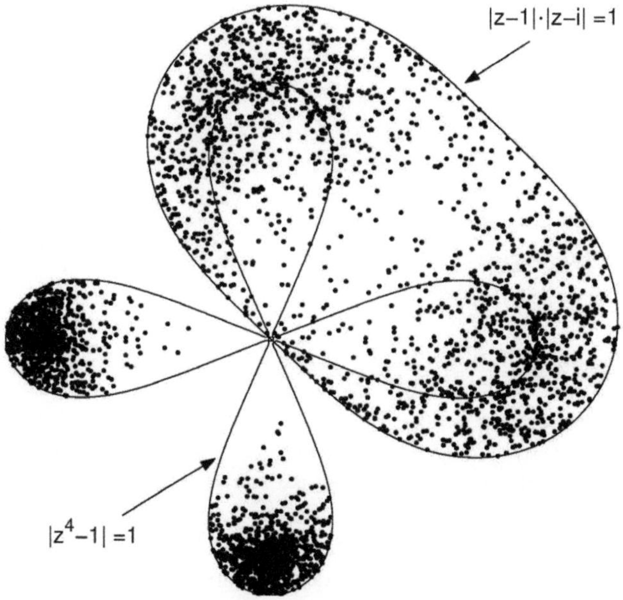

Fig. 2.10. Random Eigenvalues of $\mathcal{B}(E)$

E, while the common data of (2.59) applies to *all* matrices which lie in the closure of $\omega_{\mathcal{B}}(E)$.

Bibliography and Discussion

2.1. Ostrowski (1937b) first obtained the nonsingularity result of Theorem 2.1. Brauer (1947) later independently obtained the equivalent eigenvalue inclusion result of Theorem 2.2 by means of a direct proof. (Brauer, in the same paper, obtained Ostrowski's Theorem 2.1 by constructively deriving, from the hypothesis of (2.1), that $\min\{|\lambda| : \lambda \in \sigma(A)\} > 0$.) Though well-known today, the equivalence of these two results was not widely recognized until several years after Brauer's paper in 1947.

Brauer's name had become synonymous with the ovals of Cassini, but, for unknown reasons, they are rarely mentioned, even though they are **superior** (cf. Theorem 2.3) to the Geršgorin disks. (An exception is the book by Korganoff (1961).) Even today, they have not been widely utilized, most likely because i) there are $n(n-1)/2$ such ovals, as compared with n Geršgorin disks, and ii) these ovals are, by their definition, more complicated than disks. The author, however, hopes that this book will generate more interest and applications for this area.

For a further corroboration of Theorem 2.3, as in Fig. 2.2, where the Geršgorin set $\Gamma(A)$ with the Brauer set $\mathcal{K}(A)$ are compared, we recommend the following **interactive supplement**. For an arbitrary 3×3 complex matrix, go to the website:

http://etna.mcs.kent.edu

for the electronic journal ETNA (Electronic Transactions on Numerical Analysis), click on volume 8 (1999), and go to the paper by Varga and Krautstengl (1999), "On Geršgorin-type problems and ovals of Cassini", 15-20. Then click on the **Interactive Supplement**, which was written in Java by Dr. Bryan Lewis. There, for a 3×3 complex matrix of your choice, or a 3×3 complex matrix which is generated randomly, one gets the associated spectrum of this matrix, Geršgorin disks (in color), and the Cassini ovals (also in color)!

The fact the Brauer's ovals of Cassini do a **perfect job** in estimating the spectra of all matrices $\omega(A)$ of (2.15) for any matrix $A = [a_{i,j}] \in \mathbb{C}^{n \times n}, n \geq 3$, was shown recently by Varga and Krautstengl (1999), though the exact result of (2.19) of Theorem 2.4 appeared earlier in Theorem 6.15 of Engel (1973), with essentially the same proof. This paper of Engel (1973) is a very important contribution, as it derives results for column and row linear matrix functions, which includes determinants and permanents.

2.2 Part of the disappointment in using ovals of Cassini came possibly from the natural urge to march forward, beyond ovals of Cassini, to higher order lemniscates with the hope of superior eigenvalue inclusion results. This, unfortunately, failed in simple cases, as in the case of the 4×4 matrix in (2.33). Subsequently, Brualdi, in 1982, brilliantly showed how to "solve" this problem by using cycles from the directed graph of a matrix. In essence, Brualdi (1982) derived his related result of Theorem 2.5 for **weakly irreducible** matrices, which are matrices having only strong cycles (see Exercise 1 of Section 2.2). The Brualdi sets of (2.40), which are a new but modest generalization of Brualdi's work, include the notion of weak cycles, which permits **all** matrices to be analyzed from a knowledge of their directed graphs. (Our proof, given in Theorem 2.5, is perhaps simpler than Brualdi's, as it avoids Brualdi's use of partial orderings.

We also mention that Karow (2003), in his Ph.D. thesis, also effectively developed the analog of **weak cycles** in his definition of $\sigma_0(A)$, used in his Theorem 4.6.4,which is also an extension of Brualdi's work. This extension is done without using the normal reduced form of (2.35), which, however, is needed for the sharpness in Theorem 2.11 of this chapter.

As in Chapter 1, we remark that (2.45) of Theorem 2.8 gives, for an irreducible matrix A in $\mathbb{C}^{n \times n}$, a *necessary condition* for an eigenvalue λ of A to lie on the boundary of the Brualdi set $\mathcal{B}(A)$. In this regard, important related necessary *and* sufficient for this to happen, for a given **fixed** matrix, have been given in Li and Tsatsomeros (1997) and Kolotolina (2001), (2003a) and (2003b). Our interest here, as is apparent, has been in the related, but diametrically opposed, problem of seeing if *each* point of an eigenvalue inclusion set is an eigenvalue of *some* matrix associated with that eigenvalue inclusion set.

2.3 The comparison of the Brualdi sets with the Brauer sets was carried out recently in Theorem 2.6 in Varga (2001b).

The notion of a **reduced cycle set**, in this section, is new. The result of Theorem 2.10, which gives sufficient conditions for replacing a cycle of an irreducible matrix by lower-order cycles, is an unpublished consequence of an exciting exchange of e-mails with Richard Brualdi, for which the author is most thankful.

It is interesting to mention that Brauer (1952) gave in his Theorem 22 an **erroneous** result, patterned after Taussky's Theorem 1.12, which stated that if $A = [a_{i,j}] \in \mathbb{C}^{n \times n}$, $n \geq 2$, is irreducible, then λ, a boundary point of the union of its associated Cassini ovals of (2.6), can be an eigenvalue of A only if λ is a boundary point of **each** of the $n(n-1)/2$ ovals of Cassini $K_{i,j}(A)$ of (2.5). This error was undetected, in the current literature of widely available journals, until Zhang and Gu (1994) gave a simple 3×3 matrix counterexample, which is given in Exercise 6 of Section 2.2. But, it was kindly pointed out to me recently by Ludwig Elsner that such a counterexample was *earlier* published by his student Rein (1967) in the more obscure Kleine Mitteilungen of the journal Zeit. Angew. Math. Mech. What is also interesting is that the counterexample of Zhang and Gu is a 3×3 matrix whose first and third rows are identical to the earlier 3×3 matrix of Rein, while the second row of the Zhang/Gu matrix is exactly a multiple of 2 of the corresponding Rein matrix!

It should also be quietly mentioned that Feingold and Varga (1962) used Brauer's incorrect result to obtain in a "generalization" to partitioned matrices, which is also incorrect, but now easily corrected in Chapter 6.

2.4 The sharpness of the Brualdi lemniscate sets, as given in Theorem 2.11, comes from Varga (2001a).

3. More Eigenvalue Inclusion Results

3.1 The Parodi-Schneider Eigenvalue Inclusion Sets

In this chapter, the emphasis is again on eigenvalue inclusion sets, for general matrices in $\mathbb{C}^{n\times n}$, which differ from the previously studied eigenvalue inclusion sets of Chapters 1 and 2. We also consider the computational effort involved in determining these new eigenvalue inclusion sets, as compared with those of Chapters 1 and 2.

We begin with the original results of Parodi (1952) and Schneider (1954). Given any $A = [a_{i,j}] \in \mathbb{C}^{n\times n}$ and given an $\mathbf{x} = [x_1, x_2, \cdots, x_n]^T > \mathbf{0}$ in $\mathbb{R}^n$, let $X := \text{diag}\,[x_1, x_2, \cdots, x_n]$, so that X is a nonsingular matrix in $\mathbb{R}^{n\times n}$. If ϕ is a fixed permutation on the elements of $N := \{1, 2, \cdots, n\}$, let

$$P_\phi := [\delta_{i,\phi(j)}]$$

denote its associated permutation matrix in $\mathbb{R}^{n\times n}$. Then for any $z \in \mathbb{C}$, consider the matrix

$$B(z) = [b_{i,j}(z)] := (X^{-1}AX - zI_n) \cdot P_\phi \in \mathbb{C}^{n\times n}, \tag{3.1}$$

whose entries are given by

$$b_{i,j}(z) := \left(a_{i,\phi(j)}x_{\phi(j)}/x_i\right) - z\delta_{i,\phi(j)} \qquad (i, j \in N). \tag{3.2}$$

Following Parodi (1952) and Schneider (1954), if $z \in \sigma(A)$, then the matrix $B(z)$ of (3.1) is evidently singular, and thus cannot be strictly diagonally dominant. Hence (cf. Theorem 1.4), there exists an $i \in N$ for which

$$|b_{i,i}(z)| \leq \sum_{j\in N\backslash\{i\}} |b_{i,j}(z)| =: r_i(B(z)). \tag{3.3}$$

The inequality in (3.3) is then used to define the associated Geršgorin-type set

$$\Gamma^{r^{\mathbf{x}}}_{i,\phi}(A) := \{z \in \mathbb{C} : |b_{i,i}(z)| \leq r_i(B(z))\} \quad (i \in N), \tag{3.4}$$

which can be equivalently expressed, from (3.2), in terms of the familiar row sums $r_i^{\mathbf{x}}(A)$, of (1.13), as

$$(3.5)\begin{cases} \Gamma^{r^{\mathbf{x}}}_{i,\phi}(A) := \{z \in \mathbb{C} : |z - a_{i,i}| \leq r_i^{\mathbf{x}}(A)\} & \text{if } \phi(i) = i, \\ \text{or} & \\ \Gamma^{r^{\mathbf{x}}}_{i,\phi}(A) := \{z \in \mathbb{C} : |z - a_{i,i}| \geq -r_i^{\mathbf{x}}(A) + 2|a_{i,\phi(i)}|x_{\phi(i)}/x_i\} & \text{if } \phi(i) \neq i, \end{cases}$$

which shows the dependence of this set on ϕ and $\mathbf{x}$.

Note from (3.5) that $\Gamma^{r^{\mathbf{x}}}_{i,\phi}(A)$ is just **a closed disk** when $\phi(i) = i$. But when $\phi(i) \neq i$, then on setting

$$(3.6) \qquad \mu^{\mathbf{x}}_{i,\phi} := -r_i^{\mathbf{x}}(A) + 2|a_{i,\phi(i)}|x_{\phi(i)}/x_i,$$

we see from (3.5) that

$$(3.7) \qquad \Gamma^{r^{\mathbf{x}}}_{i,\phi}(A) = \mathbb{C}_\infty := \mathbb{C} \cup \{\infty\} \text{ if } \mu^{\mathbf{x}}_{i,\phi} \leq 0,$$

i.e., $\Gamma^{r^{\mathbf{x}}}_{i,\phi}(A)$ is the **extended complex plane**. Similarly,

$$(3.8) \qquad \Gamma^{r^{\mathbf{x}}}_{i,\phi}(A) = \{z \in \mathbb{C} : |z - a_{i,i}| \geq \mu^{\mathbf{x}}_{i,\phi}\} \text{ if } \mu^{\mathbf{x}}_{i,\phi} > 0,$$

i.e., $\Gamma^{r^{\mathbf{x}}}_{i,\phi}(A)$ is **the closed exterior of a disk**. This latter type of set will have interesting consequences for us! In any event, as each eigenvalue λ of A must, from our construction, lie in some $\Gamma^{r^{\mathbf{x}}}_{i,\phi}(A)$, then with the definition of

$$(3.9) \qquad \Gamma^{r^{\mathbf{x}}}_{\phi}(A) := \bigcup_{i \in N} \Gamma^{r^{\mathbf{x}}}_{i,\phi}(A),$$

we have that $\lambda \in \Gamma^{r^{\mathbf{x}}}_{\phi}(A)$. This gives us

Theorem 3.1. *For any $A = [a_{i,j}] \in \mathbb{C}^{n \times n}$, for any permutation ϕ on N, and for any $\mathbf{x} > \mathbf{0}$ in $\mathbb{R}^n$, then* (cf. (3.9))

$$(3.10) \qquad \sigma(A) \subseteq \Gamma^{r^{\mathbf{x}}}_{\phi}(A).$$

We call $\Gamma^{r^{\mathbf{x}}}_{\phi}(A)$ the (weighted) **permuted Geršgorin set** for A, with respect to ϕ, which, from (3.10), is another eigenvalue inclusion set for any matrix A in $\mathbb{C}^{n \times n}$. Of course, if some $\Gamma^{r^{\mathbf{x}}}_{i,\phi}(A) = \mathbb{C}_\infty$, then from (3.9), we necessarily have

$$\Gamma^{r^{\mathbf{x}}}_{\phi}(A) = \mathbb{C}_\infty,$$

meaning that the eigenvalue inclusion of (3.10), while valid, adds *nothing of interest* in estimating the actual eigenvalues of A!

For a fixed ϕ and a fixed $\mathbf{x} > \mathbf{0}$ in $\mathbb{R}^n$, the (weighted) permuted Geršgorin set $\Gamma^{r^{\mathbf{x}}}_{\phi}(A)$ is as easily determined as is the weighted Geršgorin set $\Gamma^{r^{\mathbf{x}}}(A)$ of (1.14). But if the permutation ϕ on N is such that

$$(3.11) \qquad a_{i,\phi(i)} = 0 \quad \text{with } \phi(i) \neq i,$$

it follows from (3.5) that $\Gamma^{r^{\mathbf{x}}}_{i,\phi}(A) = \mathbb{C}_\infty$, implying that $\Gamma^{r^{\mathbf{x}}}_{\phi}(A) = \mathbb{C}_\infty$ for **all** $\mathbf{x} > \mathbf{0}$ in $\mathbb{R}^n$. Such a permutation ϕ is called a **trivial permutation** for A, and such permutations are to be avoided!

We give below a matrix example of the utility of the permuted Geršgorin sets. Consider the matrix

$$(3.12) \qquad C_3 = \begin{bmatrix} 2 & 0 & 1 \\ 0 & 1 & 1 \\ 1 & 1 & 2 \end{bmatrix} = [c_{i,j}] \in \mathbb{R}^{3\times 3}.$$

As $c_{1,2} = c_{2,1} = 0$ in (3.12), we discard the following trivial permutations for C_3 (where we use the standard cyclic permutation notation from (2.34)):

$$\phi_1 = (1\ \ 2)(3),\ \phi_2 = (1\ \ 2\ \ 3),\ \text{and}\ \phi_3 = (2\ \ 1\ \ 3),$$

leaving the permutations

$$(3.13) \qquad \phi_4 = (1)(2)(3),\ \phi_5 = (1\ \ 3)(2),\ \text{and}\ \phi_6 = (1)(2\ \ 3).$$

For the following positive vectors from $\mathbb{R}^3$, namely,

$$(3.14) \qquad \mathbf{x} = [1, 1, 1]^T,\ \mathbf{y} = [1.25, 1, 0.5]^T,\ \text{and}\ \mathbf{z} = [0.5, 1, 0.25]^T,$$

we have determined the eigenvalue inclusion sets of $\Gamma^{r^{\mathbf{x}}}_{\phi_4}(C_3)$, $\Gamma^{r^{\mathbf{y}}}_{\phi_5}(C_3)$, and $\Gamma^{r^{\mathbf{z}}}_{\phi_6}(C_3)$, where the set $\Gamma^{r^{\mathbf{x}}}_{\phi_4}(C_3)$ is the *bounded* union of three closed disks while $\Gamma^{r^{\mathbf{y}}}_{\phi_5}(C_3)$ and $\Gamma^{r^{\mathbf{z}}}_{\phi_6}(C_3)$ are each the *unbounded* unions of the one closed disk and closed exteriors of two disks. The intersection of these three sets, which gives an eigenvalue inclusion set for C_3 of (3.12), is shown in Fig. 3.1, where the eigenvalue of C_3 are shown as "×'s". (The dotted circles in this figure represent internal boundaries of this intersection.)

The result of Fig. 3.1, for the matrix C_3 of (3.12), suggests that taking the intersection of some weighted Geršgorin sets, in conjunction with some (nontrivial) permuted Geršgorin sets, can give *useful information* about the spectrum of a given matrix!

Though the extensions of weighted permuted Geršgorin sets to Cassini ovals and higher-order lemniscates have not, to the author's knowledge, been considered in the literature, this can be easily done. In a completely analogous way, applying Brauer's Theorem 2.2 to the matrix $B(z)$ of (3.1) gives

Theorem 3.2. *For any* $A = [a_{i,j}] \in \mathbb{C}^{n\times n}$, $n \geq 2$, *for any permutation* ϕ *on* N, *for any* $\mathbf{x} > \mathbf{0}$ *in* $\mathbb{R}^n$, *and for any* $\lambda \in \sigma(A)$, *there is a pair of distinct integers* i *and* j *in* N *such that* (cf. (3.1) and 3.3))

$$(3.15)\ \lambda \in K^{r^{\mathbf{x}}}_{i,j,\phi}(A) := \{z \in \mathbb{C} : |b_{i,i}(z)| \cdot |b_{j,j}(z)| \leq r^{\mathbf{x}}_i(B(z)) \cdot r^{\mathbf{x}}_j(B(z))\}.$$

Thus, if

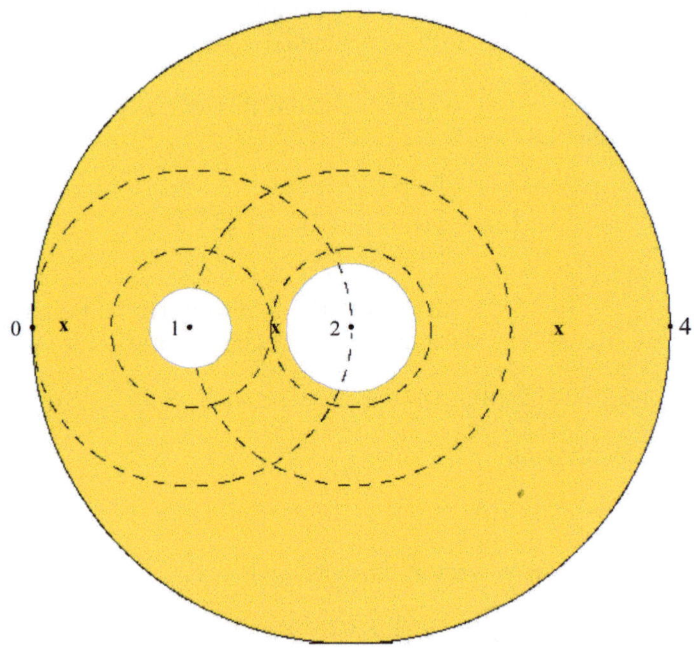

Fig. 3.1. The intersection of $\Gamma_{\phi_4}^{r^x}(C_3)$, $\Gamma_{\phi_5}^{r^y}(C_3)$, and $\Gamma_{\phi_6}^{r^z}(C_3)$ for C_3 of (3.12)

$$\mathcal{K}_{\phi}^{r^{\mathbf{x}}}(A) := \bigcup_{\substack{i,j \in N \\ i \neq j}} K_{i,j,\phi}^{r^{\mathbf{x}}}(A), \tag{3.16}$$

then

$$\sigma(A) \subseteq \mathcal{K}_{\phi}^{r^{\mathbf{x}}}(A). \tag{3.17}$$

We call $\mathcal{K}_{\phi}^{r^{\mathbf{x}}}(A)$ the (weighted) **permuted Brauer set**, with respect to ϕ, for A. We remark that the sets $K_{i,j,\phi}^{r^{\mathbf{x}}}(A)$ of (3.15) *can be* more complicated than Cassini ovals, and trivial permutations, giving $K_{i,j,\phi}^{r^{\mathbf{x}}}(A) = \mathbb{C}_{\infty}$, also arise here. As an example of this extension, consider again the matrix C_3 of (3.12) and the (nontrivial) permutation $\phi_5 = (1\ 3)(2)$ of (3.13). In this case with $\mathbf{y} = [2, 1, 1]^T$, it can be verified that

$$\mathcal{K}_{2,3,\phi_5}^{r^{\mathbf{y}}}(C_3) = \{z \in \mathbb{C} : 2 \cdot |1 - z| \leq 1 + |2 - z|\}. \tag{3.18}$$

This set is a bounded set which, from the definition in (2.5), is **not** a Cassini oval.

Continuing, it is natural to ask if the inclusion of (2.9) of Theorem 2.3, showing that the Brauer set $\mathcal{K}(A)$ is always a subset of the Geršgorin set $\Gamma(A)$ for any $A \in \mathbb{C}^{n \times n}$ with $n \geq 2$, similarly holds for the related permuted Brauer set $\mathcal{K}_{\phi}^{r^{\mathbf{x}}}(A)$ and the permuted Geršgorin set $\Gamma_{\phi}^{r^{\mathbf{x}}}$ of (3.16) and (3.9). It turns out that this *is* true, and we give this as our next result, Theorem 3.3, which is new. (As its proof follows along the lines of the proof of Theorem 2.3, we leave this as Exercise 3 in this section.)

Theorem 3.3. *For any* $A = [a_{i,j}] \in \mathbb{C}^{n \times n}$, $n \geq 2$, *for any permutation* ϕ *on* N, *and for any* $\mathbf{x} > \mathbf{0}$ *in* $\mathbb{R}^n$, *then* ((cf. (3.9) and 3.16))

$$\mathcal{K}_\phi^{r^{\mathbf{x}}}(A) \subseteq \Gamma_\phi^{r^{\mathbf{x}}}(A). \tag{3.19}$$

The previous considerations leave open similar extensions of the permuted Geršgorin sets to Brualdi sets, under a permutation ϕ on N. Again, while this has not been carried out in the literature, this can be quickly achieved.

For any matrix $A = [a_{i,j}] \in \mathbb{C}^{n \times n}$, for any permutation ϕ on N, and for any $\mathbf{x} = [x_1, x_2, \cdots, x_n]^T > \mathbf{0}$ in $\mathbb{R}^n$, consider the matrix $B(z)$, defined in (3.1), which depends on A, ϕ, z, and $\mathbf{x}$. Let $C_\phi(B(z))$ denote its associated circuit set of strong and weak cycles, as described in Section 2.2 of Chapter 2. (We note that this set may now be dependent on the variable z. We also remark that the circuit set $C_\phi(B(z))$ *can* differ from the circuit set $C(A)$, derived from the directed graph of A.) If γ is a strong cycle from $C_\phi(B(z))$, we define its associated Brualdi lemniscate $\mathcal{B}_{\gamma,\phi}^{r^{\mathbf{x}}}(A)$ as

$$\mathcal{B}_{\gamma,\phi}^{r^{\mathbf{x}}}(A) := \{z \in \mathbb{C} : \prod_{i \in \gamma} |b_{i,i}(z)| \leq \prod_{i \in \gamma} \tilde{r}_i^{\mathbf{x}}(B(z))\}, \tag{3.20}$$

and if $\gamma = \{i\}$ is a weak cycle from $C_\phi(B(z))$, we define its associated Brualdi lemniscate $\mathcal{B}_{\gamma,\phi}^{r^{\mathbf{x}}}(A)$ as

$$\mathcal{B}_{\gamma,\phi}^{r^{\mathbf{x}}}(A) := \{z \in \mathbb{C} : |b_{i,i}(z)| = \tilde{r}_i^{\mathbf{x}}(B(z)) = 0\}, \tag{3.21}$$

where this set can be empty. The set $\mathcal{B}_\phi^{r^{\mathbf{x}}}(A)$, called the (weighted) **permuted Brualdi set**, with respect to ϕ, for A, is then defined by

$$\mathcal{B}_\phi^{r^{\mathbf{x}}}(A) := \bigcup_{\gamma \in C_\phi(B(z))} \mathcal{B}_{\gamma,\phi}^{r^{\mathbf{x}}}(A). \tag{3.22}$$

With these definitions, we have the following new results, which are the analogs of Theorems 2.5 and 2.9. (Their proofs are left as Exercises 5 and 6 in this section.)

Theorem 3.4. *For any* $A = [a_{i,j}] \in \mathbb{C}^{n \times n}$, *for any permutation* ϕ *on* N, *and for any* $\mathbf{x} > \mathbf{0}$ *in* $\mathbb{R}^n$, *then* (cf.(3.22))

$$\sigma(A) \subseteq \mathcal{B}_\phi^{r^{\mathbf{x}}}(A), \tag{3.23}$$

and

Theorem 3.5. *For any $A = [a_{i,j}] \in \mathbb{C}^{n \times n}$, $n \geq 2$, for any permutations ϕ on N, and for any $\mathbf{x} > \mathbf{0}$ in $\mathbb{R}^n$, we have* (cf.(3.19))

$$\mathcal{B}_{\phi}^{r^{\mathbf{x}}}(A) \subseteq \mathcal{K}_{\phi}^{r^{\mathbf{x}}}(A) \subseteq \Gamma_{\phi}^{r^{\mathbf{x}}}(A). \tag{3.24}$$

To illustrate the inclusions of (3.24), consider the intersection of the three Brualdi sets $\mathcal{B}_{\phi_4}^{r^{\mathbf{x}}}(C_3)$. $\mathcal{B}_{\phi_5}^{r^{\mathbf{y}}}(C_3)$, and $\mathcal{B}_{\phi_6}^{r^{\mathbf{z}}}(C_3)$, determined from the permutations of (3.13) and vectors of (3.14). This intersection, which also contains the eigenvalues of C_3, is the dark set of points in Fig 3.2, where the eigenvalues of C_3 are shown as "$\times$'s". Comparing the closed bounded set of Fig. 3.2 with the corresponding intersection of Fig. 3.1, we graphically confirm that the shaded set of Fig. 3.2 is indeed a subset of the shaded set in Fig. 3.1.

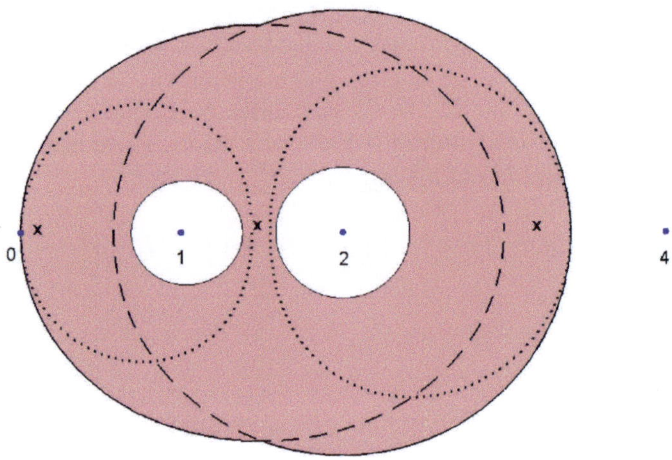

Fig. 3.2. The intersection of $\mathcal{B}_{\phi_4}^{r^{\mathbf{x}}}(C_3)$, $\mathcal{B}_{\phi_5}^{r^{\mathbf{y}}}(C_3)$, and $\mathcal{B}_{\phi_6}^{r^{\mathbf{z}}}(C_3)$ for C_3 of (3.12)

There are of course related questions of **sharpness** (cf. Theorems 2.4 and 2.11) for the permuted Brauer sets $\mathcal{K}_{\phi}^{r^{\mathbf{x}}}(A)$ and the permuted Brualdi sets $\mathcal{B}_{\phi}^{r^{\mathbf{x}}}(A)$. We leave these as **open questions** for readers.

Exercises

1. For the matrix C_3 of (3.12), determine from (3.5) the following sets: $\Gamma_{\phi_4}^{r^{\mathbf{x}}}(C_3)$, $\Gamma_{\phi_5}^{r^{\mathbf{y}}}(C_3)$, and $\Gamma_{\phi_6}^{r^{\mathbf{z}}}(C_3)$, and verify that their intersection agrees with Fig. 3.1.

2. Give a complete proof of Theorem 3.2.

3. Verify the display in equation (3.18).

4. Give a complete proof of Theorem 3.3.

5. Give a complete proof of Theorem 3.4.

6. Give a complete proof of Theorem 3.5.

3.2 The Field of Values of a Matrix

We continue here with another method which gives an eigenvalue inclusion result for general matrices in $\mathbb{C}^{n\times n}$.

Definition 3.6. For any $A = [a_{i,j}] \in \mathbb{C}^{n\times n}$, its **field of values**, $F(A)$, is given by

$$F(A) := \{\mathbf{x}^* A\mathbf{x} : \mathbf{x} \in \mathbb{C}^n \text{ with } \mathbf{x}^*\mathbf{x} = 1\}. \tag{3.25}$$

We see, by its definition, that $F(A)$ is the image of the (connected) surface of the Euclidean unit ball in $\mathbb{C}^n$, under the continuous mapping $\mathbf{x} \mapsto \mathbf{x}^* A\mathbf{x}$. Thus (cf. Royden (1988), p.182), $F(A)$ is a compact and connected set in $\mathbb{C}$. It is a well-known result (cf. Horn and Johnson (1991), p.8) that $F(A)$ is also a **convex subset** of $\mathbb{C}$, the proof of this being known as the **Toeplitz-Hausdorff Theorem**. This fact will be used below.

What interests us here is the easy result of

Theorem 3.7. *For any* $A = [a_{i,j}] \in \mathbb{C}^{n\times n}$,

$$\sigma(A) \subseteq F(A). \tag{3.26}$$

Proof. For any $\lambda \in \sigma(A)$, there is some nonzero $\mathbf{x} \in \mathbb{C}^n$ with $A\mathbf{x} = \lambda\mathbf{x}$, where we may assume that $\mathbf{x}$ is normalized so that $\mathbf{x}^*\mathbf{x} = 1$. Then, $\lambda = \lambda\mathbf{x}^*\mathbf{x} = \mathbf{x}^*(\lambda\mathbf{x}) = \mathbf{x}^* A\mathbf{x} \in F(A)$. As this holds for each $\lambda \in \sigma(A)$, (3.26) is valid. ■

We see from Theorem 3.7 that the field of values, $F(A)$, of a matrix A, also gives an eigenvalue inclusion result for any matrix A, and it is then of interest to consider how to numerically determine the field of values, and how this compares with previous eigenvalue inclusion results, both in terms of work and how well the spectrum of A is captured. For this, we need the following. Let

$$H(A) := (A + A^*)/2$$

denote the **Hermitian part** of the matrix $A \in \mathbb{C}^{n\times n}$. We then have

Lemma 3.8. *For any* $A \in \mathbb{C}^{n\times n}$, *there holds*

$$F(H(A)) = Re\ F(A) := \{Re\ s : s \in F(A)\}. \tag{3.27}$$

Proof. For any $\mathbf{x} \in \mathbb{C}^n$ with $\mathbf{x}^*\mathbf{x} = 1$, then $\mathbf{x}^* H(A)\mathbf{x} = [\mathbf{x}^* A\mathbf{x} + \mathbf{x}^* A^*\mathbf{x}]/2 = [\mathbf{x}^* A\mathbf{x} + (\mathbf{x}^* A\mathbf{x})^*]/2 = [\mathbf{x}^* A\mathbf{x} + (\overline{\mathbf{x}^* A\mathbf{x}})]/2 = Re(\mathbf{x}^* A\mathbf{x})$. Thus, each point of $F(H(A))$ is of the form $Re\ s$ for some $s \in F(A)$, and conversely. ∎

Continuing, as $F(A)$ is a compact and connected set in $\mathbb{C}$, it follows that $Re\ F(A)$ is a closed real interval, say $[\alpha, \beta]$, where α and β are the extreme eigenvalues of the Hermitian matrix $H(A)$. In particular, we have that $F(A)$ is contained in the following closed half-plane:

$$F(A) \subseteq \{z \in \mathbb{C} : Re\ z \leq \beta\}, \tag{3.28}$$

i.e., the vertical line $Re\ z = \beta$ is a **support line** for $F(A)$, where β is sharp in the sense that it cannot be reduced in (3.28), since β is an eigenvalue of $H(A)$. Moreover, if $\mathbf{y} \in \mathbb{C}^n$ is a normalized (i.e., $\mathbf{y}^*\mathbf{y} = 1$) eigenvector associated with this eigenvalue β of $H(A)$, then the number $w := \mathbf{y}^* A\mathbf{y}$ is in $F(A)$, from (3.25). In fact, it is geometrically evident that $w \in \partial F(A)$, the boundary of $F(A)$, since $Re\ w = \beta$. This is shown in Fig. 3.3.

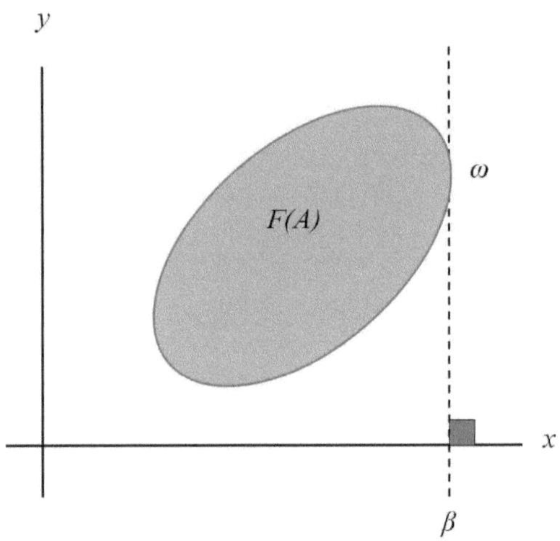

Fig. 3.3. $F(A)$ (shaded), with the vertical support line $Re\ z = \beta$.

The above construction of a specific boundary point of $F(A)$ can be extended and implemented, as follows. Select m distinct angles $\{\theta_j\}_{j=1}^m$ in $[0, 2\pi)$, and consider the Hermitian matrix

$$H(e^{i\theta_j} A) = (e^{i\theta_j} A + e^{-i\theta_j} A^*)/2. \tag{3.29}$$

Using Matlab 6, find this matrix's largest real eigenvalue $\mu(\theta_j)$ (usually in single precision), as well as its corresponding normalized eigenvector $\mathbf{y}_j$ in $\mathbb{C}^n$, i.e.,

$$\mathbf{y}_j^* H(e^{i\theta_j} A)\mathbf{y}_j = \mu(\theta_j), \text{ and } \mathbf{y}_j^*\mathbf{y}_j = 1 \quad (\text{all } 1 \le j \le m). \tag{3.30}$$

Again, the complex number $\mathbf{y}_j^* A\mathbf{y}_j$ is a point of $F(A)$, but because $\mu(\theta_j)$ is the largest eigenvalue of $H(e^{i\theta_j} A)$, it follows that

$$\mathbf{y}_j^* A\mathbf{y}_j \in \partial F(A) \quad (\text{all } 1 \le j \le m), \tag{3.31}$$

where the associated support line, through this point, now makes an angle of $(\pi/2) - \theta_j$ with the real axis. (This rotation of $-\theta_j$ is a consequence (cf. (3.27)) of

$$F(H(e^{i\theta_j} A)) = Re\ F(e^{i\theta_j} A) = Re\{e^{i\theta_j} F(A)\},$$

since $F(\alpha A) = \alpha F(A)$ for any scalar α.) Then, on plotting the values $\{\mathbf{y}_j^* A\mathbf{y}_j\}_{j=1}^m$, one obtains a discrete approximation to the boundary of $F(A)$.

As an example of the above method, consider the matrix

$$A_2 = \begin{bmatrix} 1 & i & 0 \\ \frac{1}{2} & 4 & \frac{i}{2} \\ 1 & 0 & 7 \end{bmatrix}, \tag{3.32}$$

associated with the Geršgorin set $\Gamma(A_2)$ of Fig. 1.2. Choosing 200 equally spaced θ_j's in $[0, 2\pi)$, the approximate field of values of A_2 is the shaded set in Fig. 3.4, along with the eigenvalues of A_2, given by "x's", and its three Geršgorin disks. (A Matlab 6 program for finding the 200 boundary points is given in Appendix D.)[1]

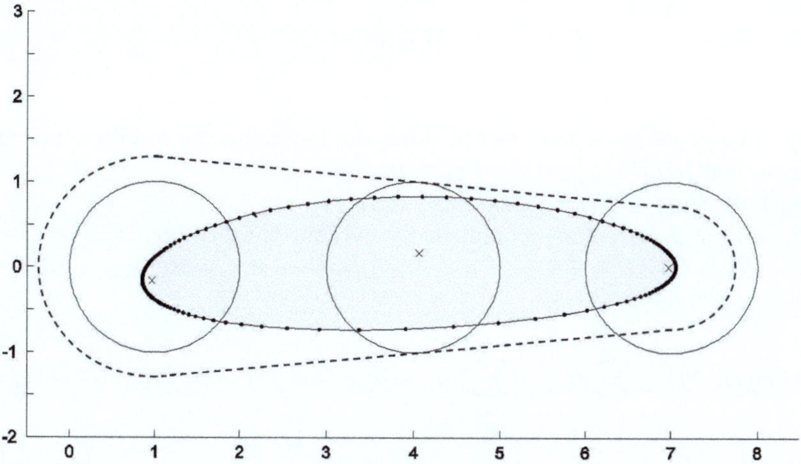

Fig. 3.4. $F(A_2)$ (shaded), $\Gamma(A_2)$ (solid boundary), and $J(A_2)$ of (3.35) (dashed boundary) for the matrix A_2 of (3.32).

[1] The author thanks Professor Anne Greenbaum for this.

We see in Fig. 3.4 that the three Geršgorin disks of $\Gamma(A_2)$ nicely separate the spectrum of A_2, while its field of values, $F(A_2)$, being convex and connected, **cannot** do this. It is also evident that the procedure we mentioned, to approximate the field of values of a matrix, requires numerically determining, for **each** θ_j, the largest real eigenvalue of the matrix in (3.29), as well as its normalized eigenvector $\mathbf{y}_j$. Thus, we see that finding the field of values of a matrix in this manner, while giving an eigenvalue inclusion set for its spectrum, is very far removed from the **simplicity** of earlier Geršgorin-type eigenvalue inclusions!

There is a simpler Geršgorin-like method, due to C.R. Johnson (1973) and given below in Theorem 3.9, which gives, for any $A \in \mathbb{C}^{n\times n}$, a set *containing* $F(A)$, and hence, $\sigma(A)$. For notation, given a set S in the complex plane, then $Co(S)$ denotes its **convex hull**, i.e.,

$$Co(S) := \bigcap_T \{T : T \text{ is convex and with } T \supseteq S\}. \tag{3.33}$$

Theorem 3.9. *Given* $A = [a_{i,j}] \in \mathbb{C}^{n\times n}$, *let* $\{r_j(A)\}_{i=1}^n$ *and* $\{c_i(A)\}_{i=1}^n$ *denote, respectively, the rows sums and column sums of* A (cf. (1.4) and (1.21)). *If*

$$g_i(A) := (r_i(A) + c_i(A))/2 \quad (\text{all } i \in N), \tag{3.34}$$

set

$$J(A) := Co\left(\bigcup_{i=1}^n \{z \in \mathbb{C} : |z - a_{i,i}| \le g_i(A)\}\right). \tag{3.35}$$

Then,

$$F(A) \subseteq J(A). \tag{3.36}$$

Proof. The proof consists of establishing the following three claims. For extra notation, let $RHP := \{z \in \mathbb{C} : Re\, z > 0\}$.

Claim 1. If $J(A) \subset RHP$, then $F(A) \subset RHP$.

If $J(A) \subset RHP$, it is geometrically evident that $Re\ a_{i,i} > g_i(A)$ for all $i \in N$. With $H(A) := (A + A^*)/2 =: [b_{i,j}]$, then $b_{i,i} = Re\ a_{i,i}$, and, by the triangle inequality,

$$r_i(H(A)) := \sum_{\substack{j=1\\ j\neq i}}^n |b_{i,j}| = \sum_{\substack{j=1\\ j\neq 1}}^n |a_{i,j} + \overline{a}_{j,i}|/2 \le (r_i(A) + c_i(A))/2 = g_i(A), \tag{3.37}$$

the last equality following from (3.34). Thus, the Geršgorin disks for $H(A)$ all lie in RHP; hence, by Geršgorin's Theorem 1.11, $\sigma(H(A)) \subset RHP$. But $H(A)$ is Hermitian, so that $H(A)$ has only real eigenvalues, and $F(H(A))$, as a convex set, is a real line segment in RHP. Thus by Lemma 3.8, $F(A) \subset RHP$, as claimed.

Claim 2. If $0 \notin J(A)$, then $0 \notin F(A)$.

Assume that $0 \notin J(A)$. By definition (cf.(3.35), $J(A)$ is convex. Then by the **separating hyperplane theorem** (Horn and Johnson (1985), Appendix B), there is a $\theta \in [0, 2\pi)$ such that $J(e^{i\theta}A) = e^{i\theta}J(A) \subset RHP$. This implies, from Claim 1, that $F(e^{i\theta}A) \subset RHP$. But as (3.35) gives $F(A) = e^{-i\theta}F(e^{i\theta}A)$, it follows that $0 \notin F(A)$ as claimed.

Claim 3. If $\alpha \notin J(A)$, then $\alpha \notin F(A)$.

If $\alpha \notin J(A)$, then $0 \notin J(A - \alpha I)$ follows directly from (3.35). Applying Claim 2 gives $0 \notin F(A - \alpha I) = F(\alpha) - \alpha$, which gives that $\alpha \notin F(\alpha)$, as claimed.

Putting all these claims together gives the desired result of Theorem 3.9. ■

It is clear, for n not too large, that forming the disks in (3.35) of $J(A)$ and then the convex hull of the union of these disks, is not difficult. This would seem to be far simpler than approximating the boundary of $F(A)$, using Matlab 6, as described earlier in this section. On the other hand, the set $J(A)$ in (3.35) may **not** be a particularly close approximation of $F(A)$, as can be seen in Fig. 3.4, where the set $J(A)$ of (3.35), for the matrix A_2 of (3.32), is also shown in Fig. 3.4, with a dashed boundary.

Exercises

1. For arbitrary matrices A and B in $\mathbb{C}^{n\times n}$ and an arbitrary scalar $\alpha \in \mathbb{C}$, verify from (3.25) that:
   a. $F(A)$ is a compact set in $\mathbb{C}$;
   b. $F(\alpha A) = \alpha F(A)$, for any $\alpha \in \mathbb{C}$;
   c. $F(\alpha I + A) = \alpha + F(A)$, for any $\alpha \in \mathbb{C}$;
   d. $F(A + B) \subseteq F(A) + F(B)$;
   e. if U is a unitary matrix in $\mathbb{C}^{n\times n}$, i.e. if $U^*U = I$, then $F(U^*AU) = F(A)$.

2. For the matrix $A = \begin{bmatrix} 1 & 2 \\ 0 & 1 \end{bmatrix} \in \mathbb{C}^{2\times 2}$, show that $F(A)$ is given by $\{z \in \mathbb{C} : |z - 1| \leq 1\}$.

3. For the 3×3 matrix A_2 of (3.32), show, with (3.34), that the ovals of Cassini variant of (3.35), defined by

$$J_K(A) := Co\left[\bigcup_{\substack{k,\ell=1 \\ k\neq\ell}}^{3} \{z \in \mathbb{C} : |z - a_{k,k}| \cdot |z - a_{\ell,\ell}| \leq g_k(A) \cdot g_\ell(A)\}\right],$$

does **not** contain the field values $F(A_2)$. In other words, extensions of Theorem 3.5 to higher-order lemniscates are **not** in general valid.

4. Given any $A = [a_{i,j}] \in \mathbb{C}^{n \times n}, n \geq 2$, set

$$r_k(H[(e^{i\theta}A + e^{-i\theta}A^*)/2]) := \frac{1}{2} \sum_{j \in N \backslash \{k\}} |e^{i\theta}a_{k,j} + e^{-i\theta}\overline{a}_{j,k}| \ (k \in N),$$

and set

$$\hat{g}_k(A) := \max_{\theta \in [0,2\pi]} \left\{ \frac{1}{2} \sum_{j \in N \backslash \{k\}} |e^{i\theta}a_{k,j} + e^{-i\theta}\overline{a}_{j,k}| \right\} \quad (k \in N).$$

Show, with (3.34) and the triangle inequality, that
a. $\hat{g}_k(A) \leq g_k(A)$ (all $k \in N$), and
b. Theorem 3.5 can be extended to

$$F(A) \subseteq Co \left[\bigcup_{k=1}^{n} \{z \in \mathbb{C} : |z - a_{k,k}| \leq \hat{g}_k(A)\} \right],$$

so that from a.) and (3.35),

$$Co \left[\bigcup_{k=1}^{n} \{z \in \mathbb{C} : |z - a_{k,k}| \leq \hat{g}_k(A)\} \right] \subseteq J(A).$$

5. Give a complete proof of Theorem 3.4.

6. Give a complete proof of Theorem 3.5.

3.3 Newer Eigenvalue Inclusion Sets

In this section, we present new results of Cvetkovic, Kostic and Varga (2004) on nonsingularity results for matrices, and their related eigenvalue inclusion sets in the complex plane. This work extends the earlier, less well-known, result of Dashnic and Zusmanovich (1970). We also show how this work is related to the work of Huang (1995).

To begin, let S denote a nonempty subset of $N = \{1, 2, \cdots, n\}, n \geq 2$, and let $\overline{S} := N \backslash S$ denote its complement in N. Then, given any matrix $A = [a_{i,j}] \in \mathbb{C}^{n \times n}$, split each row sum, $r_i(A)$ from (1.4), into two parts, depending on S and $\overline{S}$, i.e.,

$$(3.38)\quad \begin{cases} r_i(A) := \sum_{j\in N\backslash\{i\}} |a_{i,j}| = r_i^S(A) + r_i^{\overline{S}}(A), \quad \text{where} \\ r_i^S(A) := \sum_{j\in S\backslash\{i\}} |a_{i,j}|, \text{ and } r_i^{\overline{S}}(A) := \sum_{j\in \overline{S}\backslash\{i\}} |a_{i,j}| \quad (\text{all } i \in N). \end{cases}$$

Definition 3.10. Given any matrix $A = [a_{i,j}] \in \mathbb{C}^{n\times n}, n \geq 2$, and given any nonempty subset S of N, then A is an **S-strictly diagonally dominant matrix** if

$$\begin{cases} i) \;\; |a_{i,i}| > r_i^S(A) \quad (\text{all } i \in S), \\ \quad \text{and} \\ ii) \;\; (|a_{i,i}| - r_i^S(A)) \cdot (|a_{j,j}| - r_j^{\overline{S}}(A)) > r_i^{\overline{S}}(A) \cdot r_j^S(A) \quad (\text{all } i \in S, \text{ all } j \in \overline{S}). \end{cases}$$
(3.39)

We note, from (3.39 *i*), that as $|a_{i,i}| - r_i^S(A) > 0$ for all $i \in S$, then on dividing by this term in (3.39 *ii*) gives

$$\left(|a_{j,j}| - r_j^{\overline{S}}(A)\right) > \frac{r_i^{\overline{S}}(A) \cdot r_j^S(A)}{(|a_{i,i}| - r_i^S(A))} \geq 0 \quad (\text{all } j \in \overline{S}),$$

so that we also have

$$(3.40)\qquad |a_{j,j}| - r_j^{\overline{S}}(A) > 0 \quad (\text{all } j \in \overline{S}).$$

If $S = N$, so that $\overline{S} = \emptyset$, then the conditions of (3.39 *i*) reduces to $|a_{i,i}| > r_i(A)$ (all $i \in N$), and this is just the familiar statement that A is strictly diagonally dominant, and hence is nonsingular from Theorem 1.4.

With Definition 3.10, we next establish the following nonsingularity result of Cvetkovic, Kostic and Varga (2004).

Theorem 3.11. *Let S be a nonempty subset of N, and let $A = [a_{i,j}] \in \mathbb{C}^{n\times n}$, $n \geq 2$, be S-strictly diagonally dominant. Then, A is nonsingular.*

Proof. If $S = N$, then, as we have seen, A is strictly diagonally dominant, and thus nonsingular (cf. Theorem 1.4). Hence, we assume that S is a nonempty subset of N with $\overline{S} \neq \emptyset$. The idea of the proof is to construct a positive diagonal matrix W such that AW is strictly diagonally dominant. Now, define W as $W = \text{diag}[w_1, w_2, \cdots, w_n]$, where

$$w_k := \begin{cases} \gamma, & \text{for all } k \in S, \text{ where } \gamma > 0, \text{ and} \\ 1, & \text{for all } k \in \overline{S}. \end{cases}$$

It then follows that $AW := [\alpha_{i,j}] \in \mathbb{C}^{n\times n}$ has its entries given by

$$\alpha_{i,j} := \begin{cases} \gamma a_{i,j}, & \text{if } j \in S, \text{ all } i \in N, \text{ and} \\ a_{i,j}, & \text{if } j \in \overline{S}, \text{ all } i \in N. \end{cases}$$

Then, the row sums of AW are, from (3.38), just

$$r_\ell(AW) = r_\ell^S(AW) + r_\ell^{\overline{S}}(AW) = \gamma r_\ell^S(A) + r_\ell^{\overline{S}}(A) \quad (\text{all } \ell \in N),$$

and AW is then strictly diagonally dominant if

$$\begin{cases} \gamma|a_{i,i}| > \gamma r_i^S(A) + r_i^{\overline{S}}(A) & (\text{all } i \in S), \text{ and} \\ |a_{j,j}| > \gamma r_j^S(A) + r_j^{\overline{S}}(A) & (\text{all } j \in \overline{S}). \end{cases}$$

The above inequalities can be also expressed as

$$(3.41) \qquad \begin{cases} i)\ \gamma(|a_{i,i}| - r_i^S(A)) > r_i^{\overline{S}}(A) & (\text{all } i \in S), \text{ and} \\ ii)\ |a_{j,j}| - r_j^{\overline{S}}(A) > \gamma r_j^S(A) & (\text{all } j \in \overline{S}), \end{cases}$$

which, upon division, can be further reduced to

$$(3.42) \quad \frac{r_i^{\overline{S}}(A)}{(|a_{i,i}| - r_i^S(A))} < \gamma \ (\text{all } i \in S), \text{ and } \gamma < \frac{(|a_{j,j}| - r_j^{\overline{S}}(A))}{r_j^S(A)} \ (\text{all } j \in \overline{S}),$$

where the final fraction in (3.42) is defined to be $+\infty$ if $r_j^S(A) = 0$ for some $j \in \overline{S}$. The inequalities of (3.41) will be satisfied if there is a $\gamma > 0$ for which

$$(3.43) \quad 0 \leq B_1 := \max_{i \in S} \frac{r_i^{\overline{S}}(A)}{(|a_{i,i}| - r_i^S(A))} < \gamma < \min_{j \in \overline{S}} \frac{(|a_{j,j}| - r_j^{\overline{S}}(A))}{r_j^S(A)} =: B_2.$$

But since (3.39 *ii*) exactly gives that $B_2 > B_1$, then, for any $\gamma > 0$ with $B_1 < \gamma < B_2$, AW is strictly diagonally dominant and hence nonsingular. Then, as W is nonsingular, so is A. ■

We remark that the result of Dashnic and Zusmanovich (1970) is the special case of Theorem 3.11 when S is a singelton, i.e., $S := \{i\}$ for some $i \in N$. More recently, Huang (1995) has also worked on this problem, and similarly breaks N into disjoint subsets S and $\overline{S}$, but assumes a variant of the inequalities of (3.39). Now, if $S = \{i_1, i_2, \cdots, i_k\}$, then $A_{S,S} := [a_{i_j,i_k}]$ (all i_j, i_k in S) is a $k \times k$ principal submatrix of A, whose associated **comparison matrix** (see (C.4) of Appendix C) is given by

$$\mathcal{M}(A_{S,S}) := \begin{bmatrix} +|a_{i_1,i_1}| & -|a_{i_1,i_2}| & \cdots & -|a_{i_1,i_k}| \\ -|a_{i_2,i_1}| & +|a_{i_2,i_2}| & \cdots & -|a_{i_2,i_k}| \\ \vdots & & & \vdots \\ -|a_{i_k,i_1}| & -|a_{i_k,i_2}| & \cdots & +|a_{i_k,i_k}| \end{bmatrix},$$

and it is assumed that $\mathcal{M}(A_{S,S})$ is a **nonsingular M-Matrix** (see Definition C.3 of Appendix C), with the additional assumption (in analogy with (3.43)) that the k components, of the following matrix-vector product, satisfy

$$(3.44) \qquad 0 \leq \max_{i_j \in S} \left(\mathcal{M}^{-1}(A_{S,S}) \cdot \begin{bmatrix} r_{i_1}^{\overline{S}}(A) \\ \vdots \\ r_{i_k}^{\overline{S}}(A) \end{bmatrix} \right)_{i_j} < B_2,$$

where B_2 is defined in (3.43). Huang's result is more general than the result of Theorem 3.11, but it comes with the added expense of having to explicitly determine $\mathcal{M}^{-1}(A_{S,S})$ for use in (3.44). (The relationship of Theorem 3.11 and Huang's result is studied in more detail in Cvetkovic, Kostic and Varga (2004).)

As is now familiar, the nonsingularity in Theorem 3.11 immediately gives, by negation, the following equivalent eigenvalue inclusion set in the complex plane.

Theorem 3.12. *Let S be any nonempty subset of $N := \{1, 2, \cdots, n\}$, $n \geq 2$, with $\overline{S} := N \backslash S$. Then, for any $A = [a_{i,j}] \in \mathbb{C}^{n \times n}$, define the Geršgorin-type disks*

$$(3.45) \qquad \Gamma_i^S(A) := \{z \in \mathbb{C} : |z - a_{i,i}| \leq r_i^S(A)\} \text{ (any } i \in S),$$

and the sets

$$(3.46) \quad V_{i,j}^S(A) := \{z \in \mathbb{C} : (|z - a_{i,i}| - r_i^S(A)) \cdot (|z - a_{j,j}| - r_j^{\overline{S}}(A)) \leq r_i^{\overline{S}}(A) \cdot r_j^S(A)\},$$

(any $i \in S$, any $j \in \overline{S}$). Then,

$$(3.47) \qquad \sigma(A) \subseteq C^S(A) =: \left(\bigcup_{i \in S} \Gamma_i^S(A) \right) \cup \left(\bigcup_{i \in S, j \in \overline{S}} V_{i,j}^S(A) \right).$$

We remark that each set $V_{i,j}^S(A)$ of (3.46), where $\overline{S} \neq \emptyset$, resembles the Brauer Cassini oval $K_{i,j}(A)$ of (2.5). Moreover, the numerical work involved in determining the boundary of *either* of these two sets, $V_{i,j}^S(A)$ or $K_{i,j}(A)$, is roughly the same. This suggests comparisons of $C^S(A)$ in (3.47) with $\Gamma(A)$ of (1.5) and with $\mathcal{K}(A)$ of (2.6), both in terms of how all these compare as sets in the complex plane, and the numerical effort associated with obtaining these sets. We note, of course, that, given a nonempty subset S of N, $n \geq 2$, with $\overline{S} := N \backslash S$, the data needed in Theorem 3.12 are (cf. (3.38))

$$(3.48) \qquad \{a_{i,i}\}_{i \in N}, \{r_j^S(A)\}_{i \in N}, \text{ and } \{r_i^{\overline{S}}(A)\}_{i \in N},$$

whereas less data, namely

$$(3.49) \qquad \{a_{i,i}\}_{i \in N}, \text{ and } \{r_j^S(A)\}_{i \in N},$$

are needed for determining $\Gamma(A)$ and $\mathcal{K}(A)$.

To begin these comparisons, consider the singleton case of Dashnic and Zusmanovich (1970), where $S_i := \{i\}$ for some $i \in N$. In this case, it follows from (3.47) that, on defining $\mathcal{D}_i(A) := C^{S_i}(A)$, we have, for any given $A = [a_{i,j}] \in \mathbb{C}^{n\times n}$, $n \geq 2$, that

$$\mathcal{D}_i(A) = \Gamma_i^{S_i}(A) \cup \left(\bigcup_{j\in N\backslash\{i\}} V_{i,j}^{S_i}(A) \right). \tag{3.50}$$

Now, $r_i^{S_i}(A) = 0$ from (3.38) so that $\Gamma_i^{S_i}(A) = \{a_{i,i}\}$ from (3.45). Moreover, we also have, from (3.46) in this case that, for all $j \neq i$ in N,

$$V_{i,j}^{S_i}(A) = \{z \in \mathbb{C} : |z - a_{i,i}| \cdot (|z - a_{j,j}| - r_j(A) + |a_{j,i}|\ ) \leq r_i(A) \cdot |a_{j,i}|\}.$$

But as $z = a_{i,i}$ is necessarily contained in $V_{i,j}^{S_i}(A)$ for all $j \neq i$, then we can simply write from (3.50) that

$$\mathcal{D}_i(A) = \bigcup_{j\in N\backslash\{i\}} V_{i,j}^{S_i}(A) \quad (\text{any } i \in N). \tag{3.51}$$

This shows that $D_i(A)$ is determined from $(n-1)$ sets $V_{i,j}^{S_i}(A)$, (which can be disks or oval-like sets), plus the added information from (3.48) on the partial row sums of A, while the associated Geršgorin set $\Gamma(A)$, from (1.5), is determined from n disks and the associated Brauer set $\mathcal{K}(A)$, from (2.6) is determined from $\binom{n}{2}$ Cassini ovals. These sets are compared in

Theorem 3.13. *For any* $A = [a_{i,j}] \in \mathbb{C}^{n\times n}, n \geq 2$, *and for any* $i \in N$, *consider* $\mathcal{D}_i(A)$ *of* (3.51). *Then* (cf. (1.5)),

$$\mathcal{D}_i(A) \subseteq \Gamma(A), \tag{3.52}$$

but there are matrices E *in* $\mathbb{C}^{n\times n}$, *where* (cf.(2.6)), *the following hold simultaneously:*

$$\mathcal{D}_i(E) \nsubseteq \mathcal{K}(E) \text{ and } \mathcal{K}(E) \nsubseteq \mathcal{D}_i(E). \tag{3.53}$$

Proof. To establish (3.52), fix some $i \in N$ and consider any $z \in \mathcal{D}_i(A)$. Then from (3.51), there is a $j \neq i$ such that $z \in V_{i,j}^{S_i}(A)$, i.e.,

$$|z - a_{i,i}| \cdot (|z - a_{j,j}| - r_j(A) + |a_{j,i}|) \leq r_i(A) \cdot |a_{j,i}|. \tag{3.54}$$

If $z \notin \Gamma(A)$, then $|z - a_{k,k}| > r_k(A)$ for *all* $k \in N$, so that $|z - a_{i,i}| > r_i(A) \geq 0$, and $|z - a_{j,j}| > r_j(A) \geq 0$. Then, the left part of (3.54) satisfies

$$|z - a_{i,i}| \cdot (|z - a_{j,j}| - r_j(A) + |a_{j,i}|) > r_i(A) \cdot |a_{j,i}|,$$

which contradicts the inequality in (3.54). Thus, $z \in \Gamma(A)$ for each $z \in \mathcal{D}_i(A)$, which establishes (3.52).

Finally, to show (3.53), a specific 3×3 matrix E is given in Exercise 3, where $\mathcal{D}_i(E) \nsubseteq \mathcal{K}(E)$ and $\mathcal{K}(E) \nsubseteq \mathcal{D}_i(E)$. ■

Next, it is evident from (3.47) of Theorem 3.12 that, for any $A = [a_{i,j}] \in \mathbb{C}^{n\times n}$,

$$\sigma(A) \subseteq \mathcal{D}_i(A) \quad (\text{all } i \in N),$$

so that

$$\sigma(A) \subseteq \mathcal{D}(A) := \bigcap_{i\in N} \mathcal{D}_i(A). \tag{3.55}$$

Now, as each $\mathcal{D}_i(A)$, from (3.51), depends on $(n-1)$ oval-like sets $V_{i,j}^{S_i}(A)$, it follows that $\mathcal{D}(A)$ of (3.55) is determined from $n(n-1)$ oval-like sets $V_{i,j}^{S_i}(A)$, which is *twice* the number of Cassini ovals, namely $\binom{n}{2}$, which determine the Brauer set $\mathcal{K}(A)$. This suggests, perhaps, that $\mathcal{D}(A) \subseteq \mathcal{K}(A)$. This is true, and is established in

Theorem 3.14. *For any $A = [a_{i,j}] \in \mathbb{C}^{n\times n}, n \geq 2$, then the associated sets $\mathcal{D}(A)$, of (3.55), and $\mathcal{K}(A)$, of (2.6), satisfy*

$$\mathcal{D}(A) \subseteq \mathcal{K}(A). \tag{3.56}$$

Proof. First, we observe, from (3.52), that as $\mathcal{D}_i(A) \subseteq \Gamma(A)$ for each $i \in N$, then $\mathcal{D}(A)$, as defined in (3.55), evidently satisfies

$$\mathcal{D}(A) \subseteq \Gamma(A). \tag{3.57}$$

To establish (3.56), consider any $z \in \mathcal{D}(A)$ so that, for each $i \in N$, $z \in \mathcal{D}_i(A)$. Hence, from (3.51), $z \in V_{i,j}^{S_i}(A)$ for each $i \in N$, and some $j \in N\backslash\{i\}$, where the inequality of (3.54) is valid. But from (3.57), $\mathcal{D}(A) \subseteq \Gamma(A)$ implies that there is a $k \in N$ with $|z-a_{k,k}| \leq r_k(A)$. For this index k, there is a $t \in N\backslash\{k\}$ such that $z \in V_{k,t}^{S_i}(A)$, i.e.,

$$|z - a_{k,k}|\left(|z - a_{t,t}| - r_t(A) + |a_{t,k}|\right) \leq r_k(A) \cdot |a_{t,k}|.$$

This can be rewritten as

$$\begin{aligned}|z - a_{k,k}| \cdot |z - a_{t,t}| &\leq |z - a_{k,k}| \cdot (r_t(A) - |a_{t,k}|) + r_k(A) \cdot |a_{t,k}| \\ &\leq r_k(A)(r_t(A) - |a_{t,k}|) + r_k(A) \cdot |a_{t,k}| = r_k(A) \cdot r_t(A),\end{aligned}$$

that is,

$$|z - a_{k,k}| \cdot |z - a_{t,t}| \leq r_k(A) \cdot r_t(A).$$

Hence, from (2.5), $z \in K_{k,t}(A) \subseteq \mathcal{K}(A)$. As this is true for each $z \in \mathcal{D}(A)$, then $\mathcal{D}(A) \subseteq \mathcal{K}(A)$. ■

To illustrate the result of Theorem 3.14, let us consider again the 4×4 matrix B of (2.14), i.e.,

$$B = \begin{bmatrix} 1 & 1 & 0 & 0 \\ \frac{1}{2} & i & \frac{1}{2} & 0 \\ 0 & 0 & -1 & 1 \\ 1 & 0 & 0 & -i \end{bmatrix}, \tag{3.58}$$

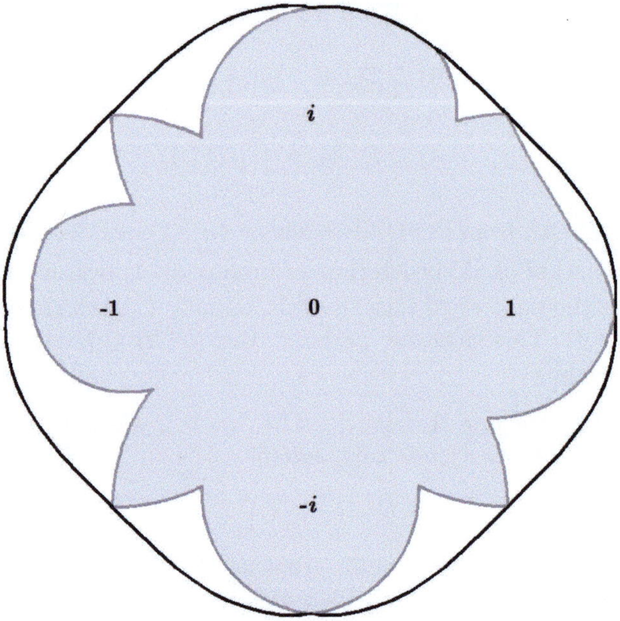

Fig. 3.5. The sets $\mathcal{D}(B)$ (shaded) and $\mathcal{K}(B)$ for the matrix B of (3.58).

whose Brauer set $\mathcal{K}(B)$ is given in Fig. 2.2. Then, the set $\mathcal{D}(B)$ for this matrix is shown as the shaded set in Fig. 3.5, along with the boundary of $\mathcal{K}(B)$, from which, the inclusion $\mathcal{D}(B) \subseteq \mathcal{K}(B)$ from Theorem 3.14 is then evident. We also remark that the Brualdi set $\mathcal{B}(B)$, for the matrix of (3.58), can be verified to be the same as the Brualdi set $\mathcal{B}(E)$ of Fig. 2.9, and the $\mathcal{B}(B)$ and $\mathcal{D}(B)$ are shown together in Fig. 3.6, from which we see that

$$\mathcal{B}(B) \nsubseteq \mathcal{D}(B) \quad \text{and} \quad \mathcal{D}(B) \nsubseteq \mathcal{B}(B).$$

This, perhaps is not unexpected, as the determination of the sets $\mathcal{D}(A)$ and $\mathcal{B}(A)$ are based on different data (i.e., one uses modified row sums, the other uses cycles).

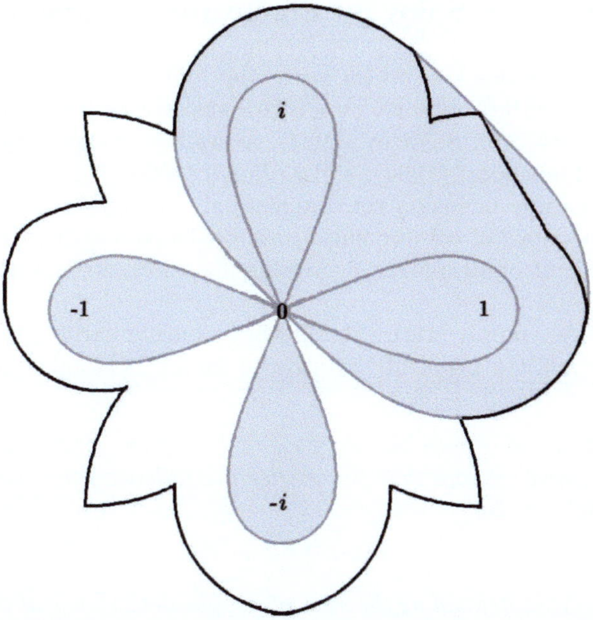

Fig. 3.6. The sets $\mathcal{B}(B)$ (shaded) and $\mathcal{D}(B)$ for the matrix B of (3.58).

Exercises

1. If $A = [a_{i,j}] \in \mathbb{C}^{n\times n}$ is a strictly diagonally dominant matrix (cf. Theorem 1.4), let $\mathcal{M}(A)$ be its comparison matrix (cf. (C.4) of Appendix C). Show that $\mathcal{M}(A)$ is a nonsingular M-matrix, using Definition C.2 of Appendix C. (Hint: if $D := \text{diag}[|a_{1.1}|, |a_{2,2}|, \cdots |a_{n,n}|]$, define the matrix B from $D^{-1}\mathcal{M}(A) = I - B$, where $B \geq O$. Using (C.2) of Appendix C, show that $\rho(B) < 1$.)

2. If $A = [a_{i,j}] \in \mathbb{C}^{n\times n}$ is an irreducibly diagonally dominant matrix (cf. Definition 1.10), and if $\mathcal{M}(A)$ is its comparison matrix, show that $\mathcal{M}(A)$ is a nonsingular M-matrix. (Hint: Follow the hint of Exercise 1, but use (C.1) of Appendix C.)

3. Consider the matrix $E = \begin{bmatrix} 1 & \frac{1}{2} & \frac{1}{2} \\ 0 & i & 1 \\ 0 & 1 & -1 \end{bmatrix}$. Show from (3.51) that
   a. $\mathcal{D}_1(E) = \{1\} \cup \{z \in \mathbb{C} : |z - i| \leq 1\} \cup \{z \in \mathbb{C} : |z + 1| \leq 1\}$; so that
   b. $\mathcal{D}_1(E) \subseteq \Gamma(E)$ where $\Gamma(E)$ is defined in (1.5), but
   c. $\mathcal{D}_1(E) \nsubseteq \mathcal{K}(E)$, and $\mathcal{K}(E) \nsubseteq \mathcal{D}_1(E)$, where $\mathcal{K}(E)$ is defined in (2.6).

3.4 The Pupkov-Solov'ev Eigenvalue Inclusions Set

We investigate in this section an intriguing matrix eigenvalue inclusion set, which we call the Pupkov-Solov'ev set.This has evolved from the initial papers of Pupkov (1984) and Solov'ev (1984), as well as the subsequent papers of Brualdi and Mellendorf (1994), and Hoffman (2000).

This eigenvalue inclusion set, interestingly enough, can depend simultaneously on portions of column sums *and* row sums, but it is, perhaps one of the most difficult of all matrix eigenvalue inclusion sets to implement, as we shall see.

To begin, we simply state the following nonsingularity result of Solov'ev (1984). (A complete proof of this can be found in Brualdi and Mellendorf (1994).)

Theorem 3.15. *For any $A = [a_{i,j}] \in \mathbb{C}^{n\times n}, n \geq 2$, let $\ell \in N := \{1, 2, \cdots, n\}$, and assume that A satisfies the following two conditions:*

i) For each $j \in N$,

$$|a_{j,j}| > c_j^{(\ell-1)}(A), \tag{3.59}$$

where $c_j^{(\ell-1)}(A)$ is defined as the sum of the absolute values of the $\ell-1$ largest off-diagonal entries in the j-th column of A (where, for $\ell = 1$, the condition in (3.59) is vacuous).

ii) For each set P of ℓ district numbers for N, say $i_1, i_2, \cdots, i_\ell$, it is assumed that

$$\sum_{k=1}^{\ell} |a_{i_k,i_k}| > \sum_{k=1}^{\ell} r_{i_k}(A) \quad (\text{where } r_i(A) := \sum_{k\in N\backslash\{i\}} |a_{i,k}|). \tag{3.60}$$

Then, A is nonsingular.

Remark 1. If $\ell = 1$, then only (3.60) applies, and in this case, the assumption of (3.60) gives that A is then strictly *row* diagonally dominant, so that A is nonsingular (cf. Theorem 1.4). If $\ell = n$, then (3.59) gives that A is strictly *column* diagonally dominant, i.e.,

$$|a_{j,j}| > \sum_{k\in N\backslash\{j\}} |a_{k,j}| \quad (\text{all } j \in N), \tag{3.61}$$

so that A is again nonsingular (cf. Corollary 1.13). (We note that adding all n inequalities of (3.59) directly gives, upon rearranging terms, the inequality of (3.60), in this case of $\ell = n$.)

Remark 2. An equivalent form for Theorem 3.15 was obtained by Pupkov (1984).

Of interest to us now is the associated matrix eigenvalue inclusion result, Theorem 3.16, which follows directly from Theorem 3.15.

Theorem 3.16. *For any matrix* $A = [a_{i,j}] \in \mathbb{C}^{n \times n}, n \geq 2$ *let* $\ell \in N := \{1, 2, \cdots, n\}$. *Then, each eigenvalue of* A, *lies either in one of the* n *disks*

$$\{z \in \mathbb{C} : |z - a_{j,j}| \leq c_j^{(\ell-1)}(A)\} \quad (\text{any } j \in N), \tag{3.62}$$

(*where* $c_j^{(\ell-1)}(A)$ *is defined in i) of Theorem 3.15*) *or in one of the sets*

$$\{z \in \mathbb{C} : \sum_{i \in \mathcal{P}} |z - a_{i,i}| \leq \sum_{i \in \mathcal{P}} r_i(A)\}, \tag{3.63}$$

where $\mathcal{P}$ *is any subset of* N *having cardinality* ℓ, *i.e.*, $|\mathcal{P}| = \ell$. *Consequently, on setting*

$$\begin{aligned} PS_\ell(A) := & \left(\bigcup_{j=1}^{n} \{z \in \mathbb{C} : |z - a_{j,j}| \leq c_j^{(\ell-1)}(A)\} \right) \cup \\ & \left(\bigcup_{|\mathcal{P}|=\ell} \{z \in \mathbb{C} : \sum_{i \in \mathcal{P}} |z - a_{i,i}| \leq \sum_{i \in \mathcal{P}} r_i(A)\} \right), \end{aligned} \tag{3.64}$$

then

$$\sigma(A) \subseteq PS_\ell(A). \tag{3.65}$$

We call the set, $PS_\ell(A)$ in (3.64), the **Pupkov-Solov'ev set** for A, which consists of n disks, from (3.62), provided that $\ell > 1$, and of $\binom{n}{\ell}$ sets, from (3.63). For example, if $n = 10$ and if $\ell = 5$, one must consider 10 disks from (3.62), and $252 = \binom{10}{5}$ sets from (3.63), to determine $PS_5(A)$, for a given matrix A in $\mathbb{C}^{10 \times 10}$. This is rather far removed from the simplicity in forming, by comparison, the 10 Geršgorin disks, or the 45 Brauer ovals of Cassini for this matrix.

There are many unanswered questions about these Pupkov-Solov'ev sets. The first might be how to select the parameter ℓ in N, when determining the set $PS_\ell(A)$. We see, from the definition of the $c_j^{(\ell-1)}(A)$'s in Theorem 3.15, that the radius of each of the n disks, in the first union in (3.64), is a *nondecreasing* function of ℓ, as is the associated union of these n disks. On the other hand, for $\ell = 2$, the second union of (3.64) consists of sets of the form

$$\{z \in \mathbb{C} : |z - a_{i,i}| + |z - a_{j,j}| \leq r_i(A) + r_j(A)\} \quad (i, j \text{ in } N, \text{ with } i \neq j),$$

and the above set is *empty* if the two disks, $\{z \in \mathbb{C} : |z - a_{i,i}| \leq r_i(A)\}$ and $\{z \in \mathbb{C} : |z - a_{j,j}| \leq r_j(A)$ are *disjoint*.

There are also unanswered questions on how these Pupkov-Solov'ev sets compare with the corresponding Brauer and Brualdi sets. These open questions are left for our attentive readers!

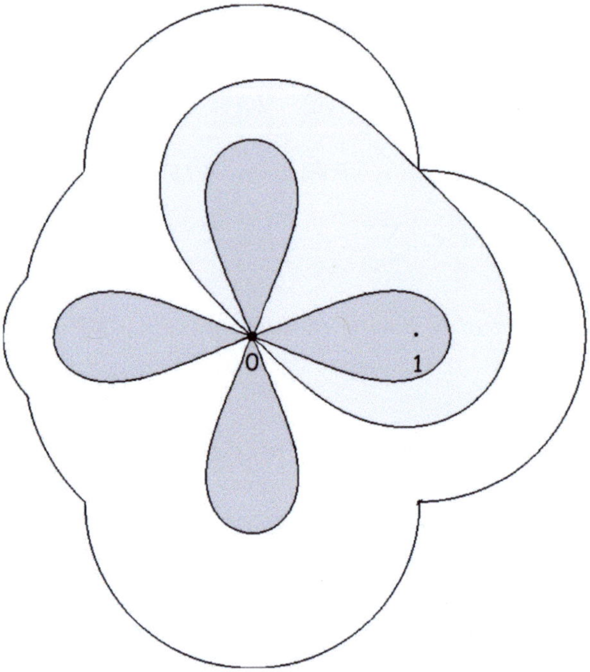

Fig. 3.7. The Pupkov-Solov'ev set $PS_2(B)$ and the Brualdi set $\mathcal{B}(B)$ (shaded) for the matrix of (3.66).

Finally, for completeness, we include the following example of the Pupkov-Solov'ev set for the matrix $B = [b_{i,j}] \in \mathbb{C}^{4\times 4}$, already considered in (2.14), i.e.,

$$B = \begin{bmatrix} 1 & 1 & 0 & 0 \\ \frac{1}{2} & i & \frac{1}{2} & 0 \\ 0 & 0 & -1 & 1 \\ 1 & 0 & 0 & -i \end{bmatrix}, \qquad (3.66)$$

for which the Brauer set $\mathcal{K}(B)$ and the Geršgorin set $\Gamma(B)$ for this matrix are shown in Fig. 2.2. We then apply the case $\ell = 2$ of Theorem 3.16 to this matrix B of (3.66), and this is shown in Fig. 3.7. Specifically, the outer boundary in Fig. 3.7 is the boundary of $PS_2(B)$, while the shaded set in this figure is the Brualdi set $\mathcal{B}(B)$ for this matrix, as in Fig. 2.9. It can be seen that $\mathcal{B}(B)$ is a **proper subset** of $PS_2(B)$. (For another example of a Pupkov-Solov'ev matrix, see Brualdi and Mellendorf (1994).)

Bibliography and Discussion

3.1 The use of permutations on a matrix to get Geršgorin-type eigenvalue inclusion results for matrices, involving the *closed* exteriors of disks, was independently discovered by Parodi (1952) and Schneider (1954), and developed further in the book of Parodi (1959). The extension of this idea, in this section, to Brauer Cassini ovals and Brualdi lemniscates is new, but follows easily from the results in Chapter 2.

Perhaps the most attractive feature of the results in this section is that they can lead to eigenvalue inclusions which *carve away* sets about diagonal entries, something which cannot be done with usual Geršgorin arguments! The results of Figs. 4.1 and 4.2 suggest that these permuted Geršgorin sets are best used in conjunction with the standard eigenvalue inclusion results of Chapters 1 and 2.

3.2 The concept of a **field of values** of a matrix (or operator), is a very useful notion in analysis, and, as it contains all the eigenvalues of a matrix, it is thus useful also in linear algebra. But, as we have seen in this section, determining numerically even reasonable approximations of this field of values of a matrix, by using Matlab 6, in general involves much effort, which is diametrically opposed to the easy calculations of Geršgorin disks. On the other hand, the Geršgorin-like result of C.R. Johnson (1973), in Theorem 3.9, is a very interesting result, as it contains the field of values of a matrix, and it is easy to implement. Our proof of Theorem 3.9 follows that of C.R. Johnson (1973).

Note that some results, related to Theorem 3.9, are discussed in Exercises 3 and 4 of this section. Exercise 3 shows that generalization of Theorem 3.9 to higher-order lemniscates are **not** in general valid, while Exercise 4 shows that there is a minor improvement in Theorem 3.9, where the triangle inequality in (3.37) is improved upon. This unpublished improvement was the result of an interesting e-mail correspondence with Anne Greenbaum, for which the author is again most thankful.

Our treatment of the field of values is brief, only because there are excellent detailed discussions of this in Horn and Johnson (1991), Ch.1 and Eiermann and Niethammer (1997), where, for example, the convexity of the field of values (i.e., the Toeplitz-Hausdorff Theorem) is established. It should also be mentioned that the field of values, as defined in (3.25), has been extended in Bauer (1962), Bauer (1968) and Zenger (1968). It should be remarked that these generalizations, unlike the field of values of (3.48), may produce sets in the complex plane which are **not** convex (cf. Nirschl and Schneider (1964)). In addition, proving the exact analog of Geršgorin's Theorem 1.6 (i.e., where each component of $\Gamma(A)$ of (1.5), consisting of r of its Geršgorin disks, contain precisely r eigenvalues of A), by means of Bauer's generalizations of the field of values of a matrix, "seems to be impossible" (cf. Zenger (1984), p.69). See also Deutsch and Zenger (1975).

3.3 This section, on newer eigenvalue inclusion sets, was added to bring to the readers' attention that there **is** on-going research in this field where extensions are based on linear algebra tools. The papers by Dashnic and Zusmanovich (1970), Huang (1995) and Cvetkovic, Kostic and Varga (2004) consider similar generalization of diagonal dominance, via a partitioning of $N = \{1, 2, \cdots, n\}$ into two disjoint subsets. The associated eigenvalue inclusion set, for a given matrix, is then compared in this section with those associated Geršgorin set and the Brauer set. Curiously, the extension, along the same lines, to partitionings of N into *more than two* disjoint subsets becomes very tedious, but it can be easily carried out, however, using the theory of matrix and vector norms as tools. This is done in Chapter 6.

3.4 This section is devoted to the results of Pupkov (1984) and Solov'ev (1984), which has inspired papers by Brualdi and Mellendorf (1994) and Hoffman (2000), on related eigenvalue inclusion results. It is an elaborate scheme, which is, perhaps, difficult to implement, but it does have a beauty of its own.

4. Minimal Geršgorin Sets and Their Sharpness

4.1 Minimal Geršgorin Sets

To begin this chapter, recall from Chapter 1 that if $N := \{1, 2, \ldots, n\}$, if $A = [a_{i,j}] \in \mathbb{C}^{n \times n}$ with $n \geq 2$, and if $\mathbf{x} = [x_1, x_2, \ldots, x_n]^T > \mathbf{0}$ in $\mathbb{R}^n$, then

$$\begin{aligned} r_i^{\mathbf{x}}(A) &:= \sum_{j \in N \setminus \{i\}} |a_{i,j}| x_j / x_i \quad (i \in N), \\ \Gamma_i^{r^{\mathbf{x}}}(A) &:= \{z \in \mathbb{C} : |z - a_{i,i}| \leq r_i^{\mathbf{x}}(A)\} \quad (i \in N), \text{ and} \\ \Gamma^{r^{\mathbf{x}}}(A) &:= \bigcup_{i \in N} \Gamma_i^{r^{\mathbf{x}}}(A). \end{aligned}$$

It then follows from Corollary 1.5 that

$$\sigma(A) \subseteq \Gamma^{r^{\mathbf{x}}}(A) \quad (\text{any } \mathbf{x} > \mathbf{0} \text{ in } \mathbb{R}^n),$$

but as this inclusion holds for any $\mathbf{x} > \mathbf{0}$ in $\mathbb{R}^n$, we immediately have

$$\sigma(A) \subseteq \bigcap_{\mathbf{x} > \mathbf{0}} \Gamma^{r^{\mathbf{x}}}(A). \tag{4.1}$$

The quantity on the right in (4.1) is given a special name in

Definition 4.1. For any $A = [a_{i,j}] \in \mathbb{C}^{n \times n}$, $n \geq 2$, then

$$\Gamma^{\mathcal{R}}(A) := \bigcap_{\mathbf{x} > \mathbf{0}} \Gamma^{r^{\mathbf{x}}}(A) \tag{4.2}$$

is the **minimal Geršgorin set for** A, relative to the collection of all weighted row sums, $r_i^{\mathbf{x}}(A)$, where $\mathbf{x} = [x_1, x_2, \ldots, x_n]^T > \mathbf{0}$ in $\mathbb{R}^n$.

The minimal Geršgorin set of (4.2) is of interest theoretically because it gives a set, containing $\sigma(A)$ in the complex plane, which is a subset of the weighted Geršgorin set $\Gamma^{r^{\mathbf{x}}}(A)$ for *any* $\mathbf{x} > \mathbf{0}$ in $\mathbb{R}^n$.

It is easily seen that the inclusion of (4.1) is valid for *each* matrix B in either of the following subsets $\Omega(A)$ and $\hat{\Omega}(A)$ of $\mathbb{C}^{n \times n}$, $n \geq 2$, where

$$
(4.3)\quad \begin{cases} \Omega(A) := \{B = [b_{i,j}] \in C^{n\times n} : b_{i,i} = a_{i,i} \text{ and } |b_{i,j}| = |a_{i,j}|, i \neq j (\text{all } i,j \in N)\}, \\ \text{and} \\ \hat{\Omega}(A) := \{B = [b_{i,j}] \in C^{n\times n} : b_{i,i} = a_{i,i} \text{ and } |b_{i,j}| \leq |a_{i,j}|, i \neq j (\text{all } i,j \in N)\}, \end{cases}
$$

The set $\Omega(A)$ is called the **equimodular set** and the set $\hat{\Omega}(A)$ is called the **extended equimodular set** for A. (The sets $\Omega(A)$ and $\hat{\Omega}(A)$ are seen to be more *restrictive* than the equiradial set $\omega(A)$ and the extended equiradial set $\hat{\omega}(A)$ of (2.15) and (2.16). See Exercise 2 of this section.) With the notation (cf. (2.17)) of

$$
(4.4)\qquad \sigma(\Omega(A)) := \bigcup_{B\in\Omega(A)} \sigma(B) \quad \text{and } \sigma(\hat{\Omega}(A)) := \bigcup_{B\in\hat{\Omega}(A)} \sigma(B),
$$

it readily follows (see Exercise 3 of this section) that

$$
(4.5)\qquad \sigma(\Omega(A)) \subseteq \sigma(\hat{\Omega}(A)) \subseteq \Gamma^R(A)
$$

We now investigate the *sharpness* of the inclusions of (4.5), for any given $A \in \mathbb{C}^{n\times n}$. The following result (which is given as Exercise 4 of this section) is the basis for the definition of the set $\Omega(A)$ of (4.3).

Lemma 4.2. *Given matrices $A = [a_{i,j}] \in \mathbb{C}^{n\times n}$ and $B = [b_{i,j}] \in \mathbb{C}^{n\times n}, n \geq 2$, then $r_i^{\mathbf{x}}(B) = r_i^{\mathbf{x}}(A)$, for all $i \in N$ and all $\mathbf{x} = [x_1, x_2, \cdots, x_n]^T > \mathbf{0}$ in $\mathbb{R}^n$, if and only if $|b_{i,j}| = |a_{i,j}|$ for all $i \neq j$ $(i,j \in N)$.*

Because of Lemma 4.2, we see that, for a given matrix $A = [a_{i,j}] \in \mathbb{C}^{n\times n}, n \geq 2$, we can only change each nonzero off-diagonal entry by a multiplicative factor $e^{i\theta}$, to define the associated entry of a matrix $B = [b_{i,j}]$ in $\Omega(A)$ of (4.3). This greater restriction gives rise to tighter eigenvalue inclusion results, but the tools for analyzing the matrices in $\Omega(A)$, as we shall see below, shift from graph theory and cycles of directed graphs, to the Perron-Frobenius theory of nonnegative matrices, our **second recurring theme** in this book.

To start, given any matrix $A = [a_{i,j}]$ in $\mathbb{C}^{n\times n}$, let z be any complex number and define the matrix $\mathcal{Q} = [q_{i,j}]$ in $\mathbb{R}^{n\times n}$ by

$$
(4.6)\qquad q_{i,i} := -|z - a_{i,i}| \text{ and } q_{i,j} := |a_{i,j}| \text{ for } i \neq j \qquad (i,j \in N),
$$

so that $\mathcal{Q}$ is dependent on A and z. It is evident from (4.6) that all off-diagonal entries of $\mathcal{Q}$ are nonnegative real numbers. On setting

$$
(4.7)\qquad \mu := \max\{|z - a_{i,i}| : i \in N\},
$$

and on defining the matrix $B = [b_{i,j}]$ in $\mathbb{R}^{n\times n}$ by

$$
(4.8)\qquad b_{i,i} := \mu - |z - a_{i,i}| \text{ and } b_{i,j} := |a_{i,j}| \text{ for } i \neq j \qquad (i,j \in N),
$$

then B has only nonnegative entries, which we write as $B \geq O$. With the above definitions of μ and B, the matrix $\mathcal{Q}$ can then be expressed as

$$\mathcal{Q} = -\mu I_n + B, \tag{4.9}$$

so that the eigenvalues of $\mathcal{Q}$ are a simple shift, by $-\mu$, of the eigenvalues of B. (The matrix $\mathcal{Q}$ is said to be an **essentially nonnegative matrix**, in the terminology of Appendix C, so that, from (C.3), $-Q \in \mathbb{Z}^{n\times n}$; see also Exercise 5 of this section.) Consequently, using the Perron-Frobenius theory of nonnegative matrices (see Theorem C.2 of Appendix C), $\mathcal{Q}$ possesses a real eigenvalue $\nu(z)$ which has the property that if λ is any eigenvalue of $\mathcal{Q}(z)$, then

$$Re\ \lambda \leq \nu(z). \tag{4.10}$$

It is also known (from Theorem C.2 of Appendix C) that to $\nu(z)$, there corresponds a nonnegative eigenvector $\mathbf{y}$ in $\mathbb{R}^n$, i.e.,

$$\mathcal{Q}\mathbf{y} = \nu(z)\mathbf{y} \text{ where } \mathbf{y} \geq \mathbf{0} \text{ with } \mathbf{y} \neq \mathbf{0}, \tag{4.11}$$

and that $\nu(z)$ can be characterized by

$$\nu(z) = \inf_{\mathbf{x}>\mathbf{0}} \left\{ \max_{i\in N} \left[(\mathcal{Q}\mathbf{x})_i / x_i \right] \right\}. \tag{4.12}$$

We also note from (4.6) that

$$(\mathcal{Q}\mathbf{x})_i / x_i = r_i^{\mathbf{x}}(A) - |z - a_{i,i}| \qquad (i \in N,\ \mathbf{x} > \mathbf{0} \text{ in } \mathbb{R}^n), \tag{4.13}$$

which will be used below. We further remark that since the entries of the matrix $\mathcal{Q}$ are, from (4.6), continuous in the variable z, then $\nu(z)$, as an eigenvalue of $\mathcal{Q}$, is also a *continuous* function of the complex variable z.

The coupling of the function $\nu(z)$ to points of the minimal Geršgorin set $\Gamma^{\mathcal{R}}(A)$ of Definition 4.1 is provided by

Proposition 4.3. *For any* $A = [a_{i,j}] \in \mathbb{C}^{n\times n}, n \geq 2$, *then* (cf. (4.2))

$$z \in \Gamma^{\mathcal{R}}(A) \ if \ and \ only \ if \ \nu(z) \geq 0. \tag{4.14}$$

Proof. Suppose that z is an arbitrary point of the minimal Geršgorin set $\Gamma^{\mathcal{R}}(A)$ for A. By Definition 4.1, $z \in \Gamma^{r^{\mathbf{x}}}(A)$ for each $\mathbf{x} > \mathbf{0}$ in $\mathbb{R}^n$. Consequently, for each $\mathbf{x} > \mathbf{0}$, there exists an $i \in N$ (with i dependent on $\mathbf{x}$) such that $|z - a_{i,i}| \leq r_i^{\mathbf{x}}(A)$, or equivalently, $r_i^{\mathbf{x}}(A) - |z - a_{i,i}| \geq 0$. But from (4.13), this means, for this value of i, that $(\mathcal{Q}\mathbf{x})_i / x_i \geq 0$. Hence, $\max_{j\in N} [(\mathcal{Q}\mathbf{x})_j / x_j] \geq 0$ for each $\mathbf{x} > \mathbf{0}$, and from (4.12), it follows that $\nu(z) \geq 0$.

Conversely, suppose that $\nu(z) \geq 0$. Then for each $\mathbf{x} > \mathbf{0}$ in $\mathbb{R}^n$, (4.12) gives us that there is an $i \in N$ (with i dependent on $\mathbf{x}$) such that

$$0 \leq \nu(z) \leq (\mathcal{Q}\mathbf{x})_i/x_i = r_i^{\mathbf{x}}(A) - |z - a_{i,i}|,$$

the last equality coming from (4.13). As the above inequalities imply that $|z - a_{i,i}| \leq r_i^{\mathbf{x}}(A)$, then $z \in \Gamma_i^{r^{\mathbf{x}}}(A)$, and thus, $z \in \Gamma^{r^{\mathbf{x}}}(A)$. But as this inclusion holds for each $\mathbf{x} > \mathbf{0}$, then $z \in \Gamma^{\mathcal{R}}(A)$ from (4.2) of Definition 4.1. ∎

With $\mathbb{C}_\infty := \mathbb{C} \cup \{\infty\}$ again denoting the extended complex plane, then $(\Gamma^{\mathcal{R}}(A))' := \mathbb{C}_\infty \backslash \Gamma^{\mathcal{R}}(A)$ denotes the *complement* of $\Gamma^{\mathcal{R}}(A)$ in the extended complex plane $\mathbb{C}_\infty$. As $\Gamma^{\mathcal{R}}(A)$ is a compact set in $\mathbb{C}$, its complement is open and unbounded. Moreover, Proposition 4.3 shows that $z \in (\Gamma^{\mathcal{R}}(A))'$ if and only if $\nu(z) < 0$. Now, the *boundary* of $\Gamma^{\mathcal{R}}(A)$, denoted by $\partial\Gamma^{\mathcal{R}}(A)$, is defined as usual by

$$\text{(4.15)} \qquad \partial\Gamma^{\mathcal{R}}(A) := \overline{\Gamma^{\mathcal{R}}(A)} \cap \overline{(\Gamma^{\mathcal{R}}(A))'} = \Gamma^{\mathcal{R}}(A) \cap \overline{(\Gamma^{\mathcal{R}}(A))'},$$

the last equality arising from the fact that $\Gamma^{\mathcal{R}}(A)$ is closed. Thus, it follows from Proposition 4.3 and the continuity of $\nu(z)$, as a function of z, that

$$\text{(4.16)} \quad z \in \partial\Gamma^{\mathcal{R}}(A) \text{ if and only if } \begin{cases} i)\ \nu(z) = 0, \text{ and} \\ ii)\ \text{there exists a sequence of complex} \\ \quad \text{numbers } \{z_j\}_{j=1}^{\infty} \text{with } \lim\limits_{j\to\infty} z_j = z, \\ \quad \text{for which } \nu(z_j) < 0 \text{ for all } j \geq 1. \end{cases}$$

We remark that $\nu(z) = 0$ in (4.16) *alone* does not in general imply that $z \in \partial\Gamma^{\mathcal{R}}(A)$; see Exercise 9 of this section.

As a first step in assessing the sharpness of the inclusions in (4.5), we establish

Theorem 4.4. *For any $A = [a_{i,j}] \in \mathbb{C}^{n\times n}$ and any $z \in \mathbb{C}$ with $\nu(z) = 0$, there is a matrix $B = [b_{i,j}] \in \Omega(A)$ (cf. (4.3)) for which z is an eigenvalue of B. In particular, each point of $\partial\Gamma^{\mathcal{R}}(A)$, from (4.16), is in $\sigma(\Omega(A))$, and*

$$\text{(4.17)} \qquad \partial\Gamma^{\mathcal{R}}(A) \subseteq \sigma(\Omega(A)) \subseteq \sigma(\hat{\Omega}(A)) \subseteq \Gamma^{\mathcal{R}}(A).$$

Proof. If $z \in \mathbb{C}$ is such that $\nu(z) = 0$, then from (4.11) there is a vector $\mathbf{y} \geq \mathbf{0}$ in $\mathbb{R}^n$ with $\mathbf{y} \neq \mathbf{0}$, such that $\mathcal{Q}\mathbf{y} = \mathbf{0}$, or equivalently, from (4.6),

$$\text{(4.18)} \qquad \sum_{j\in N\backslash\{k\}} |a_{k,j}| y_j = |z - a_{k,k}| y_k \qquad (\text{all } k \in N).$$

Next, let the real numbers $\{\psi_j\}_{j=1}^n$ satisfy

$$\text{(4.19)} \qquad z - a_{k,k} = |z - a_{k,k}| e^{i\psi_k} \qquad (\text{all } k \in N).$$

With these numbers $\{\psi_j\}_{j=1}^n$, define the matrix $B = [b_{k,j}]$ in $\mathbb{C}^{n \times n}$ by means of

$$(4.20) \qquad b_{k,k} := a_{k,k} \text{ and } b_{k,j} := |a_{k,j}|e^{i\psi_k} \text{ for } k \neq j \qquad (k,j \in N),$$

where it follows from (4.3) that $B \in \Omega(A)$. On computing $(B\mathbf{y})_k$, we find from (4.20) that

$$(B\mathbf{y})_k = \sum_{j \in N} b_{k,j} y_j = a_{k,k} y_k + e^{i\psi_k} \left(\sum_{j \in N \backslash \{k\}} |a_{k,j}| y_j \right) \qquad (k \in N).$$

But from (4.19), $a_{k,k} = z - |z - a_{k,k}|e^{i\psi_k}$ for all $k \in N$, and substituting this in the above equations gives

$$(B\mathbf{y})_k = z y_k + e^{i\psi_k} \left[-|z - a_{k,k}| y_k + \sum_{j \in N \backslash \{k\}} |a_{k,j}| y_j \right] \qquad (k \in N).$$

As the terms in the above brackets are zero from (4.18) for all $k \in N$, the above n equations can be expressed, in matrix form, simply as

$$B\mathbf{y} = z\mathbf{y}.$$

As $\mathbf{y} \neq \mathbf{0}$, then z is an eigenvalue of the particular matrix B of (4.20). Since $B \in \Omega(A)$, this shows that $\nu(z) = 0$ implies $z \in \sigma(\Omega(A))$. Finally, since each point z of $\partial\Gamma^{\mathcal{R}}(A)$ necessarily satisfies $\nu(z) = 0$ from (4.16i), then $\partial\Gamma^{\mathcal{R}}(A) \subseteq \sigma(\Omega(A))$ which, with the inclusions of (4.5), gives the final result of (4.17). ■

It may come as a surprise that the first inclusion in (4.17) of Theorem 4.4 is valid, *without* having A irreducible. To understand why this inclusion is valid also for reducible matrices A, suppose that $A = [a_{i,j}]$ in $\mathbb{C}^{n \times n}$, $n \geq 2$, is reducible, of the form (cf. (1.19))

$$A = \left[\begin{array}{c|c} A_{1,1} & A_{1,2} \\ \hline O & A_{2,2} \end{array} \right],$$

where $A_{1,1} \in \mathbb{C}^{s \times s}$ and $A_{2,2} \in \mathbb{C}^{(n-s) \times (n-s)}$, with $1 \leq s < n$. The form of the matrix A above implies that

$$\sigma(A) = \sigma(A_{1,1}) \cup \sigma(A_{2,2}),$$

so that the submatrix $A_{1,2}$ has *no* effect on the eigenvalues of A. However, the entries of $A_{1,2}$ *can* affect the weighted row sums $r_i^{\mathbf{x}}(A)$ for $1 \leq i \leq s$, since

$$r_i^{\mathbf{x}}(A) = \sum_{\substack{j=1 \\ j \neq i}}^{s} |a_{i,j}| x_j / x_i + \sum_{j=s+1}^{n} |a_{i,j}| x_j / x_i \qquad (1 \leq i \leq s).$$

But, by making the positive components $x_{s+1}, x_{s+2}, \ldots, x_n$ *small* while simultaneously making the positive components $x_1, x_2, \ldots, x_s$ *large*, the last sum above can be made arbitrarily small, where such choices are permitted because the intersection in (4.2), which defines $\Gamma^{\mathcal{R}}(A)$, is over *all* $\mathbf{x} > \mathbf{0}$ in $\mathbb{R}^n$.

What remains from Theorem 4.4 is to investigate the sharpness of the final inclusion in (4.17). This is completed in

Theorem 4.5. *For any* $A = [a_{i,j}] \in \mathbb{C}^{n \times n}$, *then* (cf.(4.3))

$$\sigma(\hat{\Omega}(A)) = \Gamma^{\mathcal{R}}(A). \tag{4.21}$$

Proof. Let z be an arbitrary point of $\Gamma^{\mathcal{R}}(A)$, so that from Proposition 4.3, $\nu(z) \geq 0$. From (4.11), there is a vector $\mathbf{y} \geq \mathbf{0}$ in $\mathbb{R}^n$ with $\mathbf{y} \neq \mathbf{0}$, such that $\mathcal{Q}\mathbf{y} = \nu(z)\mathbf{y}$. The components of this last equation can be expressed (cf. (4.6)) as

$$\sum_{j \in N \backslash \{k\}} |a_{k,j}| y_j = \{|z - a_{k,k}| + \nu(z)\} y_k \qquad (\text{all } k \in N). \tag{4.22}$$

Now, define the matrix $B = [b_{i,j}]$ in $\mathbb{C}^{n \times n}$ by

$$b_{k,k} := a_{k,k} \text{ and } b_{k,j} := \mu_k a_{k,j} \text{ for } k \neq j \qquad (k, j \in N), \tag{4.23}$$

where

$$\begin{cases} \mu_k := \dfrac{\left(\displaystyle\sum_{j \in N \backslash \{k\}} |a_{k,j}| y_j \right) - \nu(z) y_k}{\displaystyle\sum_{j \in N \backslash \{k\}} |a_{k,j}| y_j}, \text{ if } \displaystyle\sum_{j \in N \backslash \{k\}} |a_{k,j}| y_j > 0, \text{ and} \\ \\ \mu_k := 1, \text{ if } \displaystyle\sum_{j \in N \backslash \{k\}} |a_{k,j}| y_j = 0. \end{cases} \tag{4.24}$$

From (4.22), (4.24), and the fact that both $|z - a_{k,k}| y_k \geq 0$ and $\nu(z) y_k \geq 0$ hold for all $k \in N$, it readily follows that $0 \leq \mu_k \leq 1$ $(k \in N)$. Thus, from the definitions in (4.3) and (4.23), we see that $B \in \hat{\Omega}(A)$. Next, on carefully considering the two cases of (4.24), it can be verified, using equations (4.22)-(4.24), that

$$|z - b_{k,k}|y_k = |z - a_{k,k}|y_k = \left(\sum_{j \in N \backslash \{k\}} |a_{k,j}|y_j \right) - \nu(z) y_k$$

$$= \mu_k \cdot \sum_{j \in N \backslash \{k\}} |a_{k,j}|y_j = \sum_{j \in N \backslash \{k\}} |b_{k,j}|y_j \qquad (k \in N),$$

i.e.,

$$|z - b_{k,k}|y_k = \sum_{j \in N \backslash \{k\}} |b_{k,j}|y_j \qquad (k \in N).$$

Now, the above expression is exactly of the form of that in (4.18) in the proof of Theorem 4.4, and the same proof, as in Theorem 4.4, shows that there is a matrix $E = [e_{i,j}]$ in $\mathbb{C}^{n \times n}$, with $E \in \Omega(B)$, such that $z \in \sigma(E)$. But, as $E \in \Omega(B)$ and $B \in \hat{\Omega}(A)$ together imply (cf. (4.3)) that $E \in \hat{\Omega}(A)$, then $z \in \sigma(\hat{\Omega}(A))$. ∎

Theorem 4.5 states that $\sigma(\hat{\Omega}(A))$ completely **fills out** $\Gamma^{\mathcal{R}}(A)$, i.e., $\sigma(\hat{\Omega}(A)) = \Gamma^{\mathcal{R}}(A)$, and, as $\Omega(A) \subseteq \hat{\Omega}(A)$ from (4.3), this then draws our attention to determining exactly what the spectrum $\sigma(\Omega(A))$ actually is. (Note from (4.3) that $\Omega(A)$ is always a *proper subset* of $\hat{\Omega}(A)$, unless A is a diagonal matrix.) While we know from Theorem 4.4 that

$$\partial\Gamma^{\mathcal{R}}(A) \subseteq \sigma(\Omega(A)) \subseteq \Gamma^{\mathcal{R}}(A) \text{ for any } A \in \mathbb{C}^{n \times n},$$

this implies that if $\sigma(\Omega(A))$ is a proper subset of $\Gamma^{\mathcal{R}}(A)$ with $\sigma(\Omega(A)) \neq \partial\Gamma^{\mathcal{R}}(A)$, then $\sigma(\Omega(A))$ necessarily has **internal boundaries** in $\Gamma^{\mathcal{R}}(A)$! While internal boundaries of $\sigma(\Omega(A))$ are discussed in the next section, we include the following result concerning other geometric properties of $\Gamma^{\mathcal{R}}(A)$.

Theorem 4.6. *For any irreducible $A = [a_{i,j}] \in \mathbb{C}^{n \times n}$, $n \geq 2$, then $\nu(a_{i,i}) > 0$ for each $i \in N$. Moreover, for each $a_{i,i}$ and for each real θ with $0 \leq \theta \leq 2\pi$, let $\hat{\rho}_i(\theta) > 0$ be the smallest $\rho > 0$ for which*

$$(4.25) \qquad \begin{cases} \nu(a_{i,i} + \hat{\rho}_i(\theta)e^{i\theta}) = 0, \text{ and} \\ \text{there exists a sequence of complex numbers } \{z_j\}_{j=1}^{\infty} \text{ with} \\ \lim\limits_{j \to \infty} z_j = a_{i,i} + \hat{\rho}_i(\theta)e^{i\theta}, \text{ such that } \nu(z_j) < 0 \text{ for all } j \geq 1. \end{cases}$$

Then, the complex interval $[a_{i,i} + te^{i\theta}]$, for $0 \leq t \leq \hat{\rho}_i(\theta)$, is contained in $\Gamma^{\mathcal{R}}(A)$ for each real θ, and, consequently, the set

$$(4.26) \qquad \bigcup_{\theta \text{ real}} [a_{i,i} + te^{i\theta}]_{t=0}^{\hat{\rho}_i(\theta)}$$

is a **star-shaped subset** *(with respect to $a_{i,i}$) of $\Gamma^{\mathcal{R}}(A)$, for each $i \in N$.*

Proof. First, the irreducibility of A, as $n \geq 2$, gives that the matrix $\mathcal{Q}$ of (4.6) is also irreducible for any choice of z, and this irreducibility also gives (from *ii*) of Theorem C.1 of Appendix C) the following extended characterization (cf. (4.12)) of $\nu(z)$: for any $z \in \mathbb{C}$, there is a $\mathbf{y} > \mathbf{0}$ in $\mathbb{R}^n$ such that

$$\nu(z) = [(\mathcal{Q}\mathbf{y})_j / y_j] \quad \text{for all } j \in N.$$

Then for any $i \in N$, choose $z = a_{i,i}$ and let $\mathbf{x} > \mathbf{0}$ be the associated vector for which the above characterization of $\nu(a_{i,i})$ is valid. Then with $j = i$ in the above display and with (4.13), we have

$$\nu(a_{i,i}) = [(\mathcal{Q}\mathbf{x})_i / x_i] = r_i^{\mathbf{x}}(A) - |z - a_{i,i}| = r_i^{\mathbf{x}}(A) > 0,$$

the last inequality holding since A is irreducible. Thus, $\nu(a_{i,i}) > 0$.

Next, for each fixed θ with $0 \leq \theta \leq 2\pi$, consider the semi-infinite complex line

$$a_{i,i} + te^{i\theta} \quad \text{for all } t \geq 0,$$

which emanates from $a_{i,i}$. Clearly, the function $\nu(a_{i,i} + te^{i\theta})$ is positive at $t = 0$, is continuous on this line, and is negative, from (4.14), outside the compact set $\Gamma^{\mathcal{R}}(A)$. Thus, there is a smallest $\hat{\rho}_i(\theta) > 0$ which satisfies (4.25), and from (4.16), we see that $(a_{i,i} + \hat{\rho}_i(\theta)e^{i\theta}) \in \partial\Gamma^{\mathcal{R}}(A)$. This means that, for each real θ, the line segment, which joins the point $a_{i,i}$ to the point $a_{i,i} + \hat{\rho}_i(\theta)e^{i\theta}$ on $\partial\Gamma^{\mathcal{R}}(A)$, is a subset of $\Gamma^{\mathcal{R}}(A)$, as is the union (4.26) of all such line segments. For this reason, this set is **star-shaped**with respect to $a_{i,i}$. ■

We remark that star-shaped sets are a highly useful concept in function theory, and examples of the star-shaped subsets of Theorem 4.6 will be given later in this section.

Continuing now our discussion of possible internal boundaries of $\sigma(\Omega(A))$ in $\Gamma^{\mathcal{R}}(A)$, consider the irreducible matrix

$$B_3 = \begin{bmatrix} 2 & 0 & 1 \\ 0 & 1 & 1 \\ 1 & 1 & 2 \end{bmatrix} \in \mathbb{R}^{3\times 3}. \tag{4.27}$$

Then, any matrix C_3 in $\Omega(B_3)$ can be expressed, from (4.3), as

$$C_3 = \begin{bmatrix} 2 & 0 & e^{i\psi_1} \\ 0 & 1 & e^{i\psi_2} \\ e^{i\psi_3} & e^{i\psi_4} & 2 \end{bmatrix}, \tag{4.28}$$

where $\{\psi_i\}_{i=1}^4$ are arbitrary real numbers satisfying $0 \leq \psi_i \leq 2\pi$. This suggests the following numerical experiment. Using the random number generator from Matlab 6 to generate a certain number S of strings $\{\beta_j(k)\}_{j=1,k=1}^{4,S}$

of four successive random numbers for the interval $[0, 1]$, then the numbers $\{\psi_j(k) := 2\pi\beta_j(k)\}_{j=1,k=1}^{4,S}$ determine S matrices C_3 in $\Omega(B_3)$. With $S = 1,000$, the subroutine "eig" of Matlab 6 then numerically determined and plotted all 3,000 eigenvalues of these 1,000 matrices C_3, and this is shown in Fig. 4.1, where each "+" is an eigenvalue of some one of these 1,000 matrices. Next, note from (4.3) that diag $(A) = \text{diag}[a_{1,1}, a_{2,2}, \cdots, a_{n,n}]$ is clearly a matrix in $\hat{\Omega}(A)$, so that $a_{i,i} \in \sigma(\hat{\Omega}(A))$ for each $i \in N$. In particular, the diagonal entries 1 and 2 of B_3 are necessarily points of $\Gamma^{\mathcal{R}}(B_3)$, but as can be seen from Fig. 4.1, these diagonal elements do *not* appear to be eigenvalues of $\sigma(\Omega(B_3))$. This figure strongly suggests that $\sigma(\Omega(B_3))$ is a *proper subset* of $\Gamma^{\mathcal{R}}(B_3)$ with $\sigma(\Omega(B_3)) \neq \partial\Gamma^{\mathcal{R}}(B_3)$, and that *internal boundaries do occur!*

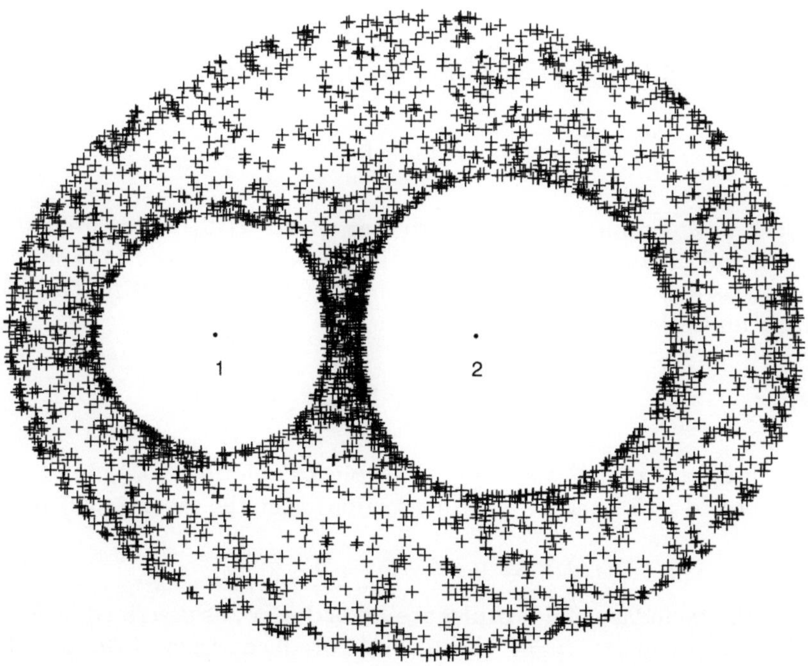

Fig. 4.1. Eigenvalues of 1000 random matrices C_3 of (4.28)

The precise analytical description of $\sigma(\Omega(B_3))$ will be given in (4.64) and (4.65) of Section 4.2.

As a final example in the section, we also consider the matrix A_n in $\mathbb{C}^{n\times n}, n \geq 2$, defined by

$$A_n := \begin{bmatrix} a_{1,1} & a_{1,2} & & & \\ & a_{2,2} & a_{2,3} & & \\ & & \ddots & \ddots & \\ & & & \ddots & a_{n-1,n} \\ a_{n,1} & & & & a_{n,n} \end{bmatrix}, \tag{4.29}$$

where

$$a_{k,k} := \exp\left(\frac{2\pi i(k-1)}{n}\right) \quad (k \in N), \text{ and } |a_{1,2}a_{2,3}\cdots a_{n,1}| = 1. \tag{4.30}$$

Note that A_n is necessarily irreducible because, by (4.30), $a_{1,2}, a_{2,3}, \cdots$, and $a_{n,1}$ are all nonzero. For any $\mathbf{x} = [x_1, x_2, \cdots, x_n]^T > \mathbf{0}$ in $\mathbb{R}^n$, it is further clear that $r_i^{\mathbf{x}}(A_n) = |a_{i,i+1}|\cdot x_{i+1}/x_i$ for $1 \leq i < n$, and that $r_n^{\mathbf{x}}(A_n) = |a_{n,1}|x_1/x_n$, so that, again with the second part of (4.30),

$$\prod_{j=1}^{n} r_j^{\mathbf{x}}(A_n) = 1 \qquad \text{for } all\ \mathbf{x} > \mathbf{0}. \tag{4.31}$$

Now, let z be any complex number such that $z \in \partial\Gamma^{\mathcal{R}}(A_n)$. Then from (4.16), $\nu(z) = 0$, and because A_n is irreducible, it can be seen from (4.11) and (4.13) that there is a $\mathbf{y} > \mathbf{0}$ such that $|z - a_{i,i}| = r_i^{\mathbf{y}}(A_n)$ for all $i \in N$. On taking products and using the result of (4.31), we see that

$$z \in \partial\Gamma^{\mathcal{R}}(A_n) \text{ only if } \prod_{i=1}^{n} |z - a_{i,i}| = 1. \tag{4.32}$$

Then writing $z = r(\theta)e^{i\theta}$ and using the explicit definition of $a_{k,k}$ in (4.30), it can be verified (see Exercise 8 of this section) that the set in (4.32) reduces to

$$(r(\theta))^n = 2\cos(n\theta), \text{ for } \theta \in [0, 2\pi],$$

which is the boundary of a **lemniscate of order n** , as described in Section 2.2. For the special case $n = 4$, the matrix A_4 reduces, from (4.29) and (4.30) with $a_{1,2} = a_{2,3} = a_{3,4} = a_{4,1} = 1$, to

$$A_4 = \begin{bmatrix} 1 & 1 & 0 & 0 \\ 0 & i & 1 & 0 \\ 0 & 0 & -1 & 1 \\ 1 & 0 & 0 & -i \end{bmatrix}, \tag{4.33}$$

and the resultant lemniscate boundary for $\partial\Gamma^{\mathcal{R}}(A_4)$ is shown in Fig. 4.2, along with the "usual" Geršgorin set (cf. (1.5)) $\Gamma(A_4)$ for A_4, whose boundary is given by the outer circular arcs.

We remark, from the special form of the matrix A_n in (4.29), that it can be verified that

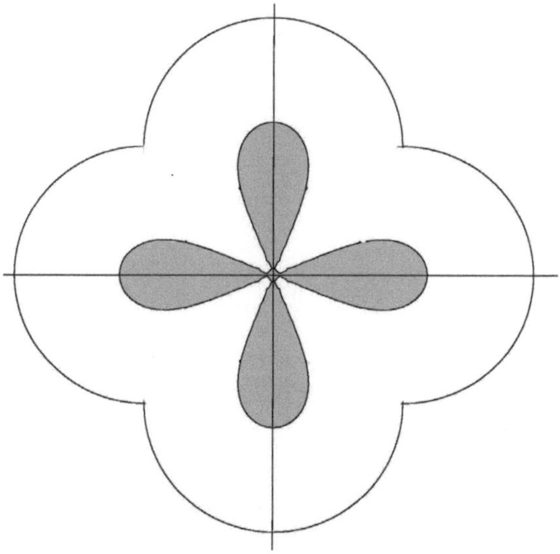

Fig. 4.2. $\partial\Gamma^{\mathcal{R}}(A_4)$ and $\Gamma(A_4)$ for A_4 of (4.33)

$$\det\,(A_n - \lambda I_n) = \prod_{i=1}^{n} (a_{i,i} - \lambda) - (-1)^n a_{1,2}a_{2,3}\cdots a_{n,1},$$

so that if λ is *any* eigenvalue of any $B \in \Omega(A_n)$, it follows from (4.30) that

$$\prod_{i=1}^{n} |a_{i,i} - \lambda| = 1.$$

But from (4.32), we know that (4.33) precisely describes the boundary, $\partial\Gamma^{\mathcal{R}}(A)$, of the minimal Geršgorin set $\Gamma^{\mathcal{R}}(A)$. Thus, for the matrix A_n of (4.29),

$$\partial\Gamma^{\mathcal{R}}(A_n) = \sigma(\Omega(A_n)),$$

which means (cf. Fig. 4.2) that while $\Gamma^{\mathcal{R}}(A_n)$ consists of n petals, where each petal has a nonempty interior, the set $\sigma(\Omega(A_n))$ is exactly the **boundary** of $\Gamma^{\mathcal{R}}(A_n)$, with $\sigma(\Omega(A_n))$ having *no intersection* with the interior of $\Gamma^{\mathcal{R}}(A_n)$. Also in this example, the *star-shaped subset* of $\Gamma^{\mathcal{R}}(A_n)$ of Theorem 4.6, for each diagonal entry, is just the entire petal containing that diagonal entry! Note further that each star-shaped subset cannot be continued beyond the point $z = 0$, as $z = 0$ is a limit point of points z_j with $\nu(z_j) < 0$ (cf. (4.16)).

In the next section, we show, using permutations, how the set $\sigma(\Omega(A))$, with its possible internal boundaries, can be precisely represented.

Exercises

1. Show, with (2.15)-(2.16) and (4.3) that

$$\Omega(A) = \omega(A) \text{ and } \hat{\Omega}(A) = \hat{\omega}(A) \text{ for } n = 2,$$

and that

$$\Omega(A) \subseteq \omega(A) \text{ and } \hat{\Omega}(A) \subseteq \hat{\omega}(A) \text{ for } n > 2.$$

2. From the definition of (4.3), verify that the inclusions of (4.5) are valid.

3. For any $A = [a_{i,j}] \in \mathbb{C}^{2\times 2}$, show (cf.(4.2) and (2.6)) that $\Gamma^{\mathcal{R}}(A) = \mathcal{K}(A)$, i.e., the minimal Geršgorin set for A and the Brauer Cassini set for A **coincide** for any $A \in \mathbb{C}^{2\times 2}$.

4. Give a proof of Lemma 4.2.

5. With the definition in (4.6) of the matrix $\mathcal{Q}$ (which is dependent on z), then $-\mathcal{Q}$ is an element of $\mathbb{Z}^{n\times n}$ of (C.3) of Appendix C, so that $-\mathcal{Q} = \mu I_n - B$ where $B \geq O$. Show that $z \notin \Gamma^{\mathcal{R}}(A)$ if and only if $-\mathcal{Q}$ is a nonsingular M**-matrix**, as defined in Appendix C.

6. With the result of the previous exercise, show that $z \notin \Gamma^{\mathcal{R}}(A)$ implies that every principal minor of $-\mathcal{Q}$ is positive, where a principal minor, of a matrix $A = [a_{i,j}]$ in $\mathbb{C}^{n\times n}$, is the determinant of the matrix which results from deleting the *same* rows and columns from A. (Hint: Apply (A_1) of Theorem 4.6 of Berman and Plemmons (1994), p.134.)

7. For a fixed matrix $A = [a_{i,j}] \in \mathbb{C}^{n\times n}$, we have from (4.12) and (4.13) that

$$\nu(z) := \inf_{\mathbf{x}>\mathbf{0}} \left\{ \max_{i\in N} (r_i^{\mathbf{x}}(A) - |z - a_{i,i}|) \right\} \quad (\text{any } z \in \mathbb{C}).$$

With this definition of $\nu(z)$, show that

$$|\nu(z) - \nu(z')| \leq |z - z'| \quad (\text{all } z, \text{ all } z' \text{ in } \mathbb{C}),$$

so that $\nu(z)$ is **uniformly continuous** on $\mathbb{C}$. (Hint: Write $|z - a_{i,i}| = |(z - z') + (z' - a_{i,i})|$, so that $|z - a_{i,i}| \leq |z - z'| + |z' - a_{i,i}|$.)

8. Verify, with the definition of $\{a_{k,k}\}_{k=1}^n$ in (4.30), that the lemniscate of (4.32) is indeed given by $(r(\theta))^n = 2\cos(n\theta)$, for $\theta \in [0, 2\pi]$.

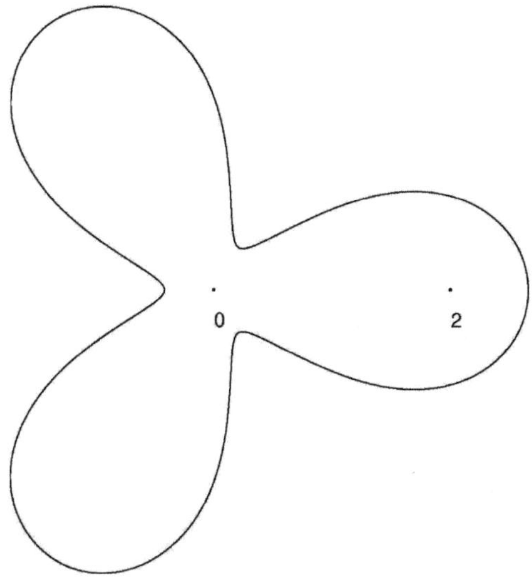

Fig. 4.3. $\Gamma^{\mathcal{R}}(A)$ for the matrix A of Exercise 7

9. Consider the matrix $A = \begin{bmatrix} 2 & 1 & 1 \\ 1 & 2e^{2\pi i/3} & 1 \\ 1 & 1 & 2e^{-2\pi i/3} \end{bmatrix}$, so that (cf. (4.6))

$$\mathcal{Q} = \begin{bmatrix} -|z-2| & 1 & 1 \\ 1 & -|z-2e^{2\pi i/3}| & 1 \\ 1 & 1 & -|z-2e^{-2\pi i/3}| \end{bmatrix}.$$

Show that $\nu(0) = 0$, but that $\nu(z) > 0$ in every sufficiently small deleted neighborhood of $z = 0$. (Hint: Since $\nu(z)$ is, from (4.11), an eigenvalue of $\mathcal{Q}$ for any complex number z, then $\nu(z) = 0$ implies that $\det \mathcal{Q} = 0$, where $\det \mathcal{Q} = 0$, for the given matrix A of this exercise, satisfies

$$\begin{aligned} &-|z-2| \cdot |z-2e^{2\pi i/3}| \cdot |z-2e^{-2\pi i/3}| + 2 \\ &+|z-2| + |z-2e^{2\pi i/3}| + |z-2e^{-2\pi i/3}| = 0. \end{aligned}$$

Then use the fact that $\{z \in \mathbb{C} : \det\mathcal{Q} = 0\}$ consists, from Fig. 4.3, of a Jordan curve, which is $\partial\Gamma^{\mathcal{R}}(A)$, and the sole point $z = 0$ in its interior, and then use Proposition 4.3.)

10. Give an example of an irreducible matrix $A = [a_{i,j}] \in \mathbb{C}^{n \times n}, n \geq 2$, for which one or more of the functions $\{\hat{\rho}_i(\theta)\}_{i=1}^n$ of Theorem 4.6 can be *discontinuous* on $[0, 2\pi]$. (Hint: Apply the result of Exercise 1 to a nonconvex Brauer Cassini set.)

4.2 Minimal Geršgorin Sets via Permutations

The goal of this section is to obtain the precise characterization, in Theorem 4.11, of the set $\sigma(\Omega(A))$ of (4.3) which will then determine any existing internal boundaries of $\sigma(\Omega(A))$ in $\Gamma^{\mathcal{R}}(A)$, as in Fig. 4.1. To obtain such interior boundaries, we make use of the technique, via permutations of Parodi (1952) and Schneider (1954) in Section 3.1, which can give rise to **eigenvalue exclusion sets** in $\mathbb{C}$. (The coupling of this technique with minimal Geršgorin sets, in this section, comes from Levinger and Varga (1966a).)

As in Section 3.1, given a matrix $A = [a_{i,j}] \in \mathbb{C}^{n \times n}$, and given an $\mathbf{x} = [x_1, x_2, \ldots, x_n]^T > \mathbf{0}$ in $\mathbb{R}^n$, let $X :=$ diag $[x_1, x_2, \ldots, x_n]$, so that X is a nonsingular diagonal matrix in $\mathbb{R}^{n \times n}$. If ϕ is any fixed permutation on the elements of $N := \{1, 2, \ldots, n\}$, let

$$P_\phi := [\delta_{i,\phi(j)}] \in \mathbb{R}^{n \times n}$$

denote its associated permutation matrix. From the definition in (4.3), assume that $B = [b_{i,j}] \in \Omega(A)$, and consider the matrix M, defined in analogy with (3.1) by

$$M := \left(X^{-1}BX - \lambda I_n\right) P_\phi = [m_{i,j}] \in \mathbb{C}^{n \times n} \qquad (\lambda \in \mathbb{C}), \tag{4.34}$$

whose entries can be verified to be

$$m_{i,j} := \left(b_{i,\phi(j)} x_{\phi(j)} / x_i\right) - \lambda \delta_{i,\phi(j)} \qquad (i, j \in N). \tag{4.35}$$

As in Section 3.1, if $\lambda \in \sigma(B)$, then the matrix M of (4.34) is evidently singular and thus *cannot* be strictly diagonally dominant (cf. Theorem 1.4). Hence, there exists an $i \in N$ for which

$$|m_{i,i}| \leq \sum_{j \in N \backslash \{i\}} |m_{i,j}|. \tag{4.36}$$

With the definition of $\Omega(A)$ in (4.3) and with the definitions of (4.35), it can be verified that the inequality of (4.36), for *any* $B \in \Omega(A)$, can be expressed, in (4.37), in terms of the familiar weighted row sums $r_i^{\mathbf{x}}(A)$ of A:

$$\begin{cases} |a_{i,i} - \lambda| \leq r_i^{\mathbf{x}}(A), & \text{if } \phi(i) = i, \\ \quad \text{or} & \\ |a_{i,i} - \lambda| \geq -r_i^{\mathbf{x}}(A) + 2|a_{i,\phi(i)}| x_{\phi(i)} / x_i, & \text{if } \phi(i) \neq i. \end{cases} \tag{4.37}$$

The above inequalities are then used, as in Section 3.1, to define the following sets in the complex plane, which are dependent on ϕ and $\mathbf{x}$:

$$(4.38)\quad \begin{cases} \Gamma^{r^{\mathbf{x}}}_{i,\phi}(A) := \{z \in \mathbb{C} : |z - a_{i,i}| \le r_i^{\mathbf{x}}(A)\}, & \text{if } \phi(i) = i, \\ \text{or} & \\ \Gamma^{r^{\mathbf{x}}}_{i,\phi}(A) := \{z \in \mathbb{C} : |z - a_{i,i}| \ge -r_i^{\mathbf{x}}(A) + 2|a_{i,\phi(i)}|x_{\phi(i)}/x_i\}, & \text{if } \phi(i) \ne i, \end{cases}$$

and

$$(4.39)\qquad \Gamma^{r^{\mathbf{x}}}_{\phi}(A) := \bigcup_{i \in N} \Gamma^{r^{\mathbf{x}}}_{i,\phi}(A).$$

Thus, from (3.5)-(3.7), the set $\Gamma^{r^{\mathbf{x}}}_{i,\phi}(A)$ is, for $\phi(i) = i$, a closed disk in $\mathbb{C}$, and is, for $\phi(i) \ne i$, either the extended complex $\mathbb{C}_\infty$ or the closed exterior of a disk. But the point here is that $\Gamma^{r^{\mathbf{x}}}_{\phi}(A)$ must contain all the eigenvalue of **each** $B \in \Gamma(A)$, i.e.,

$$(4.40)\quad \sigma(\Omega(A)) \subseteq \Gamma^{r^{\mathbf{x}}}_{\phi}(A), \text{ for any } \mathbf{x} > \mathbf{0} \text{ in } \mathbb{R}^n \text{ and any permutation } \phi.$$

As the left side of the inclusion of (4.40) is independent of the choice of $\mathbf{x} > \mathbf{0}$, it is immediate that

$$(4.41)\qquad \sigma(\Omega(A)) \subseteq \Gamma^{\mathcal{R}}_{\phi}(A) := \bigcap_{\mathbf{x} > \mathbf{0}} \Gamma^{r^{\mathbf{x}}}_{\phi}(A).$$

Then, we define $\Gamma^{\mathcal{R}}_{\phi}(A)$in (4.41) as the **minimal Geršgorin set, relative to the permutation** ϕ on N and the weighted row sums $r_i^{\mathbf{x}}(A)$. We remark that the closed set $\Gamma^{\mathcal{R}}_{\phi}(A)$ *can* be the extended complex plane $\mathbb{C}_\infty$ for certain permutations $\phi \ne I$, where I is the identity permutation. (An example will be given later in this section.) Next, observe that the inclusion in (4.41) holds for *each* permutation ϕ. If Φ denotes the set of all $n!$ permutations on $N = \{1, 2, \cdots, n\}$, then on taking intersections in (4.41) over all permutations ϕ in Φ, we further have

$$(4.42)\qquad \sigma(\Omega(A)) \subseteq \bigcap_{\phi \in \Phi} \Gamma^{\mathcal{R}}_{\phi}(A).$$

Note that the set $\bigcap_{\phi \in \Phi} \Gamma^{\mathcal{R}}_{\phi}(A)$ is always a compact set in $\mathbb{C}$, since $\Gamma^{\mathcal{R}}_{I}(A)$ is compact.

A natural question, arising from the inclusion above, is if this inclusion of (4.42) is *sharp*. The main object of this section is to show in Theorem 4.11 below that, through the use of permutations ϕ, *equality* actually holds in (4.42), which settles this sharpness question! The importance of this is that permutations ϕ, for which $\Gamma^{r^{\mathbf{x}}}_{\phi}(A)$ is not the extended complex plane, will help define the set of **all** internal and external boundaries of $\sigma(\Omega(A))$.

The pattern of our development here is similar to that of Section 3.1. Our first objective here is to characterize the set $\Gamma_{\phi}^{\mathcal{R}}(A)$ of (4.41) for each permutation ϕ on N, again using the theory on nonnegative matrices.

Fixing $A = [a_{i,j}] \in \mathbb{C}^{n \times n}$, and fixing a permutation ϕ on N, let z be any complex number, and define the real matrix $\mathcal{Q}_{\phi} = [q_{i,j}] \in \mathbb{R}^{n \times n}$ by

$$q_{i,j} := (-1)^{\delta_{i,j}} |a_{i,\phi(j)} - z\delta_{i,\phi(j)}| \qquad (i, j \in N), \tag{4.43}$$

so that $\mathcal{Q}_{\phi}$ is dependent on z, ϕ and A. Of course, if ϕ is the identity permutation on N, written $\phi = I$, then the matrix $\mathcal{Q}_I$ defined by (4.43) is just the matrix of (4.6) in Section 4.1. Note that $\mathcal{Q}_{\phi}$, from (4.43), is a real matrix with nonnegative off-diagonal entries and nonpositive diagonal entries. (In the terminology again of Appendix C, $\mathcal{Q}_{\phi}$ is an *essentially nonnegative matrix.*) As in Section 4.1, it again follows from Theorem C.2 of Appendix C that Q_{ϕ} possesses a real eigenvalue $\nu_{\phi}(z)$ which has the property that if λ is any eigenvalue of Q_{ϕ}, then

$$Re\lambda \leq \nu_{\phi}(z),$$

that to $\nu_{\phi}(z)$ there corresponds a nonnegative eigenvector $\mathbf{y} \geq \mathbf{0}$ in $\mathbb{R}^n$ such that

$$Q_{\phi}\mathbf{y} = \nu_{\phi}(z)\mathbf{y} \text{ where } \mathbf{y} \geq \mathbf{0} \text{ with } \mathbf{y} \neq \mathbf{0}, \tag{4.44}$$

and further that

$$\nu_{\phi}(z) = \inf_{\mathbf{u}>\mathbf{0}} \left\{ \max_{i \in N} \left[(Q_{\phi}\mathbf{u})_i / u_i \right] \right\}. \tag{4.45}$$

We further note that, from its definition, $\nu_{\phi}(z)$ is also a continuous function of z.

The analog of Proposition 4.3 is

Proposition 4.7. *For any $A = [a_{i,j}] \in \mathbb{C}^{n \times n}$, and for any permutation ϕ on N, then* (cf. (4.41))

$$z \in \Gamma_{\phi}^{\mathcal{R}}(A) \textit{ if and only if } \nu_{\phi}(z) \geq 0. \tag{4.46}$$

Proof. To begin, it is convenient to define the quantities

$$\begin{cases} s_{i,\phi}^{\mathbf{x}}(A; z) := r_i^{\mathbf{x}}(A) - |a_{i,i} - z| & \text{if } \phi(i) = i, \\ \text{or} & \\ s_{i,\phi}^{\mathbf{x}}(A; z) := |a_{i,i} - z| + r_i^{\mathbf{x}}(A) - 2|a_{i,\phi(i)}| x_{\phi(i)}/x_i & \text{if } \phi(i) \neq i, \end{cases} \tag{4.47}$$

so that from (4.38), $\Gamma_{i,\phi}^{r^{\mathbf{x}}}(A)$ can be described simply by

$$\Gamma_{i,\phi}^{r^{\mathbf{x}}}(A) := \{z \in \mathbb{C} : s_{i,\phi}^{\mathbf{x}}(A; z) \geq 0\} \qquad (i \in N).$$

Next, for any $\mathbf{x} > \mathbf{0}$ in $\mathbb{R}^n$ with $\mathbf{x} = [x_1, x_2, \cdots, x_n]^T$, define the vector $\mathbf{w} = [w_1, w_2, \cdots, w_n]^T$ in $\mathbb{R}^n$ by means of

$$(4.48) \qquad w_i := x_{\phi(i)} \qquad (i \in N),$$

so that $\mathbf{w} > \mathbf{0}$. Then, it can be verified from (4.43) and (4.47) that, for any $\mathbf{x} > \mathbf{0}$ in $\mathbb{R}^n$,

$$(4.49) \qquad s^{\mathbf{x}}_{i,\phi}(A, z) = \frac{x_{\phi(i)}}{x_i} \left[(\mathcal{Q}_\phi \mathbf{w})_i / w_i \right].$$

To establish (4.46), assume first that $z \in \Gamma^{\mathcal{R}}_\phi(A)$, so that (cf. (4.41)) $z \in \Gamma^{r^{\mathbf{x}}}_\phi(A)$ for every $\mathbf{x} > \mathbf{0}$. But for each $\mathbf{x} > \mathbf{0}$, this implies that there exists an i (with i dependent on $\mathbf{x}$) such that $z \in \Gamma^{r^{\mathbf{x}}}_{i,\phi}(A)$, i.e., from the above definition of $\Gamma^{r^{\mathbf{x}}}_{i,\phi}(A)$, $s^{\mathbf{x}}_{i,\phi}(A; z) \geq 0$. Since $\mathbf{x} > \mathbf{0}$, then $x_{\phi(i)}/x_i > 0$ for all $i \in N$, and therefore, using (4.49), it follows that $(\mathcal{Q}_\phi(\mathbf{w}))_i / w_i \geq 0$. Thus,

$$(4.50) \qquad \max_{j \in N} \left[(\mathcal{Q}_\phi \mathbf{w})_j / w_j \right] \geq 0 \qquad \text{for each } \mathbf{x} > \mathbf{0}.$$

Clearly, as $\mathbf{x} > \mathbf{0}$ runs over all positive vectors in $\mathbb{R}^n$, so does its corresponding vector $\mathbf{w} > \mathbf{0}$, defined in (4.48). But with (4.50), the definition in (4.45) gives us that $\nu_\phi(z) \geq 0$.

Conversely, assume that $\nu_\phi(z) \geq 0$. Again using (4.45) and (4.49), it follows that, for each $\mathbf{x} > \mathbf{0}$, there is an $i \in N$ (with i dependent on $\mathbf{x}$), such that $s^{\mathbf{x}}_{i,\phi}(A; z) \geq 0$. But this implies that $z \in \Gamma^{r^{\mathbf{x}}}_{i,\phi}(A)$, so that from (4.39), $z \in \Gamma^{r^{\mathbf{x}}}_\phi(A)$ for *each* $\mathbf{x} > \mathbf{0}$. Hence, from the definition in (4.41), $z \in \Gamma^{\mathcal{R}}_\phi(A)$. ■

It now follows from Proposition 4.7 and the continuity of $\nu_\phi(z)$, as a function z, that the boundary, $\partial\Gamma^{\mathcal{R}}_\phi(A)$ of $\Gamma^{\mathcal{R}}_\phi(A)$, defined by

$$\partial\Gamma^{\mathcal{R}}_\phi(A) := \overline{\Gamma^{\mathcal{R}}_\phi(A)} \bigcap \overline{(\Gamma^{\mathcal{R}}_\phi(A))'},$$

if it is not empty, can be analogously characterized as follows:

$$(4.51) \quad z \in \partial\Gamma^{\mathcal{R}}_\phi(A) \text{ if and only if } \begin{cases} i) \ \nu_\phi(z) = 0, \text{ and} \\ ii) \text{ there exists a sequence of complex numbers } \{z_j\}_{j=1}^\infty \text{ with } \lim_{j\to\infty} z_j = z, \text{ for which } \nu_\phi(z_j) < 0 \text{ for all } j \geq 1. \end{cases}$$

We note, of course, that the boundary $\partial\Gamma^{\mathcal{R}}_\phi(A)$ *can* be empty, as in the case when $\Gamma^{\mathcal{R}}_\phi(A) = \mathbb{C}_\infty$.

As in Theorem 4.4, we now show that if $\partial\Gamma^{\mathcal{R}}_\phi(A) \neq \emptyset$, *each* boundary point of $\Gamma^{\mathcal{R}}_\phi(A)$ is an eigenvalue of some matrix $B \in \Omega(A)$.

Theorem 4.8. *For any $A = [a_{i,j}] \in \mathbb{C}^{n \times n}$, and for any permutation ϕ on N, then for each $z \in \mathbb{C}$ with $\nu_\phi(z) = 0$, there is a matrix $B = [b_{i,j}] \in \Omega(A)$ for which z an eigenvalue of B. In particular, if $\partial\Gamma_\phi^{\mathcal{R}}(A) \neq \emptyset$, each point of $\partial\Gamma_\phi^{\mathcal{R}}(A)$ is in $\sigma(\Omega(A))$, and*

$$\partial\Gamma_\phi^{\mathcal{R}}(A) \subseteq \sigma(\Omega(A)) \subseteq \Gamma_\phi^{\mathcal{R}}(A). \tag{4.52}$$

Proof. First, assume that z is any complex number for which $\nu_\phi(z) = 0$. Then (cf. (4.44)), there exists a nonnegative eigenvector $\mathbf{y} \geq \mathbf{0}$ in $\mathbb{R}^n$ with $\mathbf{y} \neq \mathbf{0}$, such that $\mathcal{Q}_\phi \mathbf{y} = \mathbf{0}$. Next, let the real numbers $\{\psi_k\}_{k=1}^n$ satisfy

$$z - a_{k,k} = |z - a_{k,k}| e^{i\psi_k} \qquad \text{(for all } k \in N\text{)},$$

and define the matrix $B = [b_{i,j}] \in \mathbb{C}^{n \times n}$ by

$$b_{k,k} := a_{k,k} \text{ and } b_{k,j} := |a_{k,j}| \exp\{i[\psi_k + \pi(-1 + \delta_{k,\phi(k)} + \delta_{j,\phi(k)})]\}, \tag{4.53}$$

for $k \neq j$, where $k, j \in N$. It is evident from (4.53) and the definition in (4.3) that $B \in \Omega(A)$, and if

$$w_{\phi(j)} := y_j \qquad (j \in N),$$

it can be verified (upon considering separately the cases when $\phi(i) = i$ and $\phi(i) \neq i$) that $\mathcal{Q}_\phi \mathbf{y} = \mathbf{0}$ is equivalent to

$$B\mathbf{w} = z\mathbf{w}. \tag{4.54}$$

But since $\mathbf{y} \geq \mathbf{0}$ with $\mathbf{y} \neq \mathbf{0}$, then $\mathbf{w} \neq \mathbf{0}$, and (4.54) thus establishes that z is an eigenvalue of the matrix B, which is an element of $\Omega(A)$. Finally, if $\partial\Gamma_\phi^{\mathcal{R}}(A)$ is not empty, then, as each point of $\partial\Gamma_\phi^{\mathcal{R}}(A)$ necessarily satisfies $\nu_\phi(z) = 0$ from (4.51), the inclusions of (4.52) of Theorem 4.8 directly follow. ■

Next, given any $B = [b_{i,j}] \in \mathbb{C}^{n \times n}$, it is convenient to define the associated set, $\overset{\text{rot}}{\Omega}(B)$, as

$$\overset{\text{rot}}{\Omega}(B) := \left\{C = [c_{i,j}] \in \mathbb{C}^{n \times n} : |c_{i,j}| = |b_{i,j}| \quad (i, j \in N)\right\}. \tag{4.55}$$

The set $\overset{\text{rot}}{\Omega}(B)$, called the **rotated equimodular set** for A, is very much like the equimodular set $\Omega(B)$ of (4.3), but it is larger since the *diagonal entries* of B (as well as the off-diagonal entries of B) are allowed to be multiplied by factors of absolute value 1, in passing to elements of $\overset{\text{rot}}{\Omega}(B)$. This set $\overset{\text{rot}}{\Omega}(B)$ arises very naturally from the following question: Suppose $B = [b_{i,j}]$ in $\mathbb{C}^{n \times n}$ is strictly diagonally dominant, i.e. (cf. (1.11)),

$$|b_{i,i}| > \sum_{j \in N \backslash \{i\}} |b_{i,j}| \quad \text{(for all } i \in N\text{)}, \tag{4.56}$$

so that B is nonsingular from Theorem 1.4. But as (4.56) implies that *each* $C = [c_{i,j}] \in \overset{\text{rot}}{\Omega}(B)$ is also strictly diagonally dominant, then each $C \in \overset{\text{rot}}{\Omega}(B)$ is evidently nonsingular. It is natural to ask is if the **converse** of this is true: If each $C \in \overset{\text{rot}}{\Omega}(B)$ is nonsingular, is B strictly diagonally dominant? The following easy counterexample shows this to be **false**. Consider

$$B := \begin{bmatrix} 2 & 3 \\ 5 & 6 \end{bmatrix}, \text{ so that } C := \begin{bmatrix} 2e^{i\theta_1} & 3e^{i\theta_2} \\ 5e^{i\theta_3} & 6e^{i\theta_4} \end{bmatrix}$$

represents, for real θ_i's, an arbitrary element of $\overset{\text{rot}}{\Omega}(B)$. Then,

$$\det C = 12e^{i(\theta_1+\theta_4)} - 15e^{i(\theta_2+\theta_3)} \neq 0 \text{ for any choices of real } \theta_j\text{'s}.$$

Thus, each matrix in $\overset{\text{rot}}{\Omega}(B)$ is nonsingular, while B, by inspection, *fails* to be strictly diagonally dominant. But surprisingly, Camion and Hoffman (1966) have shown that the converse of the question posed above is, in a broader sense, *true*! Their result is

Theorem 4.9. *Let* $B = [b_{i,j}] \in \mathbb{C}^{n \times n}$*, and assume that each matrix in* $\overset{\text{rot}}{\Omega}(B)$ *of* (4.55) *is nonsingular. Then, there exist a positive diagonal matrix* $X := \text{diag}\,[x_1, x_2, \ldots, x_n]$ *where* $x_i > 0$ *for all* $i \in N$*, and a permutation matrix* $P_\phi := [\delta_{i,\phi(j)}]$ *in* $\mathbb{R}^{n \times n}$*, where* ϕ *is a permutation on* N*, such that* BXP_ϕ *is strictly diagonally dominant.*

As an illustration of Theorem 4.9, it was shown above that the previous matrix B in $\mathbb{R}^{2 \times 2}$ is such that each matrix in $\overset{\text{rot}}{\Omega}(B)$ is nonsingular. Then, the choices of $X = \text{diag}(1, x_2)$, with $x_2 > 0$, and $P_\phi = \begin{bmatrix} 0 & 1 \\ 1 & 0 \end{bmatrix}$ give that

$$BXP_\phi = \begin{bmatrix} 3x_2 & 2 \\ 6x_2 & 5 \end{bmatrix}.$$

It is evident that the matrix BXP_ϕ is strictly diagonally dominant for any x_2 with $2/3 < x_2 < 5/6$, thereby corroborating the result of Theorem 4.9.

We continue with

Lemma 4.10. *For any* $A = [a_{i,j}] \in \mathbb{C}^{n \times n}$*, then*

(4.57) $z \notin \sigma(\Omega(A))$ *if and only if each matrix* R *in* $\overset{\text{rot}}{\Omega}(A - zI_n)$ *is nonsingular.*

Proof. It can be verified (see Exercise 1 of this section), from the definitions of the sets $\overset{\text{rot}}{\Omega}(A - zI_n)$ and $\Omega(A)$, that each matrix R in $\overset{\text{rot}}{\Omega}(A - zI_n)$ can be uniquely expressed as $R = D(B - zI_n)$, where $D :=$ diag $[e^{i\psi_1}, e^{i\psi_2}, \cdots, e^{i\psi_n}]$ with $\{\psi_j\}_{j=1}^n$ real, and where $B \in \Omega(A)$. Now, $z \notin \sigma(\Omega(A))$ implies that $\det(B - zI_n) \neq 0$ for every $B \in \Omega(A)$. But as $|\det D| = 1$, then $R = D(B - zI_n)$ implies that

$$\det R = \det D \cdot \det (B - zI_n) \neq 0 \text{ for each } R \in \overset{\text{rot}}{\Omega}(A - zI_n),$$

and each $R \in \overset{\text{rot}}{\Omega}(A - zI)$ is thus nonsingular. The converse follows similarly. ■

With Lemma 4.10 and the Camion-Hoffman Theorem 4.9, we now establish the sought characterization of $\sigma(\Omega(A))$ in terms of the minimal Geršgorin sets $\Gamma_\phi^{\mathcal{R}}(A)$.

Theorem 4.11. *For any $A = [a_{i,j}] \in \mathbb{C}^{n \times n}$, then*

$$\sigma(\Omega(A)) = \bigcap_{\phi \in \Phi} \Gamma_\phi^{\mathcal{R}}(A), \tag{4.58}$$

where Φ is the set of all permutations ϕ on N.

Proof. Since $\sigma(\Omega(A)) \subseteq \bigcap_{\phi \in \Phi} \Gamma_\phi^{\mathcal{R}}(A)$ from (4.42), then to establish (4.58), it suffices to show that

$$(\sigma(\Omega(A)))' \subseteq \left(\bigcap_{\phi \in \Phi} \Gamma_\phi^{\mathcal{R}}(A) \right)'. \tag{4.59}$$

Consider any $z \in \mathbb{C}$ with $z \notin \sigma(\Omega(A))$. From Lemma 4.10 and Theorem 4.9, there exist a positive diagonal matrix $X =$ diag $[x_1, x_2, \cdots, x_n]$, and a permutation ψ on N, with its associated $n \times n$ permutation matrix $P_\psi = [\delta_{i,\psi(j)}]$, such that the matrix

$$T := (A - zI_n) \cdot X \cdot P_\psi =: [t_{i,j}] \in \mathbb{C}^{n \times n},$$

whose entries are given by

$$t_{i,j} := (a_{i,\psi(j)} - z\delta_{i,\psi(j)})x_{\psi(j)} \qquad (i, j \in N), \tag{4.60}$$

is strictly diagonally dominant, i.e.,

$$|t_{i,i}| > \sum_{j \in N \backslash \{i\}} |t_{i,j}| \qquad (\text{all } i \in N). \tag{4.61}$$

On comparing the definition in (4.60) with the definition in (4.43) of the $n \times n$ matrix $\mathcal{Q}_\psi$ and on setting $w_j := x_{\psi(j)}$ for all $j \in N$ so that $\mathbf{w} := [w_1, w_2, \ldots, w_n]^T > \mathbf{0}$ in $\mathbb{R}^n$, it can be verified that the inequalities in (4.61) can be equivalently expressed as

$$0 > \left(\sum_{j \in N \setminus \{i\}} |t_{i,j}| \right) - |t_{i,i}| = (\mathcal{Q}_\psi \mathbf{w})_i \qquad (\text{all } i \in N). \tag{4.62}$$

But since $\mathbf{w} > \mathbf{0}$, we see that (4.62) gives, from the definition of $\nu_\psi(z)$ in (4.54), that $\nu_\psi(z) < 0$. Hence, from Proposition 4.7, $z \notin \Gamma_\psi^{\mathcal{R}}(A)$, which further implies that $z \notin \bigcap_{\phi \in \Phi} \Gamma_\phi^{\mathcal{R}}(A)$. As z was any point not in $\sigma(\Omega(A))$, (4.59) follows. ∎

Given a matrix $A = [a_{i,j}]$ in $\mathbb{C}^{n \times n}$, it may be the case that a particular permutation ψ on N is such that, for each $\mathbf{x} > \mathbf{0}$, there is an $i \in N$ for which (cf.(4.38)) $\Gamma_{i,\psi}^{r^{\mathbf{x}}}(A) = \mathbb{C}_\infty$, so that (cf.(4.39)) $\Gamma_\psi^{r^{\mathbf{x}}}(A) = \mathbb{C}_\infty$. As this holds for all $\mathbf{x} > \mathbf{0}$, then (cf.(4.41)) $\Gamma_\psi^{\mathcal{R}}(A) = \mathbb{C}_\infty$. Obviously, such a permutation ψ plays no role at all in defining $\bigcap_{\phi \in \Phi} \Gamma_\phi^{\mathcal{R}}(A)$, so we call such a permutation ψ a **trivial permutation** for the matrix A. It turns out, however, that each **nontrivial permutation** ψ for the matrix A may determine **some** boundary of $\bigcap_{\phi \in \Phi} \Gamma_\phi^{\mathcal{R}}(A)$, which, from (4.58) of Theorem 4.11, then helps to delineate the set $\sigma(\Omega(A))$.

To show this more explicitly, consider again the matrix

$$B_3 := \begin{bmatrix} 2 & 0 & 1 \\ 0 & 1 & 1 \\ 1 & 1 & 2 \end{bmatrix} \in \mathbb{R}^{3 \times 3}. \tag{4.63}$$

Using the standard notation (see (2.34)) for representing a permutation in terms of its disjoint cycles, consider the permutation $\phi_1 = (1 \;\; 2)(3)$, i.e., $\phi_1(1) = 2$, $\phi_1(2) = 1$, $\phi_1(3) = 3$. The associated sets $\Gamma_{j,\phi_1}^{r^{\mathbf{x}}}(B_3)$ are, from (4.38), given by

$$\begin{aligned} \Gamma_{1,\phi_1}^{r^{\mathbf{x}}}(B_3) &= \{z \in \mathbb{C} : 0 \leq |z-2| + r_1^{\mathbf{x}}(B_3)\}, \\ \Gamma_{2,\phi_1}^{r^{\mathbf{x}}}(B_3) &= \{z \in \mathbb{C} : 0 \leq |z-1| + r_2^{\mathbf{x}}(B_3)\}, \\ \Gamma_{3,\phi_1}^{r^{\mathbf{x}}}(B_3) &= \{z \in \mathbb{C} : |z-2| \leq r_3^{\mathbf{x}}(B_3)\}, \end{aligned}$$

where each of the first two sets above is clearly the extended complex plane, for any $\mathbf{x} > \mathbf{0}$. Thus, the union of the above sets is the extended complex plane, and ϕ_1 is then a *trivial permutation* for B_3. It can be verified that, of the six permutations of $(1, 2, 3)$, there are only three **nontrivial permutations** for B_3, namely $(13)(2)$, $(1)(23)$, and $I = (1)(2)(3)$, and the associated minimal Geršgorin sets for these three permutations can be further verified to be

$$(4.64)\quad \begin{cases} \Gamma_I^{\mathcal{R}}(B_3) = \{z \in \mathbb{C} : |z-2|^2 \cdot |z-1| \leq |z-1| + |z-2|\}, \\ \Gamma_{(13)(2)}^{\mathcal{R}}(B_3) = \{z \in \mathbb{C} : |z-2|^2 \cdot |z-1| \geq |z-1| - |z-2|\}, \\ \Gamma_{(1)(23)}^{\mathcal{R}}(B_3) = \{z \in \mathbb{C} : |z-2|^2 \cdot |z-1| \geq -|z-1| + |z-2|\}. \end{cases}$$

The boundaries of the above minimal Geršgorin sets are obviously determined by choosing the case of equality in each of the above sets of (4.64). Specifically,

$$(4.65)\quad \begin{cases} \partial\Gamma_I^{\mathcal{R}}(B_3) = \{z \in \mathbb{C} : |z-2|^2 \cdot |z-1| = |z-1| + |z-2|\}, \\ \partial\Gamma_{(13)(2)}^{\mathcal{R}}(B_3) = \{z \in \mathbb{C} : |z-2|^2 \cdot |z-1| = |z-1| - |z-2|\}, \\ \partial\Gamma_{(1)(23)}^{\mathcal{R}}(B_3) = \{z \in \mathbb{C} : |z-2|^2 \cdot |z-1| = -|z-1| + |z-2|\}. \end{cases}$$

As a consequence of Theorem 4.11, $\sigma(\Omega(B_3))$ is the shaded multiply-connected closed set shown in Fig. 4.4. This example shows how the minimal Geršgorin sets $\Gamma_\phi^{\mathcal{R}}(B_3)$, for nontrivial permutations ϕ of B_3, *can* give boundaries defining the spectrum $\sigma(\Omega(B_3))$. Of course, we note that the boundaries, analytically given in (4.65), are what had been *suggested* from the numerical computations of the eigenvalues of C_3 in Fig. 4.1!

While it is the case that, for the matrix B_3 of (4.63), each of its three nontrivial permutations gave rise to a *different boundary* of $\sigma(\Omega(B_3))$, there are matrices A in $\mathbb{C}^{n\times n}, n \geq 2$, where two *different* nontrivial permutations for this matrix give rise to the *same* boundary of $\sigma(\Omega(A))$. (See Exercise 6 of this section.)

Finally in this section, we remark that the irreducible matrix B_3, of (4.63), is such that its cycle set is given by $C(B_3) := (1\;\;3) \cup (2\;\;3)$. Now, the cycles $(1\;\;3)$ and $(2\;\;3)$ are not permutations on the entire set $(1\;\;2\;\;3)$, but they become permutations on $(1\;\;2\;\;3)$ by annexing missing **singletons**, i.e., $(1\;\;3) \mapsto (1\;\;3)(2)$ and $(2\;\;3) \mapsto (2\;\;3)(1)$ are then permutations on $\{1\;\;2\;\;3\}$. Curiously, we see that these permutations, on $\{1,2,3\}$, plus the identity mapping, arise in (4.64) as the **exact** set of nontrivial permutations for the matrix B_3. But, as we shall see, this is no surprise.

Given any irreducible matrix $A = [a_{i,j}] \in \mathbb{C}^{n\times n}$, $n \geq 2$, let $C(A)$ be the cycle set of all (strong) cycles γ in its directed graph, $\mathbb{G}(A)$, as discussed in Section 2.2. Given any (possibly empty) collection $\{\gamma_j\}_{j=1}^m$ of *distinct* cycles of $C(A)$, (i.e., for any k and ℓ with $k \neq \ell$ where $1 \leq k,\ \ell \leq m$, then γ_k and γ_ℓ have **no** common entries), we annex any singletons necessary to form a permutation on $N := \{1, 2, \cdots, n\}$, and call $\Phi(A)$ the set of all such permutations on N. Thus, the identity mapping I is an element of $\Phi(A)$. Then,

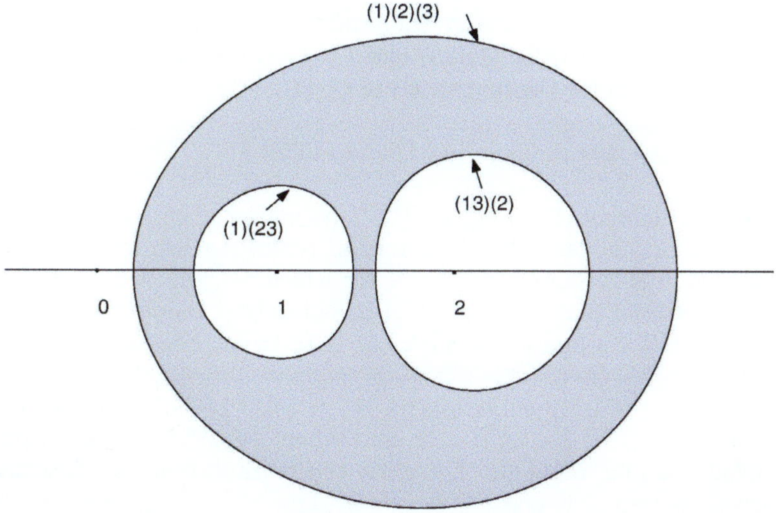

Fig. 4.4. $\sigma(\Omega(B_3)) = \bigcap_{\phi \in \Phi} \Gamma_{\phi}^{\mathcal{R}}(B_3)$ for the matrix B_3 of (4.63), showing the permutations which define the boundaries of $\sigma(\Omega(B_3))$

with the results of Exercise 4 and 5 of this section, we have the new result of Theorem 4.12, which exactly characterizes the set of nontrivial permutations for A. (Its proof is left as Exercise 7 of this section.)

Theorem 4.12. *For any irreducible* $A = [a_{i,j}] \in \mathbb{C}^{n \times n}$*,* $n \geq 2$*, let* $\Phi(A)$ *be the collection of permutations on* $N := \{1, 2, \cdots, n\}$*, derived, as defined above, from the cycle set* $C(A)$ *of* A*. Then,* $\Phi(A)$ *is the exact set of nontrivial permutations for* A*.*

Remark. If $A = [a_{i,j}] \in \mathbb{C}^{n \times n}, n \geq 2$, has all its off-diagonal entries nonzero, it follows from Theorem 4.12 that *every* permutation on N is a nontrivial permutation for A. (See Exercise 8 of this section).

Exercises

1. Given any $A = [a_{i,j}] \in \mathbb{C}^{n \times n}, n \geq 2$, verify, from the definitions in (4.3) and (4.55), that each R in $\overset{\text{rot}}{\Omega}(A)$ can be uniquely expressed as $R = DB$, where $D = \text{diag}[e^{i\psi_1}, \cdots, e^{i\psi_n}]$ with $\{\psi_j\}_{j=1}^n$ real, and where $B \in \Omega(A)$, the equimodular set for A.

2. Verify that of the six permutations on the numbers 1, 2, and 3, only the permutations (13)(2), (1)(23) and $I = (1)(2)(3)$ give rise to nontrivial permutations for the matrix B_3 of (4.63).

3. Verify the expressions in (4.64) and (4.65).

4. Given a matrix $A = [a_{i,j}] \in \mathbb{C}^{n \times n}$, $n \geq 2$, let ϕ be a permutation on N. Then, ϕ is defined to be a **trivial permutation** for A if, for each $\mathbf{y} > \mathbf{0}$ in $\mathbb{R}^n$, there is an $i \in N$ for which $\Gamma_{i,\phi}^{r^{\mathbf{y}}}(A) = \mathbb{C}_\infty$. Show that ϕ is a trivial permutation for A if and only if there is an i in N for which $\phi(i) \neq i$ and for which $a_{i,\phi(i)} = 0$. (Hint: Use (4.38).)

5. Consider the irreducible matrix $A_n = [a_{i,j}]$ of (4.29), where $n \geq 2$, where $|a_{1,2} \cdot a_{2,3} \cdots a_{n,1}| = 1$, and where $\{a_{i,i}\}_{i=1}^n$ are arbitrary but fixed complex numbers. Let ψ be the permutation on N, which, in cyclic notation, is $\psi = (1 \;\; 2 \;\; \cdots \;\; n)$, i.e.,

$$\psi(i) = i + 1 \quad (1 \leq i < n), \text{ and } \psi(n) = 1.$$

If ϕ is a permutation on N with $\phi \neq \psi$ and with $\phi \neq I$, show, using the result of the previous exercise, that $\Gamma_\phi^{r^{\mathbf{x}}}(A_n) = \mathbb{C}_\infty$ for any $\mathbf{x} > \mathbf{0}$ in $\mathbb{R}^n$, i.e., each permutation ϕ, with $\phi \neq \psi$ and $\phi \neq I$, is a trivial permutation for A_n.

6. With the definitions of the previous exercise, show (cf.(4.39)) that

$$\Gamma_I^{\mathcal{R}}(A_n) = \{z \in \mathbb{C} : \prod_{i=1}^n |z - a_{i,i}| \leq 1\}, \text{and}$$
$$\Gamma_\psi^{\mathcal{R}}(A_n) = \{z \in \mathbb{C} : \prod_{i=1}^n |z - a_{i,i}| \geq 1\},$$

so that

$$\Gamma_I^{\mathcal{R}}(A_n) \cap \Gamma_\psi^{\mathcal{R}}(A_n) = \{z \in \mathbb{C} : \prod_{i=1}^n |z - a_{i,i}| = 1\}.$$

Thus, the permutations I and ψ give, respectively, the interior and exterior of the single set

$$\{z \in \mathbb{C} : \prod_{i=1}^n |z - a_{i,i}| = 1\}.$$

7. Using the results of Exercise 4 and 5 above, prove the result of Theorem 4.12.

8. Using Theorem 4.12, show that if $A = [a_{i,j}] \in \mathbb{C}^{n \times n}, n \geq 2$ has all its off-diagonal entries nonzero, then every permutation on N is a nontrivial permutation for A.

9. Let $A = [a_{i,j}] \in \mathbb{C}^{n \times n}$ be irreducible, let ϕ be any permutation mapping of $N = \{1, 2, \cdots, n\}$ onto N, and let $P_\phi := [\delta_{i,\phi(j)}]$ be its associated permutation matrix. Show (Loewy (1971), Lemma 1) that the condition, that $a_{i,i} \neq 0$ for *each* i with $i \neq \phi(i)$, is *sufficient* for the product $A \cdot P_\phi$ to be irreducible. Also, show by means of a 3×3 matrix, that this condition is not always *necessary*.

4.3 A Comparison of Minimal Geršgorin Sets and Brualdi Sets

We have examined, in the previous sections of this chapter, the properties of the minimal Geršgorin set for a given matrix $A = [a_{i,j}]$ in $\mathbb{C}^{n \times n}$, but there is the remaining question of how the minimal Geršgorin set, $\Gamma^{\mathcal{R}}(A)$ of (4.2) for the matrix $A = [a_{i,j}]$, which is a compact set in the complex plane, compares with its Brualdi set $\mathcal{B}(A)$ of (2.40), also a compact set in the complex plane. To answer this question, we begin with

Lemma 4.13. *For any* $A = [a_{i,j}] \in \mathbb{C}^{n \times n}$, *then* (cf. (4.3) and (2.70))

$$\hat{\Omega}(A) \subseteq \overline{\omega}_{\mathcal{B}}(A). \tag{4.66}$$

Proof. First, assume that A is irreducible with $n \geq 2$, so that, $r_i(A) > 0$ for all $i \in N$. For any matrix $B = [b_{i,j}]$ in $\hat{\Omega}(A)$, we have from (4.3) that

$$b_{i,i} = a_{i,i} \text{ (all } i \in N), \text{ and } 0 \leq |b_{i,j}| \leq |a_{i,j}| \text{ (all } i \neq j \text{ in } N). \tag{4.67}$$

Define the set

$$S_i(A) := \{j \in N : j \neq i \text{ and } |a_{i,j}| > 0\} \quad (i \in N). \tag{4.68}$$

Then, $S_i(A) \neq \emptyset$ for any $i \in N$, since A is irreducible. If there is a j in $S_i(A)$ for which $b_{i,j} = 0$, we note from (4.3) and (1.4) that

$$r_i(B) < r_i(A). \tag{4.69}$$

Then, for a fixed $\epsilon > 0$, replace this zero (i,j)-th entry of B by any number having modulus ϵ, and do the same for every k in $S_i(A)$ for which $b_{i,k} = 0$, leaving the remaining entries in this i-th row, of B, unchanged. On carrying out this procedure for all rows of the matrix B, a matrix $B(\epsilon)$, in $\mathbb{C}^{n\times n}$, is created, whose entries are continuous in the parameter ϵ, and for which the cycle set $\mathcal{C}(B(\epsilon))$ of $B(\epsilon)$ is identical with the cycle set $\mathcal{C}(A)$ of A, for each $\epsilon > 0$. In addition, because of the strict inequality in (4.69) whenever $b_{i,j} = 0$ with $j \in S_i(A)$, it follows, for all $\epsilon > 0$ sufficiently small, that

$$r_i(B(\epsilon)) \leq r_i(A) \text{ (all } i \in N). \tag{4.70}$$

Hence from (2.61), $B(\epsilon) \in \hat{\omega}_{\mathcal{B}}(A)$ for all sufficiently small $\epsilon > 0$. As such, we see from the definition in (2.69) that $B(0) = B \in \overline{\hat{\omega}}_{\mathcal{B}}(A)$, and, as this holds for any $B \in \hat{\Omega}(A)$, the inclusion of (4.66) is valid in this irreducible case.

For the case $n = 1$, we see by definition that equality holds in (4.66). Finally, assume that $n \geq 2$ and that A is reducible. Without loss of generality, further assume that A is in the normal reduced form of (2.35). Then, our previous construction, in the irreducible case, can be applied to each irreducible submatrices $R_{j,j}$ of case (2.36i), while the remaining case of (2.36*ii*), as in the case $n = 1$ above, is immediate, completing the proof. ∎

This brings us to the following new, but not unexpected, result which extends that of Varga (2001a). Its proof is remarkably simple.

Theorem 4.14. *For any $A = [a_{i,j}] \in \mathbb{C}^{n\times n}$, then*

$$\Gamma^{\mathcal{R}}(A) \subseteq \mathcal{B}(A). \tag{4.71}$$

Remark. The word "minimal" in the minimal Geršgorin set of (4.2) seems to be appropriate!

Proof. It is known from (4.21) of Theorem 4.5 that

$$\Gamma^{\mathcal{R}}(A) = \sigma(\hat{\Omega}(A)),$$

and as $\hat{\Omega}(A) \subseteq \overline{\hat{\omega}}_{\mathcal{B}}(A)$ from (4.66) of Lemma 4.13, then $\sigma(\hat{\Omega}(A)) \subseteq \sigma(\overline{\hat{\omega}}_{\mathcal{B}}(A))$. But as $\sigma(\overline{\hat{\omega}}_{\mathcal{B}}(A)) = \mathcal{B}(A)$ from (2.71) of Theorem 2.11, then

$$\Gamma^{\mathcal{R}}(A) \subseteq \mathcal{B}(A),$$

the desired result of (4.71). ∎

We remark that equality, of the inclusion in (4.71) of Theorem 4.14, *can* hold, as the following simple example shows. Consider the matrix

$$A_4 = \begin{bmatrix} 1 & 1 & 0 & 0 \\ 0 & i & 1 & 0 \\ 0 & 0 & -1 & 1 \\ 1 & 0 & 0 & -i \end{bmatrix}, \tag{4.72}$$

which was considered in (4.33). Then, A_4 is irreducible, and $\mathcal{C}(A_4)$ consists of just one cycle, $\gamma = \{1 \;\; 2 \;\; 3 \;\; 4\}$. Thus from (2.40),

(4.73) $$\mathcal{B}(A_4) = \{z \in \mathbb{C} : |z^4 - 1| \leq 1\}.$$

As shown in Section 4.2, $\mathcal{B}(A_4)$ of (4.73) is then the set of four petals in Fig. 4.2, along with all their interior points. It can similarly be verified that $\Gamma^{\mathcal{R}}(A_4)$ is exactly the same set, i.e.,

(4.74) $$\Gamma^{\mathcal{R}}(A_4) = \mathcal{B}(A_4),$$

which gives the case of equality in (4.71).

To conclude this chapter, we note that (4.71) of Theorem 4.14 seems, at first glance, somewhat contradictory to the **reverse inclusions** of (2.9) of Theorem 2.3 and (2.48) of Theorem 2.9, namely,

(4.75) $$\mathcal{B}(A) \subseteq \mathcal{K}(A) \subseteq \Gamma(A),$$

which is valid for any $A = [a_{i,j}] \in \mathbb{C}^{n \times n}, n \geq 2$. (Note that $n \geq 2$ is now needed for the definition of the Brauer Cassini set.) But of course, the sets in (4.75) have no dependence on weighted row sums! With $r_i^{\mathbf{x}}(A)$ of (1.13), we can define, for any $\mathbf{x} = [x_1, x_2, \cdots, x_n]^T > \mathbf{0}$ in $\mathbb{R}^n$, the weighted row sum version of the Brauer Cassini oval of (2.5) as

(4.76) $$K_{i,j}^{r^{\mathbf{x}}}(A) := \{z \in \mathbb{C} : |z - a_{i,i}| \cdot |z - a_{j,j}| \leq r_i^{\mathbf{x}}(A) r_j^{\mathbf{x}}(A)\},$$

where $i \neq j;\ i, j \in N$, and the weighted row sums version of $\mathcal{K}(A)$ of (2.5) as

(4.77) $$\mathcal{K}^{r^{\mathbf{x}}}(A) := \bigcup_{\substack{i,j \in N \\ i \neq j}} K_{i,j}^{r^{\mathbf{x}}}(A).$$

Similarly, if $\gamma \in \mathcal{C}(A)$ is any cycle of $\mathbb{G}(A)$, we define

(4.78) $$\mathcal{B}_\gamma^{r^{\mathbf{x}}}(A) := \{z \in \mathbb{C} : \prod_{i \in \gamma} |z - a_{i,i}| \leq \prod_{i \in \gamma} r_i^{\mathbf{x}}(A)\}$$

and

(4.79) $$\mathcal{B}^{r^{\mathbf{x}}}(A) := \bigcup_{\gamma \in \mathcal{C}(A)} \mathcal{B}_\gamma^{r^{\mathbf{x}}}(A).$$

Now, applying the inclusions of (4.75) to the matrix $X^{-1}AX$ where $X := \mathrm{diag}[x_1, x_2, \cdots, x_n]$ with $\mathbf{x} > \mathbf{0}$, it directly follows that

(4.80) $$\mathcal{B}^{r^{\mathbf{x}}}(A) \subseteq \mathcal{K}^{r^{\mathbf{x}}}(A) \subseteq \Gamma^{r^{\mathbf{x}}}(A), \text{ for any } \mathbf{x} > \mathbf{0},$$

so that if

(4.81) $$\mathcal{B}^{\mathcal{R}}(A) := \bigcap_{\mathbf{x} > \mathbf{0}} \mathcal{B}^{r^{\mathbf{x}}}(A) \text{ and } \mathcal{K}^{\mathcal{R}}(A) := \bigcap_{\mathbf{x} > \mathbf{0}} \mathcal{K}^{r^{\mathbf{x}}}(A),$$

we have from (4.80) that

$$\mathcal{B}^{\mathcal{R}}(A) \subseteq \mathcal{K}^{\mathcal{R}}(A) \subseteq \Gamma^{\mathcal{R}}(A). \tag{4.82}$$

On the other hand, we have from (4.71) and (4.79) that

$$\Gamma^{\mathcal{R}}(X^{-1}AX) \subseteq \mathcal{B}(X^{-1}AX) = \mathcal{B}^{r^{\mathbf{x}}}(A) \text{ for any } \mathbf{x} > \mathbf{0}.$$

But as is easily verified (see Exercise 1 of this section), $\Gamma^{\mathcal{R}}(X^{-1}AX) = \Gamma^{\mathcal{R}}(A)$ for any $\mathbf{x} > \mathbf{0}$, so that

$$\Gamma^{\mathcal{R}}(A) \subseteq \mathcal{B}^{r^{\mathbf{x}}}(A) \text{ for any } \mathbf{x} > \mathbf{0}.$$

As this inclusion holds for all $\mathbf{x} > \mathbf{0}$, then with the notation of (4.81), we have

$$\Gamma^{\mathcal{R}}(A) \subseteq \mathcal{B}^{\mathcal{R}}(A). \tag{4.83}$$

Thus, on combining (4.82) and (4.83), we immediately have the somewhat surprising result of

Theorem 4.15. *For any $A = [a_{i,j}] \in \mathbb{C}^{n\times n}, n \geq 2$, then*

$$\Gamma^{\mathcal{R}}(A) = \mathcal{B}^{\mathcal{R}}(A) = \mathcal{K}^{\mathcal{R}}(A). \tag{4.84}$$

Exercises

1. Given $A = [a_{i,j}] \in \mathbb{C}^{n\times n}, n \geq 2$, let $\mathbf{x} = [x_1, x_2, \cdots, x_n]^T > \mathbf{0}$ in $\mathbb{R}^n$ be any positive vector, and set $X := \text{diag}[x_1, x_2, \cdots, x_n]$. Show, using (4.2) of Definition 4.1, that

$$\Gamma^{\mathcal{R}}(X^{-1}AX) = \Gamma^{\mathcal{R}}(A).$$

2. Complete the proof of Lemma 4.13 in the reducible case.

Bibliography and Discussion

4.1 Minimal Geršgorin sets were introduced in Varga (1965), and most of this work is reported in this section. We remark that Theorem 4.7, showing that the spectra of all matrices in $\hat{\Omega}(A)$ of (4.3) exactly fills out the minimal Geršgorin set $\Gamma^{R}(A)$, was established in Varga (1965). This result was also obtained by Engel (1973) in his Corollary 5.4. Theorem 4.6, having to do with star-shaped subsets associated with diagonal entries of the minimal Geršgorin set $\Gamma^{\mathcal{R}}(A)$, is new. We mention that the term, "minimal Geršgorin circles," appears in Elsner (1968), but in a quite different context.

It is worth mentioning that the characterization in Section 3.1, of points z in $\Gamma^{\mathcal{R}}(A)$, was through the theory of **essentially nonnegative matrices**, which are the negatives of matrices in $\mathbb{Z}^{n\times n}$, defined in C.3 of Appendix C. It could, just as well, have been determined from the theory of *M-matrices* because, as is shown in Exercise 3 of Section 4.1, the result that $z \notin \Gamma^{\mathcal{R}}(A)$ is true if and only if $-\mathcal{Q}$ is a nonsingular M-matrix.

The exact determination of the minimal Geršgroin set $\Gamma^{\mathcal{R}}(A)$ of a given matrix A in $\mathbb{C}^{n\times n}$ can be computationally challenging, for even relatively low values of n. Because of this, their value is more for *theoretical purposes.*

4.2 The tool for using permutation matrices to obtain new eigenvalue inclusion results, where **exteriors** of disks then can come into play, is due to Parodi (1952) and Schneider (1954). See also Parodi (1959). The material in this section, based on minimal Geršgorin sets under permutations, comes from Levinger and Varga (1966a), where the aim was to completely specify the boundaries of $\sigma(\Omega(A))$ of (4.4). This was achieved in Theorem 4.11 which was also obtained later by Engel (1973) in his Theorem 5.14. This paper by Engel also contains, in his Corollary 5.11, the result of Camion and Hoffman (1966), mentioned in the proof of our Theorem 4.9.

We note that the exact characterization in Theorem 4.12 of all nontrivial permutations for the matrix A, is new.

4.3 The idea of comparing the minimal Geršgorin set, of a given matrix $A = [a_{i,j}] \in \mathbb{C}^{n\times n}$, with its Brualdi set in Theorem 4.14 is quite new; see also Varga (2001a). Perhaps the final result of Theorem 4.15 (which also appears in Varga (2001a)) is unexpected.

5. G-Functions

5.1 The Sets $\mathcal{F}_n$ and $\mathcal{G}_n$

We have seen in Chapter 1 that if $A = [a_{i,j}] \in \mathbb{C}^{n \times n}$, if $\mathbf{x} = [x_1, x_2, \cdots, x_n]^T \in \mathbb{R}^n$ with $\mathbf{x} > \mathbf{0}$, and if $0 < \alpha < 1$, then, with the definitions of $r_i^{\mathbf{x}}(A)$ and $c_i^{\mathbf{x}}(A) := r_i^{\mathbf{x}}(A^T)$ from (1.13) and (1.35), and with $N := \{1, 2, \cdots, n\}$, each of the following statements:

$$
(5.1) \qquad \begin{array}{ll}
i)\ |a_{j,j}| > r_j^{\mathbf{x}}(A) & (\text{all } j \in N), \\
ii)\ |a_{j,j}| > c_j^{\mathbf{x}}(A) & (\text{all } j \in N), \\
ii)\ |a_{j,j}| > \left(r_j^{\mathbf{x}}(A)\right)^{\alpha} \cdot \left(c_j^{\mathbf{x}}(A)\right)^{1-\alpha} & (\text{all } j \in N),
\end{array}
$$

separately implies that A is nonsingular (where, as in Chapter 1 (cf. (1.4)), we continue the convention that $r_j^{\mathbf{x}}(A) := 0 =: c_j^{\mathbf{x}}(A)$ when $n = 1$). Note that the above statements are *similar*, in that each compares the absolute values of the diagonal entries of A with various combinations of the absolute values of *only* off-diagonal entries of A. It is natural then to study the set of *all* such relations which establish the nonsingularity of matrices in $\mathbb{C}^{n \times n}$. The material in this chapter was inspired by the works of Nowosad (1965) and Hoffman (1969) where G-functions were introduced, but as we shall see, it definitely had its roots in the earlier work of Fan (1958), covered in Theorem 1.21. Our exposition in this section largely follows that of Carlson and Varga (1973a).

We begin with

Definition 5.1. For any positive integer $n, \mathcal{F}_n$ is the collection of all functions $f = [f_1, f_2, \cdots, f_n]$ such that

i) $f : \mathbb{C}^{n \times n} \to \mathbb{R}_+^n$, i.e., $0 \leq f_j(A) < +\infty$, for any $j \in N$ and for any matrix $A = [a_{i,j}] \in \mathbb{C}^{n \times n}$;

ii) for each $k \in N$, $f_k(A)$ depends only on the absolute values of the off-diagonal entries of $A = [a_{i,j}] \in \mathbb{C}^{n \times n}$, i.e., for any $B = [b_{i,j}] \in \mathbb{C}^{n \times n}$ with $|b_{i,j}| = |a_{i,j}|$ for all $i, j \in N$ with $i \neq j$, then $f_k(A) = f_k(B)$ for all $k \in N$, (with the convention, for $n = 1$, that $f = [f_1] \in \mathcal{F}_1$ implies $f_1(A) := 0$ for all $A \in \mathbb{C}^{1 \times 1}$).

Definition 5.2. Let $f = [f_1, f_2, \cdots, f_n] \in \mathcal{F}_n$. Then, f is a **G-function** if, for any $A = [a_{i,j}] \in \mathbb{C}^{n\times n}$, the relations

$$|a_{i,i}| > f_i(A) \quad (\text{all } i \in N), \tag{5.2}$$

imply that A is nonsingular. The set of all G-functions in $\mathcal{F}_n$ is denoted by $\mathcal{G}_n$.

From Definitions 5.1 and 5.2, we see, for $n = 1$, that $\mathcal{F}_1 = \mathcal{G}_1$, where the unique element in $\mathcal{F}_1$ is the null function. We next claim that $\mathcal{G}_n$ is nonempty for any n with $n \geq 2$. To show this, fix any $\mathbf{x} = [x_1, x_2, \cdots, x_n]^T$ in $\mathbb{R}^n$ with $\mathbf{x} > \mathbf{0}$ and fix any α with $0 \leq \alpha \leq 1$. Then, define $f = [f_1, f_2, \cdots, f_n]$ by

$$f_j(A) := (r_j^{\mathbf{x}}(A))^{\alpha} \cdot (c_j^{\mathbf{x}}(A))^{1-\alpha} \qquad (j \in N, A \in \mathbb{C}^{n\times n}).$$

Clearly, $f \in \mathcal{F}_n$ from Definition 5.1, and $f \in \mathcal{G}_n$ from Corollary 1.17. Later, we shall see, from Remark 4 following Theorem 5.5, that $\mathcal{G}_n$ is a *proper* subset of $\mathcal{F}_n$ for any $n \geq 2$.

Previously, we have seen that Geršgorin's Theorem 1.1, which, for any $A \in \mathbb{C}^{n\times n}$, defines a closed and bounded set in the complex plane containing all the eigenvalues of A, is *equivalent* to the Diagonal Dominance Theorem 1.4, which gives a sufficient condition for the nonsingularity for matrices in $\mathbb{C}^{n\times n}$. The analogous equivalence to the nonsingularity condition of (5.2) is given in Theorem 5.3 below, which generalizes the results of Corollaries 1.5 and 1.18. For notation, if $f = [f_1, f_2, \cdots, f_n] \in \mathcal{F}_n$, then

$$\begin{cases} \Gamma_i^f(A) := \{z \in \mathbb{C} : |z - a_{i,i}| \leq f_i(A)\} \quad (i \in N), \\ \Gamma^f(A) := \bigcup_{i \in N} \Gamma_i^f(A). \end{cases} \tag{5.3}$$

Theorem 5.3. *Let $f = [f_1, f_2, \cdots, f_n] \in \mathcal{F}_n$. Then, $f \in \mathcal{G}_n$ if and only if, for every $A = [a_{i,j}] \in \mathbb{C}^{n\times n}$,*

$$\sigma(A) \subseteq \Gamma^f(A). \tag{5.4}$$

Proof. First, suppose that $f \in \mathcal{G}_n$. For any $A = [a_{i,j}] \in \mathbb{C}^{n\times n}$, consider any $\lambda \in \sigma(A)$, so that $A - \lambda I_n$ is singular. If $|a_{i,i} - \lambda| > f_i(A - \lambda I_n)$ for *all* $i \in N$, then $A - \lambda I_n$ would be nonsingular from Definition 5.2, a contradiction. Hence, there exists a $k \in N$ for which

$$|a_{k,k} - \lambda| \leq f_k(A - \lambda I_n) = f_k(A),$$

the last equality following from property *ii)* of Definition 5.1. Thus from (5.3), $\lambda \in \Gamma_k^f(A)$, so that $\lambda \in \Gamma^f(A)$. But as λ was any eigenvalue of A, then the eigenvalue inclusion of (5.4) is valid.

Conversely, suppose that $f \in \mathcal{F}_n$, and that (5.4) is valid for all $A \in \mathbb{C}^{n\times n}$. If $f \notin \mathcal{G}_n$, there is an $A = [a_{i,j}] \in \mathbb{C}^{n\times n}$ with A singular and with A

satisfying (5.2). As (5.4) is by hypothesis valid, then $0 \in \sigma(A) \subseteq \Gamma^f(A)$. But from (5.2) and (5.3), $0 \notin \Gamma_i^f(A)$ for any $i \in N$, so that $0 \notin \Gamma^f(A)$, a contradiction. Thus, $f \in \mathcal{G}_n$. ■

Of course, the result of Theorem 5.3 is our **first recurring theme** on the equivalence of matrix nonsingularity results and matrix eigenvalue inclusion results. Its proof was given here just to show the relevance of the parts of Definition 5.1.

From Theorem 5.3, we see that, to obtain more general matrix nonsingularity results as in (5.2), or equivalently, more general matrix eigenvalue inclusion results as in (5.4), we must precisely determine *which* functions in $\mathcal{F}_n$ are elements of $\mathcal{G}_n$. This determination, given below in Theorem 5.5, depends nicely on the theory of M-matrices, our **second recurring theme**, which is summarized below. (See also Appendix C for a fuller discussion.)

To define M-matrices, we first adopt the traditional notation of

$$(5.5) \qquad \mathbb{Z}^{n\times n} := \{A = [a_{i,j}] \in \mathbb{R}^{n\times n} : a_{i,j} \leq 0 \text{ for all } i \neq j\}.$$

Then, if $A = [a_{i,j}] \in \mathbb{Z}^{n\times n}$, we set $\mu = \mu(A) := \max\{a_{i,i} : i \in N\}$, so that μ is a real number. We can then express A as $A = \mu I_n - B$, where $B = [b_{i,j}] \in \mathbb{R}^{n\times n}$ is defined by

$$(5.6) \quad b_{i,i} := \mu - a_{i,i} \geq 0 \text{ and } b_{i,j} := -a_{i,j} \geq 0 \text{ for } i \neq j \qquad (i,j \in N).$$

As B has all its entries nonnegative, we can write $B \geq O$. Recalling from (1.8) that $\rho(A)$ denotes the spectral radius of $A \in \mathbb{C}^{n\times n}$, this brings us to

Definition 5.4. Given $A = [a_{i,j}] \in \mathbb{Z}^{n\times n}$ (cf.(5.5)), express A as $A = \mu I_n - B$, where μ is a real number, and $B \geq O$. Then, A is an **M-matrix** if $\rho(B) \leq \mu$. More precisely, A is a **nonsingular M-matrix** if $\rho(B) < \mu$, and a **singular M-matrix** if $\rho(B) = \mu$.

We remark that a singular M-matrix *is* singular, thereby justifying the above terminology, because, by the Perron-Frobenius theory of nonnegative square matrices, $\rho(B)$ is an eigenvalue of B when $B \geq O$ (cf. Appendix C). We also warn the reader that, in the literature, an M-matrix often means a *nonsingular* M-matrix; here, an M-matrix *can* be singular.

To connect the theory of M-matrices with properties of the set $\mathcal{G}_n$, we need the following additional notation. Given any $f = [f_1, f_2, \cdots, f_n] \in \mathcal{F}_n$, of Definition 5.1, and given any $A = [a_{i,j}] \in \mathbb{C}^{n\times n}$, we define the associated matrix $\mathcal{M}^f(A) := [\alpha_{i,j}] \in \mathbb{R}^{n\times n}$ as

$$\alpha_{i,i} := f_i(A) \text{ and } \alpha_{i,j} := -|a_{i,j}| \text{ for } i \neq j \qquad (i,j \in N),$$

i.e.,

$$(5.7) \qquad \mathcal{M}^f(A) := \begin{bmatrix} f_1(A) & -|a_{1,2}| & \cdots & -|a_{1,n}| \\ -|a_{2,1}| & f_2(A) & \cdots & -|a_{2,n}| \\ \vdots & & & \vdots \\ -|a_{n,1}| & -|a_{n,2}| & \cdots & f_n(A) \end{bmatrix}.$$

We note from (5.7) that the real $n \times n$ matrix $\mathcal{M}^f(A) = [\alpha_{i,j}]$ is necessarily an element of $\mathbb{Z}^{n\times n}$ of (5.5).

The next result, one of our main results, connects elements of $\mathcal{G}_n$ with M-matrices and with the now-familiar weighted row sums $r_i^{\mathbf{x}}(A)$ of (1.13).

Theorem 5.5. *Let $f = [f_1, f_2, \cdots, f_n] \in \mathcal{F}_n$. Then, $f \in \mathcal{G}_n$ if and only if $\mathcal{M}^f(A)$, as defined in* (5.7), *is an M-matrix for every $A = [a_{i,j}] \in \mathbb{C}^{n\times n}$. For $n \geq 2$, if $f \in \mathcal{G}_n$ and if $A \in \mathbb{C}^{n\times n}$ is irreducible, then $\mathcal{M}^f(A)$ is an irreducible M-matrix and there exists an $\mathbf{x} = [\mathbf{x}_1, \mathbf{x}_2, \cdots, \mathbf{x}_n]^T > \mathbf{0}$ in $\mathbb{R}^n$ (where $\mathbf{x}$ is dependent on A) for which*

$$(5.8) \qquad f_i(A) \geq r_i^{\mathbf{x}}(A) \qquad (\text{all } i \in N).$$

If, moreover, $\mathcal{M}^f(A)$ is a singular M-matrix, then equality holds in (5.8) *for all $i \in N$, i.e.,*

$$(5.9) \qquad f_i(A) = r_i^{\mathbf{x}}(A) \qquad (\text{all } i \in N).$$

Remark 1. From equation (5.8), we directly see the connection of Theorem 5.5 to Fan's Theorem 1.21.

Proof. To begin, Theorem 5.5 can be seen to be trivially true when $n = 1$, since $\mathcal{F}_1 = \mathcal{G}_1$ and the unique element f in these sets is the null function. Hence, $\mathcal{M}^f(A) = [0]$ is a singular M-matrix for any $A \in \mathbb{C}^{1\times 1}$. Next, assume that $n \geq 2$. Then, suppose that $f \in \mathcal{G}_n$ and suppose that $A = [a_{i,j}] \in \mathbb{C}^{n\times n}$ is such that $\mathcal{M}^f(A)$, defined in (5.7), is *not* an M-matrix. Defining $\mu := \max\{f_i(A) : i \in N\}$, we can write that $\mathcal{M}^f(A) = \mu I_n - B$, where $B \in \mathbb{R}^{n\times n}$ satisfies $B \geq O$. Since $\mathcal{M}^f(A)$ is not an M-matrix, it follows from Definition 5.4 that $\rho(B) > \mu$. Then, define the matrix $C = [c_{i,j}] := \rho(B) \cdot I_n - B$ in $\mathbb{R}^{n\times n}$, whose entries can be verified from (5.7) to be

$$(5.10) \; c_{i,i} := f_i(A) + (\rho(B) - \mu) \text{ and } c_{i,j} := -|a_{i,j}| \text{ for } i \neq j \qquad (i,\, j \in N).$$

Because $B \geq O$, the Perron-Frobenius theory of nonnegative matrices (cf. Theorem C.2 of Appendix C) gives us that $\rho(B) \in \sigma(B)$, so that C is singular. On the other hand, as $\rho(B) > \mu$, the first equality in (5.10) gives that $c_{i,i} > 0$ with $|c_{i,i}| > f_i(A)$ for all $i \in N$, while the second equality in (5.10) gives, from *ii*) of Definition 5.1, that $f_i(A) = f_i(C)$ for all $i \in N$, i.e.,

$$|c_{i,i}| > f_i(C) \quad (i \in N).$$

But as $f \in \mathcal{G}_n$, then C is nonsingular from Definition 5.2, a contradiction. Thus, $\mathcal{M}^f(A)$ is an M-matrix for every $A \in \mathbb{C}^{n\times n}$.

Conversely, suppose that $f \in \mathcal{F}_n$ is such that $\mathcal{M}^f(A)$ is an M-matrix for any $A \in \mathbb{C}^{n\times n}$. To deduce that $f \in \mathcal{G}_n$, we must show (cf. Definition 5.2) that, for any matrix $A = [a_{i,j}] \in \mathbb{C}^{n\times n}$ satisfying

$$|a_{i,i}| > f_i(A) \quad (\text{all } i \in N),$$

the matrix A is nonsingular. Now, the above inequalities give us that $d_i := |a_{i,i}| - f_i(A) > 0$ for all $i \in N$, and we set $\mathbf{d} := [d_1, d_2, \cdots, d_n]^T \in \mathbb{R}^n$. Then define the matrix $E = [e_{i,j}] \in \mathbb{R}^{n\times n}$ by

$$E := [e_{i,j}] = \mathcal{M}^f(A) + \text{ diag}[\mathbf{d}],$$

so that $E \in \mathbb{Z}^{n\times n}$. Because E is the sum of an M-matrix and a *positive* diagonal matrix, then (cf. Propositions C.4 and C.5 of Appendix C.) E is a nonsingular M-matrix with $E^{-1} \geq O$. Moreover, by the above construction, it can be verified that $|e_{i,j}| = |a_{i,j}|$ for *all* $i, j \in N$. This implies, by the reverse triangle inequality and the sign properties (cf. (5.7)) of off-diagonal entries of $\mathcal{M}^f(A)$, that

$$|(A\mathbf{y})_i| = \left|\sum_{j\in N} a_{i,j}y_j\right| \geq |a_{i,i}| \cdot |y_i| - \sum_{j\in N\backslash\{i\}} |a_{i,j}| \cdot |y_j| = \sum_{j\in N} e_{i,j} \cdot |y_j|,$$

for any $i \in N$ and for *any* $\mathbf{y} = [y_1, y_2, \cdots, y_n]^T \in \mathbb{C}^n$. Now, it is convenient to define the particular **vectorial norm p(u)** by

$$\mathbf{p}(\mathbf{u}) := [|u_1|, |u_2|, \cdots, |u_n|]^T \quad (\text{any } \mathbf{u} = [u_1, u_2, \cdots, u_n]^T \in \mathbb{C}^n),$$

so that $\mathbf{p} : \mathbb{C}^n \to \mathbb{R}^n_+$. With this definition, the inequalities above can be expressed concisely as

$$\mathbf{p}(A\mathbf{y}) \geq E\mathbf{p}(\mathbf{y}) \quad (\text{all } \mathbf{y} \in \mathbb{C}^n). \tag{5.11}$$

Thus, E (in the terminology of Appendix C) is a **lower bound matrix** for A with respect to the vectorial norm $\mathbf{p}$. Because $E^{-1} \geq O$, then multiplying (on the left) by E^{-1} in (5.11) preserves these inequalities, giving

$$E^{-1}\mathbf{p}(A\mathbf{y}) \geq \mathbf{p}(\mathbf{y}) \quad (\text{all } \mathbf{y} \in \mathbb{C}^n). \tag{5.12}$$

But these inequalities imply that A is *nonsingular*, for if A were singular, there would be a $\mathbf{y} \in \mathbb{C}^n$ with $A\mathbf{y} = \mathbf{0}$ and with $\mathbf{y} \neq \mathbf{0}$. Then the vector $\mathbf{p}(\mathbf{y})$, on the right side of (5.12), has at least one positive component, while the left side of (5.12) would be the zero column vector in $\mathbb{R}^n$, which contradicts (5.12). The nonsingularity of A thus establishes, from Definition 5.2, that $f \in \mathcal{G}_n$.

To conclude the proof, suppose that $f \in \mathcal{G}_n$ and that $A \in \mathbb{C}^{n\times n}$ is irreducible. Because $n \geq 2$, we see, using the first part of this proposition, that $\mathcal{M}^f(A)$ of (5.7) is an (irreducible) M-matrix. As such, (cf. Theorem C.1

of Appendix C), there is an $\mathbf{x} = [x_1, x_2, \cdots, x_n]^T \in \mathbb{R}^n$ with $\mathbf{x} > \mathbf{0}$ such that $\mathcal{M}^f(A)\mathbf{x} \geq \mathbf{0}$, or equivalently from (5.7),

$$f_i(A)x_i - \sum_{j \in N \setminus \{i\}} |a_{i,j}| \cdot x_j \geq 0 \quad (\text{all } i \in N).$$

Upon dividing the above expression by $x_i > 0$ and using the definition of $r_i^{\mathbf{x}}(A)$ of (1.13), we obtain $f_i(A) \geq r_i^{\mathbf{x}}(A)$ for all $i \in N$, which is the desired result of (5.8). Similarly, if $\mathcal{M}^f(A)$ is an irreducible singular M-matrix, then there is an $\mathbf{x} > \mathbf{0}$ such that $\mathcal{M}^f(A)\mathbf{x} = \mathbf{0}$, which analogously implies (5.9). ■

We next list some additional remarks arising from Theorem 5.5.

Remark 2. The results of (5.8) and (5.9) of Theorem 5.5 apply equally well to the weighted column sums $c_i^{\mathbf{y}}(A) := r_i^{\mathbf{y}}(A^T)$ of (1.35), where $\mathbf{y} > \mathbf{0}$ in $\mathbb{R}^n$.

Remark 3. Theorem 5.5 generalizes the results of Ostrowski (1937b) and Fan (1958). In particular, the inequalities (5.8) of Theorem 5.5, coupled with the equivalences of Theorem 5.3, reproduce the main results of Fan (1958).

Remark 4. Theorem 5.5 also gives us that, for any $n \geq 2$, $\mathcal{G}_n$ is a *proper* subset of $\mathcal{F}_n$. To see this, just consider the null function $f = [f_1, \cdots, f_n]$ in $\mathcal{F}_n$, i.e., $f_i(A) := 0$ for all $i \in N$ and all $A \in \mathbb{C}^{n \times n}$ for $n \geq 2$. It is easily seen that $\mathcal{M}^f(A)$ of (5.7) cannot be an M-matrix for *every* $A \in \mathbb{C}^{n \times n}$, so that, from Theorem 5.5, this $f \notin \mathcal{G}_n$.

Exercises

1. If $f = [f_1, f_2, \cdots, f_n] \in \mathcal{G}_n$ and if $g = [g_1, g_2, \cdots, g_n] \in \mathcal{F}_n$, show that $f + g := [f_1 + g_1, f_2 + g_2, \cdots, f_n + g_n] \in \mathcal{G}_n$. Similarly, if $\boldsymbol{\tau} \in \mathbb{R}^n$ with $\boldsymbol{\tau} \geq \mathbf{0}$, show that $f + \boldsymbol{\tau} := [f_1 + \tau_1, f_2 + \tau_2, \cdots, f_n + \tau_n] \in \mathcal{G}_n$.

2. If $C = [c_{i,j}] \in \mathbb{R}^{n \times n}$ is a nonsingular M-matrix, show that $C^{-1} \geq O$. Similarly, if $C = [c_{i,j}] \in \mathbb{R}^{n \times n}$ is an irreducible nonsingular M-matrix, show that $C^{-1} > O$. (Hint: Write $C = \mu(I - B/\mu)$ where $\mu > \rho(B)$, so that $C^{-1} = \frac{1}{\mu}[I + \frac{B}{\mu} + (\frac{B}{\mu})^2 + \cdots].$)

3. Let $C = [c_{i,j}] \in \mathbb{R}^{n \times n}$, $n \geq 2$, be an irreducible singular M-matrix. Show that if *any* entry of C is reduced, the resulting matrix is not an M-matrix. Also, give an example that this last result is not in general true if C is reducible.

4. Let $A = [a_{i,j}] \in \mathbb{C}^{n\times n}$ be any (cf. (1.11)) strictly diagonally dominant matrix. Show that its **comparison matrix** $E = [e_{i,j}] \in \mathbb{C}^{n\times n}$, defined by $e_{i,i} := |a_{i,i}|$ and $e_{i,j} := -|a_{i,j}|, i \neq j$, for $i, j \in N$, is a nonsingular M-matrix. If the vectorial norm $\mathbf{p}(\mathbf{u})$ is defined by $\mathbf{p}(\mathbf{u}) := [|u_1|, |u_2|, \cdots, |u_n|]^T$ for $\mathbf{u} = [u_1, u_2, \cdots, u_n]^T$ in $\mathbb{C}^n$, show that

$$\mathbf{p}(A\mathbf{y}) \geq E\mathbf{p}(\mathbf{y}) \text{ (all } \mathbf{y} \in \mathbb{C}^n),$$

i.e., E is a *lower bound* matrix for A with respect to $\mathbf{p}$.

5. For $n = 2$, if $f_1(A) := |a_{1,2}|^\beta \cdot |a_{2,1}|^\gamma$ and $f_2(A) := |a_{1,2}|^{1-\beta} \cdot |a_{2,1}|^{1-\gamma}$, for all β and γ with $0 \leq \beta,\ \gamma \leq 1$ and all $A = [a_{i,j}] \in \mathbb{C}^{2\times 2}$, show that $f := [f_1, f_2]$ is an element of $\mathcal{G}_2$. Show moreover, using Theorem 5.5, that any $g = [g_1, g_2]$ in $\mathcal{G}_2$ for which (cf.(5.7)) $\mathcal{M}^g(A)$ is a *singular M-matrix* for any $A = [a_{i,j}] \in \mathbb{C}^{2\times 2}$, is necessarily of the form $f = [f_1, f_2]$, for suitable choices of β and γ in $[0, 1]$.

6. For each $f = [f_1, f_2, \cdots, f_n]$ in $\mathcal{F}_n$, show that $f \in \mathcal{G}_n$ if and only if each principal minor (i.e., the determinant of each principal submatrix) of $\mathcal{M}^f(A)$ of (5.7) is nonnegative for any $A = [a_{i,j}] \in \mathbb{C}^{n\times n}$. Also, show that the condition, that each principal minor of $\mathcal{M}^f(A)$ is nonnegative, gives rise to $2^n - 1$ conditions on $f = [f_1, f_2, \cdots, f_n]$. (Hint: See A_1 of Theorem 4.6 of Berman and Plemmons (1994).)

5.2 Structural Properties of $\mathcal{G}_n$ and $\mathcal{G}_n^c$

The set $\mathcal{G}_n$ of G-functions has two interesting structural properties which we wish to exploit. As the first structural property, we can **partially order** the elements of $\mathcal{F}_n$ as follows: given any $f = [f_1, f_2, \cdots, f_n] \in \mathcal{F}_n$ and any $g = [g_1, g_2, \cdots, g_n] \in \mathcal{F}_n$, we write

$$(5.13) \qquad f \succ g \text{ if } f_i(A) \geq g_i(A) \ \ (\text{all } i \in N, \text{ all } A \in \mathbb{C}^{n\times n}).$$

It can be verified that this is indeed a partial order on $\mathcal{F}_n$, i.e., for all f, g and h in $\mathcal{F}_n$,

i) (transitive) $f \succ g$ and $g \succ h$ imply $f \succ h$;
ii) (anti-symmetric) $f \succ g$ and $g \succ f$ imply $f = g$;
ii) (reflexive) $f \succ f$.

With the partial ordering (5.13) for elements of $\mathcal{F}_n$, we next give the easy, but useful, result of

Lemma 5.6. *If $g \in \mathcal{G}_n$ and if $f \in \mathcal{F}_n$ satisfy $f \succ g$, then $f \in \mathcal{G}_n$.*

Proof. For any $A = [a_{i,j}] \in \mathbb{C}^{n\times n}$ with $|a_{i,i}| > f_i(A)$ for all $i \in N$, the hypothesis $f \succ g$ gives $f_i(A) \geq g_i(A)$ for all $i \in N$, so that

$$|a_{i,i}| > g_i(A) \quad (i \in N).$$

As $g \in \mathcal{G}_n$, A is thus nonsingular from Definition 5.2, and hence, $f \in \mathcal{G}_n$. ■

Our interest in the partial order of (5.13) for $\mathcal{F}_n$ and $\mathcal{G}_n$ is based on the following simple observation. Given f and g in $\mathcal{G}_n$ and given any $A \in \mathbb{C}^{n\times n}$, we know from Theorem 5.3 that $\sigma(A)$ is contained in both sets, $\Gamma^f(A)$ and $\Gamma^g(A)$, defined in (5.3), but it is not apparent which of the sets, $\Gamma^f(A)$ or $\Gamma^g(A)$ in the complex plane, gives a *tighter* estimation of $\sigma(A)$. But if $f \succ g$, then from (5.13), $f_i(A) \geq g_i(A)$ for all $i \in N$ and all $A \in \mathbb{C}^{n\times n}$, so that (cf. (5.3)), $\Gamma^g(A) \subseteq \Gamma^f(A)$, i.e., $\Gamma^g(A)$ is always *at least as good* as $\Gamma^f(A)$ in estimating $\sigma(A)$ for *any* $A \in \mathbb{C}^{n\times n}$.

Next, it is convenient to make the following definition. Recall from Definition 5.1 that, for $n \geq 2, f = [f_1, f_2, \cdots, f_n]$ in $\mathcal{F}_n$ implies that, for any $A = [a_{i,j}] \in \mathbb{C}^{n\times n}$, each $f_i(A)$ is a function of the $n(n-1)$ nonnegative numbers $|a_{i,j}|$ for all $i, j \in N$ with $i \neq j$. Then, each f_i can be viewed as a mapping from $\mathbb{R}_+^{n(n-1)}$ to $\mathbb{R}_+$.

Definition 5.7. For $n \geq 2, f = [f_1, f_2, \cdots, f_n] \in \mathcal{F}_n$ is **continuous** if f_i is continuous from $\mathbb{R}_+^{n(n-1)}$ to $\mathbb{R}_+$, for each $i \in N$. The sets of all continuous elements in $\mathcal{F}_n$ and in $\mathcal{G}_n$ are respectively denoted by $\mathcal{F}_n^c$ and $\mathcal{G}_n^c$.

We remark that for $n = 1$, the unique null function of $\mathcal{F}_1 = \mathcal{G}_1$ can also be regarded as continuous. We also remark that our previous examples of G-functions, such as those arising from (5.1), are all *continuous*. Discontinuous G-functions *do* exist, and they play a special role in Section 5.3, as we shall see.

The next proposition gives a useful sufficient condition for the elements of $\mathcal{F}_n^c$ to be elements of $\mathcal{G}_n^c$.

Proposition 5.8. *Let $f = [f_1, f_2, \cdots, f_n] \in \mathcal{F}_n^c$, and assume that every* **irreducible** *$A = [a_{i,j}] \in \mathbb{C}^{n\times n}$, for which $|a_{i,i}| > f_i(A)$ for all $i \in N$, is nonsingular. Then, $f \in \mathcal{G}_n^c$.*

Proof. As the case $n = 1$ is again trivial, assume that $n \geq 2$. For any $A = [a_{i,j}] \in \mathbb{C}^{n\times n}$ with $|a_{i,i}| > f_i(A)$ for all $i \in N$, we must show that A is nonsingular. For every $\epsilon > 0$, define the matrix $E(\epsilon) = [e_{i,j}(\epsilon)] \in \mathbb{Z}^{n\times n}$ by means of

$$e_{i,j}(\epsilon) := \begin{cases} -|a_{i,j}| & \text{if } i \neq j \text{ and if } a_{i,j} \neq 0; \\ -\epsilon & \text{if } i \neq j \text{ and if } a_{i,j} = 0; \\ |a_{i,i}| & \text{if } i = j. \end{cases} \tag{5.14}$$

Since each off-diagonal entry of $E(\epsilon)$ is nonzero, then $E(\epsilon)$ is necessarily irreducible for every $\epsilon > 0$. For $\epsilon > 0$ chosen sufficiently *small*, it is clear, from the assumed continuity of the f_i's, that the hypothesis $|a_{i,i}| > f_i(A)$ for all $i \in N$ implies

$$|e_{i,i}(\epsilon)| > f_i(E(\epsilon)) \quad (\text{all } i \in N).$$

Since $E(\epsilon)$ is irreducible, the above inequalities give, from our hypotheses, that $E(\epsilon)$ is nonsingular. Note, moreover, that adding any nonnegative diagonal matrix to $E(\epsilon)$ produces an irreducible matrix which *also* satisfies the above inequalities, so that the resulting matrix sum is again nonsingular. This can be used as follows. The sign pattern, induced by (5.14) in the entries of the matrix $E(\epsilon)$, is such that $E(\epsilon)$ admits the representation

$$E(\epsilon) = \mu I_n - B(\epsilon), \text{ where } B(\epsilon) \geq O \text{ and where } \mu := \max_{i \in N} |a_{i,i}|.$$

Since $E(\epsilon)$ is nonsingular and since $B(\epsilon) \geq O$, the Perron-Frobenius Theorem (see Theorem C.2 of Appendix C) gives that $\mu \neq \rho(B(\epsilon))$, so that either $\mu < \rho(B(\epsilon))$ or $\mu > \rho(B(\epsilon))$. If $\mu < \rho(B(\epsilon))$, then, with $\tau := \rho(B(\epsilon)) - \mu > 0$, adding τI_n to $E(\epsilon)$ would make $E(\epsilon) + \tau I_n$ singular, which contradicts our previous statement above. Hence, $\mu > \rho(B(\epsilon))$, which implies that $E(\epsilon)$ is a nonsingular M-matrix, which also implies (cf. Proposition C.4 of Appendix C) that $E^{-1}(\epsilon) \geq O$. Then for any $\mathbf{y} = [y_1, y_2, \cdots, y_n]^T \in \mathbb{C}^n$, it follows as in (5.11), from the definitions in (5.14), that $|(A\mathbf{y})_i| \geq (E(\epsilon)|\mathbf{y}|)_i$ for all $i \in N$, or, in terms of the vectorial norm $\mathbf{p}(\mathbf{y}) := [|y_1|, |y_2|, \cdots, |y_n|]^T$,

$$\mathbf{p}(A\mathbf{y}) \geq E(\epsilon)\mathbf{p}(\mathbf{y}) \quad (\text{all } \mathbf{y} \in \mathbb{C}^n).$$

But since $E^{-1}(\epsilon) \geq O$, the above inequalities imply, as in the proof of Theorem 5.5, that

$$E^{-1}(\epsilon)\mathbf{p}(A\mathbf{y}) \geq \mathbf{p}(\mathbf{y}) \quad (\text{all } \mathbf{y} \in \mathbb{C}^n),$$

which, as in the proof of Theorem 5.5, shows that A is nonsingular. Thus, $f \in \mathcal{G}_n^c$. ∎

With the above proposition, we next establish a result of Hoffman (1969), which is a generalization of Ostrowski's Theorem 1.16. For notation, let $f = [f_1, f_2, \cdots, f_n]$ and $g = [g_1, g_2, \cdots, g_n]$ be elements of $\mathcal{G}_n^c$, and let α satisfy $0 \leq \alpha \leq 1$. Then $h = [h_1, h_2, \cdots, h_n]$, defined by

$$(5.15) \quad h_i(A) := f_i^{\alpha}(A) \cdot g_i^{1-\alpha}(A) \qquad (\text{all } i \in N, \text{ all } A \in \mathbb{C}^{n \times n}),$$

is called the **α-convolution** of f and g.

Theorem 5.9. *If $f = [f_1, f_2, \cdots, f_n]$ and $g = [g_1, g_2, \cdots, g_n]$ are elements of $\mathcal{G}_n^c$, and if $0 \leq \alpha \leq 1$, then the* **α-convolution** *$h = [h_1, h_2, \cdots, h_n]$ of (5.15) is also in $\mathcal{G}_n^c$*

Proof. Clearly, h, as defined in (5.15), is an element in $\mathcal{F}_n^c$ from Definitions 5.1 and 5.7. Also, since the result of Theorem 5.9 is obvious if $\alpha = 0$ or if $\alpha = 1$, assume that $0 < \alpha < 1$. To show that $h \in \mathcal{G}_n^c$, it suffices, from Proposition 5.8, to consider any *irreducible* $A = [a_{i,j}] \in \mathbb{C}^{n \times n}$ for which

$$(5.16) \qquad |a_{i,i}| > h_i(A) \quad (\text{all } i \in N).$$

From Theorem 5.5, there exist $\mathbf{x}$ and $\mathbf{y}$ in $\mathbb{R}^n$, with $\mathbf{x} > \mathbf{0}$ and $\mathbf{y} > \mathbf{0}$, such that $f_i(A) \geq r_i^{\mathbf{x}}(A)$ and $g_i(A) \geq r_i^{\mathbf{y}}(A)$ for all $i \in N$. Then, using (5.15) and (5.16),

$$(5.17) \qquad |a_{i,i}| > h_i(A) \geq (r_i^{\mathbf{x}}(A))^{\alpha} \cdot (r_i^{\mathbf{y}}(A))^{1-\alpha} \quad (\text{all } i \in N).$$

But with the definitions of $r_i^{\mathbf{x}}(A)$ and $r_i^{\mathbf{y}}(A)$, the product on the right in (5.17) is just the upper bound obtained from applying Hölder's inequality, with $p := 1/\alpha$ and $q := 1/(1-\alpha)$, to the sum $\sum_{j \in N \setminus \{i\}} (|a_{i,j}|x_j/x_i)^{\alpha}(|a_{i,j}|y_j/y_i)^{1-\alpha}$, where this last sum can be written as

$$\sum_{j \in N \setminus \{i\}} |a_{i,j}| x_j^{\alpha} y_j^{1-\alpha} / x_i^{\alpha} y_i^{1-\alpha}.$$

Hence, the inequalities (5.17) give rise to the inequalities of

$$(5.18) \qquad |a_{i,i}| > h_i(A) \geq (r_i^{\mathbf{x}}(A))^{\alpha} \; (r_i^{\mathbf{y}}(A))^{1-\alpha} \geq r_i^{\mathbf{z}}(A) \quad (\, i \in N),$$

where $\mathbf{z} = [z_1, z_2, \cdots, z_n]^T \in \mathbb{R}^n$ is defined by $z_i := x_i^{\alpha} y_i^{1-\alpha}$, $i \in N$, so that $\mathbf{z} > \mathbf{0}$. But as it was remarked after Definition 5.7 that $r^{\mathbf{x}}$ is an element of $\mathcal{G}_n^c$, for *any* $\mathbf{x} > \mathbf{0}$, then $r^{\mathbf{z}} \in \mathcal{G}_n^c$. Hence, (5.18) implies from Definition 5.2 that A is nonsingular. Thus from Proposition 5.8, $h \in \mathcal{G}_n^c$. ■

As a consequence of Theorem 5.9, we derive below in Corollary 5.10 the second structural property of $\mathcal{G}_n^c$, namely, **convexity**. In general, a subset T of a linear space is **convex** if, for every f and g in T and for every α with $0 \leq \alpha \leq 1$, then $k := \alpha f + (1-\alpha)g$ is also in T.

Corollary 5.10. *$\mathcal{G}_n^c$ is a convex set.*

Proof. For any $f = [f_1, f_2, \cdots, f_n]$ in $\mathcal{G}_n^c$, for any $g = [g_1, g_2, \cdots, g_n]$ in $\mathcal{G}_n^c$ and for any $0 \leq \alpha \leq 1$, define $k = [k_1, k_2, \cdots, k_n] \in \mathcal{F}_n^c$ by

$$(5.19) \qquad k_i(A) := \alpha f_i(A) + (1-\alpha) g_i(A), \quad (\text{all } i \in N, \text{all } A \in \mathbb{C}^{n \times n}).$$

Recall the generalized arithmetic-geometric mean inequality (cf. Beckenbach and Bellman (1961), p. 15), namely, that for $a \geq 0$, $b \geq 0$, and $0 \leq \alpha \leq 1$, there holds

$$\alpha a + (1-\alpha) b \geq a^{\alpha} b^{1-\alpha} \quad (\text{with the convention that } 0^0 := 1),$$

and when $a > 0, b > 0$, and $0 < \alpha < 1$, there holds

$$\alpha a + (1-\alpha)b = a^{\alpha}b^{1-\alpha} \quad \text{if and only if } a = b.$$

(This latter case of equality will be used below in the proof of Theorem 5.12.) Applying the first inequality above to (5.19) gives

$$(5.20) \quad k_i(A) \geq f_i^{\alpha}(A) g_i^{1-\alpha}(A) =: h_i(A) \quad (\text{all } i \in N, \ \text{all } A \in \mathbb{C}^{n\times n}),$$

the last equality coming from the definition in (5.15). Hence, (cf. (5.13)), $k \succ h$. But as $h \in \mathcal{G}_n^c$ from Theorem 5.9, then so is k, from the inequalities of (5.20) and Lemma 5.6. This proves that $\mathcal{G}_n^c$ is convex. ■

For a convex set, its **extreme points** play a special role. Similarly, for a set which is partially ordered, its **minimal points** are of special interest.

Definition 5.11.

i) $f \in T$ is an **extreme point** of a convex set T if $f = \alpha g + (1-\alpha)h$, for some $0 < \alpha < 1$ and $g, h \in T$, implies $f = g = h$;

ii) $f \in U$ is a **minimal point** of a set U, which is partially ordered with respect to " $\succ$" (cf. (5.13)), if, for every $g \in U$ with $f \succ g$, implies $f = g$.

As previously remarked, if f and g are elements in $\mathcal{G}_n^c$ with $f \succ g$, then the associated Geršgorin sets $\Gamma^g(A)$ and $\Gamma^f(A)$, from (5.3), satisfy $\Gamma^g(A) \subseteq \Gamma^f(A)$ for *each* $A \in \mathbb{C}^{n\times n}$. Hence, there is considerable interest in finding the minimal points of $\mathcal{G}_n^c$, as they give the tightest Geršgorin inclusions for *each* $A \in \mathbb{C}^{n\times n}$, as compared with any other element of $\mathcal{G}_n^c$. As one of the main results in this section, we next show that the extreme points of $\mathcal{G}_n^c$ **coincide** with the minimal points of $\mathcal{G}_n^c$, and, moreover, that these points precisely give the case of equality in (5.9) of Theorem 5.5.

Theorem 5.12. *Let $f \in \mathcal{G}_n^c$. Then, the following are equivalent:*

i) *f is an extreme point of the convex set $\mathcal{G}_n^c$;*

ii) *f is a minimal point of $\mathcal{G}_n^c$, partially ordered by " $\succ$" of (5.13);*

ii) *for every $A \in \mathbb{C}^{n\times n}$, the matrix $\mathcal{M}^f(A)$, defined in (5.7), is a singular M-matrix;*

iv) *for every irreducible $A \in \mathbb{C}^{n\times n}$, there exists an $\mathbf{x} > \mathbf{0}$ in $\mathbb{R}^n$ (where $\mathbf{x}$ is dependent on A) such that*

$$(5.21) \quad f_i(A) = r_i^{\mathbf{x}}(A) \quad (\text{all } i \in N).$$

Proof. Assume *i)*, i.e., let f be any extreme point of $\mathcal{G}_n^c$, and assume *ii)* is false, i.e., f is not a minimal point in $\mathcal{G}_n^c$. Then, there exists a g in $\mathcal{G}_n^c$ with $f \succ g$ and with $f \neq g$. Defining $h := 2f - g = f + (f-g)$, it follows that $h \in \mathcal{F}_n^c$ with $h \succ f$ and $h \neq f$. Since $f \in \mathcal{G}_n^c$, so is h from Lemma 5.6.

But then, by definition, $f = (h+g)/2$ with f, g and h not coinciding, which contradicts the assumption that f is an extreme point in $\mathcal{G}_n^c$. Thus, *i)* implies *ii)*.

Next, we need the following construction. Consider any $f \in \mathcal{G}_n^c$. From Theorem 5.5, we know that $\mathcal{M}^f(A)$ is an M-matrix for *any* $A \in \mathbb{C}^{n\times n}$. Thus, with the notation of Definition 5.4, we can write that

$$\mathcal{M}^f(A) = \mu I_n - B \text{ with } \mu \geq \rho(B), \tag{5.22}$$

where the matrix $B = [b_{i,j}]$ in $\mathbb{R}^{n\times n}$ has its entries defined, for $\mu := \max_{i\in N} f_i(A)$, by

$$b_{i,i} := \mu - f_i(A) \text{ and } b_{i,j} := |a_{i,j}| \text{ for } i \neq j \quad (i, j \in N),$$

so that $B \geq O$. Because $f \in \mathcal{G}_n^c$, note from (5.7) that *all* entries of $\mathcal{M}^f(A)$ and B depend continuously on the absolute values of *only* off-diagonal entries of A. Thus, if we define the mapping $g = [g_1, g_2, \cdots, g_n] : \mathbb{C}^{n\times n} \to \mathbb{R}^n$ by

$$g_i(A) := f_i(A) - [\mu - \rho(B)] \quad (\text{all } i \in N, \text{ all } A \in \mathbb{C}^{n\times n}), \tag{5.23}$$

it can be verified, using (5.23) and the previous definitions for μ and the entries $b_{i,j}$ of B, that $\mathcal{M}^g(A)$ of (5.7) can be expressed as

$$\mathcal{M}^g(A) = \rho(B)I_n - B,$$

so that $\mathcal{M}^g(A)$ is, by definition, a *singular* M-matrix for *any* $A \in \mathbb{C}^{n\times n}$. Also, since all entries of B depend continuously on the absolute values of only off-diagonal entries of A, the same is true for $\rho(B)$ and also $\mathcal{M}^g(A)$. Moreover, since $\mathcal{M}^g(A)$ is a (singular) M-matrix, its diagonal entries, $g_i(A)$ of (5.23), are all necessarily nonnegative (see Exercise 1 of this section). This implies from Definition 5.1 that $g \in \mathcal{F}_n^c$. Also from Theorem 5.5, we have that $g \in \mathcal{G}_n^c$. But the important consequence of this construction is that since $\mu \geq \rho(B)$ from (5.22), then from (5.23), $f_i(A) \geq g_i(A)$ for all $i \in N$ and all $A \in \mathbb{C}^{n\times n}$, i.e., (cf. (5.13))

$$\text{for any } f \in \mathcal{G}_n^c, \text{ there is a } g \in \mathcal{G}_n^c \text{ with } f \succ g.$$

To show that *ii)* implies *ii)*, let f be any minimal point of $\mathcal{G}_n^c$, and assume that *ii)* is not valid. Then, there is an $A \in \mathbb{C}^{n\times n}$ for which $\mathcal{M}^f(A)$ is not singular. However, as $\mathcal{M}^f(A)$ is an M-matrix from Theorem 5.5, which can be expressed as $\mu I_n - B$ with $B \geq O$, then as $\mathcal{M}^f(A)$ is not singular, it follows from (5.22) that $\mu > \rho(B)$. Hence, from (5.23), $f_i(A) > g_i(A)$ for all $i \in N$. Then, from the construction in the paragraph above, $f \succ g$ with $f \neq g$, which contradicts that f is a minimal point of $\mathcal{G}_n^c$. Hence, *ii)* implies *ii)*.

Next, assuming *ii)*, then *iv)* follows directly from Theorem 5.5. Conversely, if *iv)* is valid, then by direct computation, $\mathcal{M}^{r^{\mathbf{x}}}(A)\mathbf{x} = \mathbf{0}$ for any $A \in \mathbb{C}^{n\times n}$, so that *iv)* implies *ii)*. Thus, *ii)* and *iv)* are equivalent.

To complete the proof, we show that *ii)* implies *i)*. Assuming *ii)*, i.e. $\mathcal{M}^f(A)$ is a singular M-matrix for every $A \in \mathbb{C}^{n\times n}$, suppose that *i)* is false, i.e., that f is not an extreme point of $\mathcal{G}_n^c$. Hence, there exist g and h in $\mathcal{G}_n^c$ and an α with $0 < \alpha < 1$ such that $f = \alpha g + (1-\alpha)h$ with $g \neq h$. Thus, there is an $A \in \mathbb{C}^{n\times n}$ and an $i \in N$ for which $g_i(A) \neq h_i(A)$. But as g and h are both continuous, we may, following the construction of (5.14) in the proof of Proposition 5.8, assume without loss of generality that A is irreducible. Next, using the generalized arithmetic-geometric mean inequality, we have that

$$f_j(A) := \alpha g_j(A) + (1-\alpha)h_j(\alpha) \geq g_j^{\alpha}(A) \cdot h_j^{1-\alpha}(A) \quad (\text{all } j \in N). \tag{5.24}$$

But as g and h are elements of $\mathcal{G}_n^c$, Theorem 5.5 gives us that $\mathcal{M}^g(A)$ and $\mathcal{M}^h(A)$ are both M-matrices, and in addition as A is also irreducible, it can then be verified that $g_j(A) > 0$ and $h_j(A) > 0$ for all $j \in N$. In particular, we have

$$g_i(A) > 0, h_i(A) > 0 \text{ with } g_i(A) \neq h_i(A), \text{ and } 0 < \alpha < 1.$$

Hence, from our discussion (in the proof of Corollary 5.10) of the case of equality in the generalized arithmetic-geometric mean inequality, we then deduce that the final inequality in (5.24) must be a *strict inequality* when $j = i$. Set $k := g^{\alpha}h^{1-\alpha}$, so that $k \in \mathcal{G}_n^c$ by Theorem 5.9. Then, $\mathcal{M}^f(A)$, an irreducible singular M-matrix from *ii)* and Theorem 5.5, satisfies $\mathcal{M}^f(A) \geq \mathcal{M}^k(A)$, where the i-th diagonal entry of $\mathcal{M}^f(A)$ is strictly greater than the corresponding one of $\mathcal{M}^k(A)$. But then, $\mathcal{M}^k(A)$ is not an M-matrix (see Exercise 3 of this section), which contradicts the fact, from Theorem 5.5, that $k \in \mathcal{G}_n^c$. Thus, *ii)* implies *i)*. ■

As an immediate consequence of Theorem 5.12, we have

Corollary 5.13. *For any* $\mathbf{x} = [x_1, x_2, \cdots, x_n]^T \in \mathbb{R}^n$ *with* $\mathbf{x} > \mathbf{0}$, *then* $r^{\mathbf{x}} = [r_1^{\mathbf{x}}, r_2^{\mathbf{x}}, \cdots, r_n^{\mathbf{x}}]$ *of* (1.13) *and* $c^{\mathbf{x}} = [c_1^{\mathbf{x}}, c_2^{\mathbf{x}}, \cdots, c_n^{\mathbf{x}}]$ *of* (1.35) *are minimal continuous G-functions.*

We conclude this section with the curious result that the α-convolution (5.15) of two G-functions in $\mathcal{G}_n^c$ is **rarely minimal**! More precisely, we have

Theorem 5.14. *For* $n > 2$ *and* $0 < \alpha < 1$, *the* α- *convolution,* h, *as defined in* (5.15), *of* f *and* g *in* $\mathcal{G}_n^c$ *is a minimal point of* $\mathcal{G}_n^c$ *if and only if* $f = g$ *and* f *is a minimal point in* $\mathcal{G}_n^c$.

Proof. If $f = g$ and if f is a minimal point of $\mathcal{G}_n^c$, then, since h reduces to f in this case from (5.15), h is obviously a minimal point in $\mathcal{G}_n^c$. Conversely, with h the α-convolution of f and g in $\mathcal{G}_n^c$ we see, from the second inequality of (5.18) and the equivalence of *iv)* and *ii)* of Theorem 5.12, that h is a minimal point of $\mathcal{G}_n^c$ only if f and g are both minimal points of $\mathcal{G}_n^c$ and only if equality holds throughout in the final inequality of (5.18) for every irreducible $A \in \mathbb{C}^{n \times n}$. Now, suppose that f and g are *distinct* minimal points in $\mathcal{G}_n^c$. Then, we can find an $A \in \mathbb{C}^{n \times n}$ and a $k \in N$ for which $f_k(A) \neq g_k(A)$. Moreover, from the continuity of f and g, we may assume that A has all nonzero off-diagonal entries, so that A is irreducible. Now, for equality to hold for all $i \in N$ in the final inequality of (5.18), i.e., with $f_i(A) = r_i^{\mathbf{x}}(A)$ and $g_i(A) = r_i^{\mathbf{y}}(A)$, where $\mathbf{x} > \mathbf{0}$ and $\mathbf{y} > \mathbf{0}$, we must have $(r_i^{\mathbf{x}}(A))^{\alpha}\ (r_i^{\mathbf{y}}(A))^{1-\alpha} = r_i^{\mathbf{z}}(A)$, or equivalently,

$$\left(\sum_{\substack{j=1\\ j\neq i}}^{n} \frac{|a_{i,j}|x_j}{x_i}\right)^{\alpha} \cdot \left(\sum_{\substack{j=1\\ j\neq i}}^{n} \frac{|a_{i,j}|y_j}{y_i}\right)^{1-\alpha} = \sum_{\substack{j=1\\ j\neq i}}^{n} \left(\frac{|a_{i,j}|x_j}{x_i}\right)^{\alpha} \cdot \left(\frac{|a_{i,j}|y_j}{y_i}\right)^{1-\alpha}$$

for *all* $i \in N$. It then follows from the case of equality in the Hölder inequality (cf. Beckenbach and Bellman (1961), p. 19) that the vectors

$$\left[\frac{|a_{i,1}|x_1}{x_i}, \ldots, \frac{|a_{i,i-1}|x_{i-1}}{x_i}, \frac{|a_{i,i+1}|x_{i+1}}{x_i}, \cdots\right]^T \quad \text{and}$$

$$\left[\frac{|a_{i,1}|y_1}{y_i}, \ldots, \frac{|a_{i,i-1}|y_{i-1}}{y_i}, \frac{|a_{i,i+1}|y_{i+1}}{y_i}, \cdots\right]^T$$

in $\mathbb{R}^{n-1}$ are proportional for *each* $i \in N$. Using this and the fact that A has *all* its off-diagonal entries non-zero for $n > 2$, this proportionality can occur, as is readily verified, only when the vectors $\mathbf{x} > \mathbf{0}$ and $\mathbf{y} > \mathbf{0}$ are themselves proportional, i.e., for some $\mu > 0, x_i = \mu y_i$ for all $i \in N$. This, however, implies that

$$f_i(A) = r_i^{\mathbf{x}}(A) = \sum_{j \in N\backslash\{i\}} |a_{i,j}|x_j/x_i = r_i^{\mathbf{y}}(A) = \sum_{j \in N\backslash\{i\}} |a_{i,j}|y_j/y_i = g_i(A)$$

for all $i \in N$, which contradicts the fact that $f_k(A) \neq g_k(A)$. Thus, if f and g are distinct minimal points of $\mathcal{G}_n^c$, their α-convolution is *not* a minimal point in $\mathcal{G}_n^c$. ■

The result of Theorem 5.14, in conjunction with the equivalences of Theorem 5.12, shows that the minimal elements $f = r^{\mathbf{x}}$ and $g = c^{\mathbf{x}}$, for $\mathbf{x} \in \mathbb{C}^n$ with $\mathbf{x} > \mathbf{0}$, in $\mathcal{G}_n^c$ again play a **central**, if not dominant, role in the Geršgorin theory. It is also interesting to note that the combinatorial result of Ostrowski (1951a) in (1.30) of Theorem 1.16 is, from (5.15), just the α-convolution of $r^{\mathbf{x}}$ and $c^{\mathbf{x}}$, each of which is a minimal continuous G-function from Corollary 5.13. But from Theorem 5.14, this α-convolution is a minimal point of $\mathcal{G}_n^c$ only if $r^{\mathbf{x}} = c^{\mathbf{x}}$, i.e., if $r_i^{\mathbf{x}}(A) = c_i^{\mathbf{x}}(A)$ for $i \in N$, and all $A \in \mathbb{C}^{n \times n}$, which is **extremely restrictive**. (See Exercise 4 of this section.)

Exercises

1. If $E = [e_{i,j}] \in \mathbb{R}^{n\times n}$ is an M-matrix, show that $e_{i,i} \geq 0$ for all $i \in N$. (This is a special case of Exercise 6 of Section 5.1.)

2. Show that Theorem 5.9 is valid if $\mathcal{G}_n^c$ is replaced by $\mathcal{G}_n$. (Hoffman (1969)).

3. If $D = [d_{i,j}] \in \mathbb{R}^{n\times n}$ is an irreducible singular M-matrix, show that decreasing any diagonal entry of D gives a matrix which is not an M-matrix. (Hint: Use *ii*) of Theorem C.1 of Appendix C.)

4. Given $A = [a_{i,j}] \in \mathbb{C}^{n\times n}, n > 2$, and given $\mathbf{x} > \mathbf{0}$ in $\mathbb{R}^n$, find sufficient conditions on the moduli of the off-diagonal entries of A so that

$$r_i^{\mathbf{x}}(A) = c_i^{\mathbf{x}}(A) \text{ for all } i \in N.$$

(Hint: Start with $|a_{i,j}| = |a_{j,i}|$ for all $i \neq j$, and make small changes in these entries.)

5.3 Minimal G-Functions

The object of this section is to obtain an analog of Theorem 5.12 for the set $\mathcal{G}_n$, which is larger than $\mathcal{G}_n^c$ for $n \geq 2$. This requires us to review some graph-theoretic facts about reducible matrices, as discussed in Section 2.2. Given any reducible $A = [a_{i,j}] \in \mathbb{C}^{n\times n}$ with $n \geq 2$, then there is a permutation matrix $P \in \mathbb{R}^{n\times n}$ for which PAP^T has the form (2.35), i.e.

$$PAP^T = \begin{bmatrix} R_{1,1} & R_{1,2} & \cdots & R_{1,m} \\ O & R_{2,2} & \cdots & R_{2,m} \\ \vdots & & \ddots & \vdots \\ O & O & \cdots & R_{m,m} \end{bmatrix}, \tag{5.25}$$

where each submatrix $R_{j,j}$, $1 \leq j \leq m$, is square and is either irreducible or a 1×1 matrix, and we say that (5.25) is a **normal reduced form** of the reducible matrix A. (As examples easily show, this reduced normal form of a reducible matrix is not in general unique.)

The form of PAP^T in (5.25) gives rise to a **partitioning** of $N = \{1, 2, \cdots, n\}$ into m disjoint nonempty sets S_k of N, where this partitioning, from (5.25), clearly depends on the matrix A. If we *define* here all 1×1 null matrices to also be irreducible, then these m sets S_k exactly correspond to the distinct *connected* components of the directed graph for A. For each $i \in N$, let $\langle i \rangle$ denote the *unique* subset S_k of N which contains i, and let

$|S_k|$ denote the cardinality of S_k (i.e., the number of elements in S_k). It follows that $\sum_{k=1}^{m} |S_k| = n$. If the matrix A is irreducible, we define $m := 1$ and $\langle i \rangle := N$ for all $i \in N$. In this way, for any given $A \in \mathbb{C}^{n \times n}$, the set $\langle i \rangle$ is defined for all $i \in N$, whether A is reducible or irreducible.

For any $f \in \mathcal{F}_n$ and for any matrix $A \in \mathbb{C}^{n \times n}$, the matrix $\mathcal{M}^f(A)$ in $\mathbb{R}^{n \times n}$ is defined by (5.7). If A is reducible and has the normal reduced form of (5.25), then $\mathcal{M}^f(A)$ inherits the special form of

$$(5.26) \qquad \mathcal{M}^f(PAP^T) = \begin{bmatrix} \mathcal{M}^f(R_{1,1}) & -|R_{1,2}| & \cdots & -|R_{1,m}| \\ O & \mathcal{M}^f(R_{2,2}) & \cdots & -|R_{2,m}| \\ \vdots & & \ddots & \vdots \\ O & O & \cdots & \mathcal{M}^f(R_{m,m}) \end{bmatrix},$$

where $|R_{i,j}|$ (for $1 \leq i < j \leq m$) denotes the nonnegative matrix whose entries are the absolute values of the corresponding entries of $R_{i,j}$ in (5.25). Because of the upper-triangular block structure of the matrix in (5.26), only the diagonal submatrices, $\mathcal{M}^f(R_{k,k})$, with $1 \leq k \leq m$, determine the eigenvalues of $\mathcal{M}^f(PAP^T)$, or equivalently, $\mathcal{M}^f(A)$. Consequently, it follows from Definition 5.4 and (5.26) that

(5.27) $\quad \mathcal{M}^f(A)$ is an $M-$matrix if and only if $\mathcal{M}^f(R_{k,k})$ is an $M-$matrix for each k with $1 \leq k \leq m$.

Continuing, given any $A \in \mathbb{C}^{n \times n}$, from which the set $\langle i \rangle$ is defined for each $i \in N$, and given any $\mathbf{x} = [x_1, x_2, \cdots, x_n]^T$ in $\mathbb{R}^n$ with $\mathbf{x} > \mathbf{0}$, we set

$$(5.28)\ \hat{r}_i^{\mathbf{x}}(A) := \sum_{j \in \langle i \rangle \backslash \{i\}} |a_{i,j}| x_j / x_i; \ \hat{c}_i^{\mathbf{x}}(A) := \sum_{j \in \langle i \rangle \backslash \{i\}} |a_{j,i}| x_j / x_i \ (i \in N),$$

where $\hat{r}_i^{\mathbf{x}}(A) := \hat{c}_i^{\mathbf{x}}(A) := 0$ if $\langle i \rangle = \{i\}$. Note from Definition 5.1 that $\hat{r}^{\mathbf{x}} = [\hat{r}_1^{\mathbf{x}}, \hat{r}_2^{\mathbf{x}}, \cdots, \hat{r}_n^{\mathbf{x}}]$ and $\hat{c}^{\mathbf{x}} = [\hat{c}_1^{\mathbf{x}}, \hat{c}_2^{\mathbf{x}}, \cdots \hat{c}_n^{\mathbf{x}}]$ are certainly elements of $\mathcal{F}_n$, and that $\hat{r}^{\mathbf{x}}(A)$ and $\hat{c}^{\mathbf{x}}(A)$ reduce to the previously defined familiar functions $r^{\mathbf{x}}(A)$ and $c^{\mathbf{x}}(A)$, when A is irreducible.. It is important to note that the functions $\hat{r}^{\mathbf{x}}$ and $\hat{c}^{\mathbf{x}}$ in $\mathcal{F}_n$, as defined in (5.28), can be *discontinuous* on $\mathbb{C}^{n \times n}$. To see this, consider the 2×2 matrix

$$A(t) := \begin{bmatrix} 1 & 1 \\ t & 1 \end{bmatrix} \quad \text{for } t \in [0, 1].$$

Choosing $\boldsymbol{\xi} := [1, 1]^T > \mathbf{0}$, it follows that $\hat{r}^{\boldsymbol{\xi}}(A(t)) = [\hat{r}_1^{\boldsymbol{\xi}}(A(t)), \hat{r}_2^{\boldsymbol{\xi}}(A(t))]$, where

$$\hat{r}_1^{\boldsymbol{\xi}}(A(t)) = \left\{ \begin{array}{ll} 1 & \text{for } t \in (0, 1]; \\ 0 & \text{for } t = 0; \end{array} \right\}, \text{ and } \hat{r}_2^{\boldsymbol{\xi}}(A(t)) = t \text{ for } t \in [0, 1].$$

Thus, $\hat{r}_1^{\boldsymbol{\xi}}(A(t))$ is discontinuous at $t = 0$; this is a consequence of the fact that $A(t)$ is irreducible for all $t \in (0,1]$, but $A(0)$ is reducible.

To deduce further properties of the functions $\hat{r}^{\mathbf{x}}$ and $\hat{c}^{\mathbf{x}}$ in $\mathcal{F}_n$, observe that because $\langle i \rangle$ is a subset of N, we directly have from (5.28) that

$$r_i^{\mathbf{x}}(A) \geq \hat{r}_i^{\mathbf{x}}(A) \text{ and } c_i^{\mathbf{x}}(A) \geq \hat{c}_i^{\mathbf{x}}(A) \quad (i \in N, A \in \mathbb{C}^{n\times n}, \mathbf{x} > \mathbf{0}).$$

Hence, with the partial order defined in (5.13), this gives

$$(5.29) \qquad r^{\mathbf{x}} \succ \hat{r}^{\mathbf{x}} \text{ and } c^{\mathbf{x}} \succ \hat{c}^{\mathbf{x}} \quad (\mathbf{x} > \mathbf{0} \text{ in } \mathbb{R}^n),$$

with $r^{\mathbf{x}} \not\equiv \hat{r}^{\mathbf{x}}$, as reducible matrices show. Next, for any $A \in \mathbb{C}^{n\times n}$, we have seen that the sets $\langle i \rangle$ partition N into m disjoint sets S_k, as i runs through N. On choosing $f := \hat{r}^{\mathbf{x}}$ (for $\mathbf{x} > \mathbf{0}$ in $\mathbb{R}^n$) and on utilizing these sets S_k, it can be verified from (5.28) that $\mathcal{M}^{\hat{r}^{\mathbf{x}}}(R_{k,k})$ in (5.26) is a *singular* M-matrix for each k with $1 \leq k \leq m$, the same being true for $f := \hat{c}^{\mathbf{x}}$. Consequently, from (5.27) and Theorem 5.5, we deduce that

$$(5.30) \qquad \hat{r}^{\mathbf{x}} \text{ and } \hat{c}^{\mathbf{x}} \text{ are elements of } \mathcal{G}_n \text{ for any } \mathbf{x} > \mathbf{0} \text{ in } \mathbb{R}^n.$$

As further preparations for the main result in this section, note that given any $f \in \mathcal{G}_n$ and given any $A \in \mathbb{C}^{n\times n}$, then $\mathcal{M}^f(A)$ is an M-matrix from Theorem 5.5, so that from (5.27), $\mathcal{M}^f(R_{k,k})$ is an M-matrix for each k with $1 \leq k \leq m$. As such, there is a vector $\mathbf{x}_k > \mathbf{0}$ in $\mathbb{R}^{|S_k|}$ such that

$$(5.31) \qquad \mathcal{M}^f(R_{k,k})\mathbf{x}_k \geq \mathbf{0} \quad (1 \leq k \leq m).$$

Because the sets $\{S_k\}_{k=1}^m$ are nonempty and have as their union N, we can appropriately compose the positive components of the vectors $\mathbf{x}_k$ from (5.31) to *define* the vector $\mathbf{x} > \mathbf{0}$ in $\mathbb{R}^n$ such that (5.31) is equivalent to

$$(5.32) \qquad f_i(A) \geq \hat{r}_i^{\mathbf{x}}(A) \quad (i \in N).$$

The inequalities of (5.32) are of course a sharper form of the inequalities of (5.8) of Theorem 5.5. Moreover, by simply replacing $r^{\mathbf{x}}$ by $\hat{r}^{\mathbf{x}}$ in the proof of Theorem 5.9, we see that the α-convolution, (5.15), of any two elements in $\mathcal{G}_n$ is again an element of $\mathcal{G}_n$ so that, as in Corollary 5.10, $\mathcal{G}_n$ is *also* a convex set.

This now brings us to the following analog of Theorem 5.12.

Theorem 5.15. *Let $f \in \mathcal{G}_n$. Then, the following are equivalent:*

i) *f is an extreme point of the convex set $\mathcal{G}_n$;*
ii) *f is a minimal point of $\mathcal{G}_n$, partially ordered by* (5.13)*;*
iii) *for every $A \in \mathbb{C}^{n\times n}$, $\mathcal{M}^f(R_{k,k})$ (cf.* (5.26)*) is a singular M-matrix for each k with $1 \leq k \leq m$;*
iv) *for every $A \in \mathbb{C}^{n\times n}$, there exists an $\mathbf{x} > \mathbf{0}$ in $\mathbb{R}^n$ (where $\mathbf{x}$ is dependent on A) such that $f_i(A) = \hat{r}_i^{\mathbf{x}}(A)$, for all $i \in N$.*

Proof. That *i)* implies *ii)* follows word-for-word from the proof of Theorem 5.12. Next, assuming *ii)*, define $g \in \mathcal{F}_n$, for any $A \in \mathbb{C}^{n\times n}$, by means of

$$g_i(A) := f_i(A) - \lambda_k(A), \tag{5.33}$$

where $i \in S_k$ and where $\lambda_k(A)$ denotes the minimal nonnegative real eigenvalue of the M-matrix $\mathcal{M}^f(R_{k,k})$, i.e., if $\mathcal{M}^f(R_{k,k}) = \mu I - B$, where $B \geq O$ and $\mu \geq \rho(B)$, then $\lambda_k(A) := \mu - \rho(B)$. As is readily verified, $g \in \mathcal{G}_n$ and $f \succ g$. Now, suppose that *iii)* is not valid. Then, there is an $A \in \mathbb{C}^{n\times n}$ for which some $\mathcal{M}^f(R_{k,k})$ is nonsingular. Of course, since $\mathcal{M}^f(R_{k,k})$ is an M-matrix from Theorem 5.5, then $\mathcal{M}^f(R_{k,k})$ is necessarily a nonsingular M-matrix. As such, $\lambda_k(A) > 0$ so that there is a $j \in N$ for which $f_j(A) > g_j(A)$, i.e., $f \succ g$ and $f \neq g$, which contradicts the minimality of f. Thus, *ii)* implies *iii)*.

Next, assuming *iii)*, the singularity of each M-matrix $\mathcal{M}^f(R_{k,k}), 1 \leq k \leq m$, implies that equality only holds in (5.31) and (5.32), which implies *iv)*. Conversely, if *iv)* holds, direct computation shows that (5.31) is valid with equality, so that each $\mathcal{M}^f(R_{k,k})$ is a singular M-matrix. Thus, *iii)* and *iv)* are equivalent.

Finally, we wish to show that *iii)* implies *i)*. As in the proof of Theorem 5.12, assuming *iii)* holds with f not an extreme point of $\mathcal{G}_n$ leads us to the construction of a $k \in \mathcal{G}_n$ and an $A \in \mathbb{C}^{n\times n}$ for which $\mathcal{M}^k(R_{k,k})$ is not an M-matrix, which contradicts Theorem 5.5. Thus, *iii)* implies *i)*. ∎

The following is then an obvious consequence of Theorem 5.15.

Corollary 5.16. *For any* $\mathbf{x} \in \mathbb{R}^n$ *with* $\mathbf{x} > \mathbf{0}$*, then* $\hat{r}^{\mathbf{x}}$ *and* $\hat{c}^{\mathbf{x}}$ *of* (5.30) *are minimal G-functions in* $\mathcal{G}_n$.

Exercises

1. Give an example of a reducible matrix $A = [a_{i,j}] \in \mathbb{C}^{n\times n}, n > 2$, for which its reduced normal form of (5.25) is not unique.

2. Consider matrix A $\begin{bmatrix} a_{1,1} & 0 & 0 & 0 & 0 \\ a_{2,1} & a_{2,2} & 0 & a_{2,4} & 0 \\ a_{3,1} & a_{3,2} & a_{3,3} & a_{3,4} & a_{3,5} \\ a_{4,1} & a_{4,2} & 0 & a_{4,4} & 0 \\ a_{5,1} & a_{5,2} & a_{5,3} & a_{5,4} & a_{5,5} \end{bmatrix}$,

   where the entries with numbered subscripts are all nonzero. Then,

   a. Determine a normal reduced form (cf. (5.25)) for A;
   b. Determine the sets $< i >$ for $i = 1, 2, 3, 4, 5$;
   c. Determine $\hat{r}_i^{\mathbf{x}}(A)$ for $i = 1, 2, 3, 4, 5$, where $\mathbf{x} > \mathbf{0}$ in $\mathbb{R}^5$.

5.4 Minimal G-Functions with Small Domains of Dependence

The material for this section comes from Hoffman and Varga (1970).

To begin, for any $i, j \in N$, let $E_{i,j} = [a_{k,\ell}] \in \mathbb{R}^{n\times n}$ be defined by $a_{k,\ell} := \delta_{k,i}\delta_{\ell,j}$, for all $k, \ell \in N$, so that each entry of the matrix $E_{i,j}$ is zero, except for unity in the (i, j)-th entry.

Definition 5.17. Given $f = [f_1, f_2, \cdots, f_n] \in \mathcal{F}_n$, then f_k is **independent of the ordered pair** (i, j), with $i \neq j$ $(i, j \in N)$, if

$$(5.34) \qquad f_k(A + \tau E_{i,j}) = f_k(A) \qquad (\text{all } A \in \mathbb{C}^{n\times n}, \text{all } \tau \in \mathbb{C}).$$

Otherwise, f_k **depends on** (i, j). The set

$$(5.35)\ \mathcal{D}(f_k) := \{(i,j)\ :\ i,j \in N \text{ with } i \neq j \text{ and } f_k \text{ depends on } (i,j)\}$$

is called the **domain of dependence of** f_k.

To illustrate the above concept, for any $\mathbf{x} = [x_1, x_2, \cdots, x_n]^T > \mathbf{0}$ in $\mathbb{R}^n$, consider $r^{\mathbf{x}} = [r_1^{\mathbf{x}}, r_2^{\mathbf{x}}, \cdots, r_n^{\mathbf{x}}]$, which is certainly an element of $\mathcal{G}_n^c$, from Corollary 5.13. Because $r_i^{\mathbf{x}}(A) := \sum_{j\in N\backslash\{i\}} |a_{i,j}|x_j/x_i$ for any $A = [a_{i,j}] \in \mathbb{C}^{n\times n}$, we see that the domain of dependence of $r_i^{\mathbf{x}}$ is given by

$$(5.36) \qquad \mathcal{D}(r_i^{\mathbf{x}}) = \{(i,j)\ :\ j \in N \text{ and } j \neq i\}, \quad (i \in N).$$

In contrast to the above example, **maximal domains** of dependence can be obtained from the following construction. For any $A = [a_{i,j}] \in \mathbb{C}^{n\times n}$, let $B_A = [b_{i,j}] \in \mathbb{R}^{n\times n}$ be defined by

$$(5.37) \qquad b_{i,i} := 0,\ \ b_{i,j} := |a_{i,j}| \text{ for } i \neq j \quad (i,j \in N),$$

so that $B_A \geq O$. Next, define $g := [g_1, g_2, \cdots, g_n]$ by

$$(5.38) \qquad g_i(A) := \rho(B_A) \quad (i \in N, A \in \mathbb{C}^{n\times n}),$$

so that, from Definitions 5.1 and 5.7, it is easily seen that $g \in \mathcal{F}_n^c$. Since, from (5.7), $\mathcal{M}^g(A) := \rho(B_A)I_n - B_A$ is evidently a singular M-matrix for any $A \in \mathbb{C}^{n\times n}$, then $g \in \mathcal{G}_n^c$ from Theorem 5.5. In this case, each g_i depends on **every** ordered pair (k, ℓ) with $k \neq \ell$ $(k, \ell \in N)$, so that

$$(5.39) \qquad \mathcal{D}(g_i) = \{(k,\ell)\ :\ k \neq \ell, \text{ with } k,\ell \in N\}, \qquad \text{for all } i \in N.$$

The next example shows that some domains of dependence can be **empty**. In the special case $n = 2$, for any $A = [a_{i,j}] \in \mathbb{C}^{2\times 2}$, define

$h = [h_1, h_2] \in \mathcal{F}_2^c$ by

$$h_1(A) := 1, \quad h_2(A) := |a_{1,2}| \cdot |a_{2,1}|. \tag{5.40}$$

In this case, we have (cf. (5.7)) that

$$\mathcal{M}^h(A) = \begin{bmatrix} 1 & -|a_{1,2}| \\ -|a_{2,1}| & |a_{1,2}| \cdot |a_{2,1}| \end{bmatrix},$$

so that $\mathcal{M}^h(A)$ is a singular M-matrix for every $A \in \mathbb{C}^{2\times 2}$. Hence, with Theorem 5.5, we see that $h \in \mathcal{G}_n^c$, and in this case,

$$\mathcal{D}(h_1) = \emptyset \text{ and } \mathcal{D}(h_2) = \{(1,2) \text{ and } (2,1)\}. \tag{5.41}$$

In each of the examples above, it has been the case that for $f \in \mathcal{G}_n$, each ordered pair (i,j), with $i \neq j$ $(i,j \in N)$, is in $\mathcal{D}(f_k)$ for *some* $k \in N$. That this is in general true is established in

Theorem 5.18. *If $f = [f_1, f_2, \cdots, f_n] \in \mathcal{G}_n$, then*

$$\bigcup_{k\in N} \mathcal{D}(f_k) = \{(i,j) \ : i \neq j \text{ with } i,j \in N\}. \tag{5.42}$$

Proof. For $n = 1$, (5.42) is vacuously true from Definition 5.1. For $n \geq 2$, suppose, on the contrary, that there is an ordered pair (i,j) with $i \neq j$ $(i,j \in N)$, such that $(i,j) \notin \mathcal{D}(f_k)$ for *any* $k \in N$. To simplify notation, we assume without loss of generality that $(i,j) = (1,2)$. By definition, since each f_k is independent of the ordered pair $(1,2)$, then (cf. (5.34))

$$f_k(A) = f_k(A + \tau E_{1,2}) \text{ for all } k \in N, \text{ all } A \in \mathbb{C}^{n\times n}, \text{ all } \tau \in \mathbb{C}. \tag{5.43}$$

Now, define $B \in \mathbb{R}^{n\times n}$ by

$$B = \left[\begin{array}{cc|c} 0 & 1 & O \\ 1 & 0 & \\ \hline & O & O \end{array}\right],$$

where the null submatrices of B are not present if $n = 2$. Then, fix μ such that $\mu > \max\{f_k(B) \ : \ k \in N\}$. Defining $C := \mu I_n - B = [c_{i,j}] \in \mathbb{R}^{n\times n}$, we have that

$$\mu = |c_{i,i}| > f_i(C) = f_i(B) \qquad (\text{all } i \in N), \tag{5.44}$$

the last equality following from ii) of Definition 5.1. But, as f is by hypothesis a G-function, (5.44) implies that the matrix C is nonsingular. On the other hand, if, in place of C, we consider $C-\tau E_{1,2}$, then from (5.43), the inequalities of (5.44) similarly imply that $C-\tau E_{1,2}$ is nonsingular for *any* τ, in $\mathbb{C}$, i.e., that

$$C-\tau E_{1,2} = \left[\begin{array}{cc|c} \mu & -1-\tau & \\ -1 & \mu & O \\ \hline & O & \mu I_{n-2} \end{array}\right]$$

is nonsingular. But on choosing $\tau = \mu^2 - 1$, it can be seen that $C-\tau E_{1,2}$ is *singular*, which is a contradiction. ∎

For any $\mathbf{x} > \mathbf{0}$ in $\mathbb{R}^n$, the G-function $r^{\mathbf{x}} = [r_1^{\mathbf{x}}, r_2^{\mathbf{x}}, \cdots, r_n^{\mathbf{x}}]$, with its domains of dependence given by (5.36), has played a *central* role in Geršgorin results for matrices, as we have seen. Our final result in this section gives yet another reason for this.

Theorem 5.19. *For $n \geq 2$, let $f = [f_1, f_2, \cdots, f_n]$ be a minimal point in $\mathcal{G}_n^c$, partially ordered by "$\succ$" of* (5.13). *If, for some k,*

$$\mathcal{D}(f_k) = \{(k,\ell) \ : \ \ell \in N \text{ and } \ell \neq k\}, \tag{5.45}$$

and if

$$\left\{ \bigcup_{i \in N \backslash \{k\}} \mathcal{D}(f_i) \right\} \bigcap \mathcal{D}(f_k) = \emptyset, \tag{5.46}$$

then

$$\mathcal{D}(f_i) = \{(i,j) \ : \ j \in N \text{ and } j \neq i\} \quad \text{for all } i \in N, \tag{5.47}$$

and there exists an $\mathbf{x} \in \mathbb{R}^n$ with $\mathbf{x} > \mathbf{0}$, which is independent of the matrix A in $\mathbb{C}^{n\times n}$, such that

$$f = r^{\mathbf{x}}. \tag{5.48}$$

Proof. There is no loss of generality in assuming that $k=1$. Since f is by hypothesis a minimal point in $\mathcal{G}_n^c$, it follows from *iii*) of Theorem 5.12, that $\mathcal{M}^f(A)$ is a singular M-matrix for every $A \in \mathbb{C}^{n\times n}$. If $A \in \mathbb{C}^{n\times n}$ is, in addition, irreducible, it also follows from *iv*) of Theorem 5.12 that there is an $\mathbf{x} > \mathbf{0}$ in $\mathbb{R}^n$ (where $\mathbf{x}$ is dependent on A) such that $\mathcal{M}^f(A)\mathbf{x} = \mathbf{0}$; whence,

$$f_i(A) := \sum_{j \in N \backslash \{i\}} |a_{i,j}| x_j / x_i \quad (\text{all } i \in N). \tag{5.49}$$

Assume that all off-diagonal entries of A are non-zero. Because the x_k's occur in (5.49) as ratios, we may normalize to $x_1 = 1$. Next, as in (5.6) and (5.7), we can express $\mathcal{M}^f(A) = [\alpha_{i,j}]$ in $\mathbb{R}^{n\times n}$ as $\mu I_n - B$ where $\mu := \max\{f_i(A) : i \in N\} > 0$, and where $B = [b_{i,j}] \in \mathbb{R}^{n\times n}$ has its entries defined by

$$b_{i,i} := \mu - f_i(A) \geq 0 \text{ and } b_{i,j} := |a_{i,j}| \geq 0,\ i \neq j, \quad i,j \in N.$$

Thus, $B \geq O$ and B is irreducible. But since $\mathcal{M}^f(A) = \mu I_n - B$ is a singular M-matrix, then $\rho(B) = \mu$. Recalling that $n \geq 2$, delete both the first row and column of the matrix $\mathcal{M}^f(A)$, calling the resulting matrix $\tilde{C} \in \mathbb{R}^{(n-1)\times(n-1)}$. Then, $\tilde{C} = \rho(B)I_{n-1} - \tilde{B}$, where $\tilde{B} \in \mathbb{R}^{(n-1)\times(n-1)}$ is similarly the result of deleting the first row and first column of B. Because A and B are both irreducible, it follows from the Perron-Frobenius theory of nonnegative matrices (cf. *iii*) of Theorem C.1 of Appendix C) that $\rho(\tilde{B}) < \rho(B)$; whence, $\tilde{C}$ is a *nonsingular* M-matrix. Now, the equation $\mathcal{M}^f(A)\mathbf{x} = \mathbf{0}$ can be expressed (on recalling that $x_1 = 1$) as the pair of equations

$$\text{(5.50)} \qquad \begin{cases} f_1(A) = \sum\limits_{j>1} |a_{1,j}|x_j, \text{ and} \\ \tilde{C}[x_2, x_3, \cdots, x_n]^T = [|a_{2,1}|, |a_{3,1}|, \cdots, |a_{n,1}|]^T. \end{cases}$$

Since $\tilde{C}$ is nonsingular, then the second equation above implies that

$$\text{(5.51)} \qquad [x_2, x_3, \cdots, x_n]^T = (\tilde{C})^{-1}[|a_{2,1}|, |a_{3,1}|, \cdots, |a_{n,1}|]^T.$$

From the hypothesis of (5.46) with $k = 1$, $f_2(A), \cdots, f_n(A)$ are all independent of the ordered pairs $(1,2), (1,3), \cdots, (1,n)$; hence, from the structure of $\tilde{C}$, so is $(\tilde{C})^{-1}$. Thus, we see from (5.51) that the components $x_2, x_3, \cdots, x_n$ are also all independent of the ordered pairs $(1,2), (1,3), \cdots, (1,n)$. This means that if we now continuously *vary* the matrix A only in the entries $|a_{1,2}|, \cdots, |a_{1,n}|$, while keeping A irreducible, the first equation of (5.50) remains valid with $x_2, x_3, \cdots, x_n$ fixed, i.e.,

$$\text{(5.52)} \qquad f_1(A) = \sum_{j>1} |a_{1,j}|x_j = r_1^{\mathbf{x}}(A).$$

But, as f_1 is continuous by hypothesis, the above expression also remains valid for *any* matrix A in $\mathbb{C}^{n\times n}$, obtained by continuously varying the original matrix A only in the entries $|a_{1,2}|, \cdots, |a_{1,n}|$.

For a fixed k with $2 \leq k \leq n$, let us now, in particular, continuously vary only the entries $|a_{1,2}|, \cdots, |a_{1,n}|$ of the original matrix A, so that the first row of A becomes $[a_{1,1}, \delta_{2,k}, \cdots, \delta_{n,k}]^T$, where $\delta_{j,k}$ denotes the Kronecker delta function. Because all off-diagonal entries of the original matrix A were non-zero, the new matrix A is also irreducible. For this new matrix A, (5.50) gives us that $f_1(A) = x_k = x_k(A)$. On the other hand, since f_1, by hypothesis, depends only on the ordered pairs $\{(1,\ell) : 2 \leq \ell \leq n\}$, while x_k, from (5.51), is independent of the first row of A, then $f_1(A) = x_k$ where x_k is a positive constant which is *independent* of A. Repeating this argument for other values of k with $2 \leq k \leq n$, it similarly follows that $x_2, x_3, \cdots, x_n$ from (5.51) are positive constants which are independent of A. Thus, from (5.49), $f = r^{\mathbf{x}}$ where $\mathbf{x}$ is independent of A. ∎

We remark that an analogous result of Theorem 5.19 can similarly be formulated in terms of the weighted column sums $c^{\mathbf{x}} := [c_1^{\mathbf{x}}, \cdots, c_n^{\mathbf{x}}]$ in $\mathcal{G}_n^c$.

Exercises

1. For any $n \geq 3$, show that it is not possible to find an $h = [h_1, h_2, \cdots, h_n] \in \mathcal{G}_n^c$, for which $h_1(A) := h_2(A) := 1$ for all $A \in \mathbb{C}^{n \times n}$, so that $\mathcal{D}(h_1) = \mathcal{D}(h_2) = \emptyset$. (Hint: If $h \in \mathcal{G}_n^c$, then $\mathcal{M}^h(A)$ must be an M-matrix for any $A \in \mathbb{C}^{n \times n}$, from Theorem 5.5. As such, every principal submatrix of $\mathcal{M}^h(A)$ also must be an M-matrix for any $A \in \mathbb{C}^{n \times n}$. Then, consider the upper 2×2 principal submatrix of $\mathcal{M}^h(A)$).

2. Consider any matrix $A = [a_{i,j}] \in \mathbb{C}^{n \times n}$, $n \geq 2$, where all off-diagonal entries of A are nonzero, so that A is irreducible. If all the nondiagonal entries, except one, in the first row of A are replaced by zero, show that the resulting matrix is still irreducible.

5.5 Connections with Brauer Sets and Generalized Brualdi Sets

On reviewing the developments in Section 5.1, the idea there was to see if one could go beyond the standard weighted row or weighted column sums, of a general matrix A in $\mathbb{C}^{n \times n}$, to develop new eigenvalue inclusion results, still based on the absolute values of off-diagonal entries of A. This led to the concept of a G-function in Definition 5.2, with the associated eigenvalue inclusion of (5.4) of Theorem 5.3. But (cf. (5.3)), the new eigenvalue inclusions were still the union of n **disks** in the complex plane, with the centers of these disks again being the diagonal entries of A, but now with new definitions for the radii of these disks.

It is then natural to ask if there is a similar extension, but based now on the union of Brauer Cassini ovals, as in (2.6). This can be done as follows. With the definition of $\mathcal{F}_n$ in Definition 5.1, we now make

Definition 5.20. For any $n \geq 2$, let $f = [f_1, f_2, \cdots, f_n] \in \mathcal{F}_n$. Then, f is a **K-function** if, for any $A = [a_{i,j}] \in \mathbb{C}^{n \times n}$, the relations

$$|a_{i,i}| \cdot |a_{j,j}| > f_i(A) \cdot f_j(A) \text{ (all } i \neq j,\ i,j \in N), \tag{5.53}$$

imply that A is nonsingular. The set of all K-functions in $\mathcal{F}_n$ is denoted by $\mathcal{K}_n$.

Remark. Note that if (5.2) of Definition 5.2 is valid, then (5.53) is also valid. Conversely, if (5.53) is valid, then all but at most one of the inequalities of (5.2) must hold.

Next, for any $n \geq 2$, for any $f \in \mathcal{F}_n$ and for any $A = [a_{i,j}] \in \mathbb{C}^{n\times n}$, we set

$$\begin{cases} K_{i,j}^f(A) := \{z \in \mathbb{C} : |z - a_{i,i}| \cdot |z - a_{j,j}| \leq f_i(A) \cdot f_j(A), i \neq j, (i,j \in N)\}, \\ \text{and} \\ K^f(A) := \bigcup\limits_{\substack{i,j \in N \\ i \neq j}} K_{i,j}^f(A). \end{cases} \tag{5.54}$$

Using the method of proof of Theorem 2.3, we easily establish (see Exercise 1 of this section)

Lemma 5.21. *For any $f = [f_1, f_2, \cdots, f_n] \in \mathcal{F}_n$, $n \geq 2$, and for any $A = [a_{i,j}] \in \mathbb{C}^{n\times n}$,*

$$K^f(A) \subseteq \Gamma^f(A). \tag{5.55}$$

Next, from Definition 5.20, our first recurring theme gives the following analog of Theorem 5.3.

Theorem 5.22. *Let $f = [f_1, f_2, \cdots, f_n] \in \mathcal{F}_n$, $n \geq 2$. Then, $f \in \mathcal{K}_n$ if and only if for every $A = [a_{i,j}] \in \mathbb{C}^{n\times n}$,*

$$\sigma(A) \subseteq K^f(A). \tag{5.56}$$

Then, on combining Lemma 5.21 and Theorem 5.22, we have

Corollary 5.23. *If $f = [f_1, f_2, \cdots f_n] \in \mathcal{K}_n$, $n \geq 2$, then for any $A = [a_{i,j}] \in \mathbb{C}^{n\times n}$,*

$$\sigma(A) \subseteq K^f(A) \subseteq \Gamma^f(A). \tag{5.57}$$

As an immediate consequence of Theorem 5.3 and the final inclusion of (5.57), we also have

Corollary 5.24. *If $f = [f_1, f_2, \cdots, f_n] \in \mathcal{K}_n$, $n \geq 2$, then $f \in \mathcal{G}_n$, so that $\mathcal{K}_n \subseteq \mathcal{G}_n$.*

A striking consequence of Corollary 5.24, is that it gives us that $\mathcal{K}_n \subseteq \mathcal{G}_n$ for any $n \geq 2$, so that the results of Sections 5.1-5.3 apply *equally well* to functions in $\mathcal{K}_n$. The important feature, however, is the final inclusion of (5.57), which shows that the inclusion region $K^f(A)$ is at least as good as the inclusion region $\Gamma^f(A)$, for any $A \in \mathbb{C}^{n\times n}$, $n \geq 2$.

Having considered, in this section, extensions of Section 5.1 to K-functions associated with Brauer sets of Section 2.2, we now consider a similar extension to the case of Brualdi sets, from Section 2.2, which depend on the additional knowledge of the cycle set of a matrix. We recall from Section 2.2 that a matrix $A = [a_{i,j}] \in \mathbb{C}^{n\times n}, n \geq 1$, has a directed graph $\mathbb{G}(A)$, from which its cycle set $C(A)$, of strong and weak cycles γ, is determined.

Definition 5.25. For any $n \geq 1$, let $f = [f_1, f_2, \cdots, f_n] \in \mathcal{F}_n$.Then, f is a **B − function** if, for every matrix $A = [a_{i,j}] \in \mathbb{C}^{n\times n}$, with $C(A)$ as its cycle set, then

$$\prod_{i\in\gamma} |a_{i,i}| > \prod_{i\in\gamma} f_i(A) \quad (\text{all } \gamma \in \mathcal{C}(A)), \tag{5.58}$$

imply that A is nonsingular. The set of all B-functions in $\mathcal{F}_n$ is denoted by $\mathcal{B}_n$.

For associated additional notation, if $A = [a_{i,j}] \in \mathbb{C}^{n\times n}, n \geq 1$, has $C(A)$ as its cycle set of strong and weak cycles, we define the associated sets (cf.(2.38) and (2.40)) of

$$\begin{cases} \mathcal{B}^f_\gamma(A) := \{z \in \mathbb{C} : \prod_{i\in\gamma} |z - a_{i,i}| \leq \prod_{i\in\gamma} f_i(A)\ \} \quad \text{for } \gamma \in \mathcal{C}(A), \\ \text{and} \\ \mathcal{B}^f(A) := \bigcup_{\gamma\in\mathcal{C}(A)} B^f_\gamma(A). \end{cases} \tag{5.59}$$

Our next result in the section is the analog of Theorems 4.3 and 4.22 (Its proof is left to Exercise 4 in this section.)

Theorem 5.26. *Let $f = [f_1, f_2, \cdots, f_n] \in \mathcal{F}_n$, $n \geq 1$. Then, $f \in \mathcal{B}_n$ of (5.59) if and only if , for every $A = [a_{i,j}] \in \mathbb{C}^{n\times n}$,*

$$\sigma(A) \subseteq \mathcal{B}^f(A). \tag{5.60}$$

Combining the results of Theorems 2.3, 2.6, and 4.26, we obtain

Corollary 5.27. *If $f = [f_1, f_2, \cdots, f_n] \in \mathcal{B}_n$, $n \geq 2$, then for any $A = [a_{i,j}] \in \mathbb{C}^{n\times n}$,*

$$\sigma(A) \subseteq \mathcal{B}^f(A) \subseteq K^f(A) \subseteq \Gamma^f(A). \tag{5.61}$$

Exercises

1. Give a proof of Lemma 5.21.

2. Give a complete proof of Theorem 5.22.

3. If $f = [f_1, f_2, \cdots, f_n] \in \mathcal{K}_n$, $n \geq 2$, then with the definition of (5.7) show, for any $A = [a_{i,j}] \in \mathbb{C}^{n \times n}$, that $\mathcal{M}^f(A)$ is an M-matrix.

4. Similarly, if $f = [f_1, f_2, \cdots, f_n] \in \mathcal{B}_n, n \geq 2$, then with the definition of (5.7), show, for any $A = [a_{i,j}] \in \mathbb{C}^{n \times n}$, that $\mathcal{M}^f(A)$ is an M-matrix.

5. Using simply the definitions of $\mathcal{K}_n$ and $\mathcal{G}_n$, show for $n \geq 2$ that $\mathcal{K}_n \subseteq \mathcal{G}_n$.

Bibliography and Discussion

5.1 As stated in the text, the material in this chapter was inspired by the works of Nowosad (1965) and Hoffman (1969), where G-functions were introduced. The material in this section basically comes from Carlson and Varga (1973a).

We note that G-functions are also considered in Hoffman (1971), Hoffman (1975) and Hoffman (2003), in Novosad and Tover (1980), and in Huang and You (1993), and Huang and Zhong (1999).

As further remarked in this section, Theorem 5.5 generalizes results of Ostrowski (1937b) and Fan (1958). We remark that **vectorial norms** appeared roughly simultaneously with the papers of Ostrowski (1961), Fiedler and Pták (1962a), and, as used in the proof of Theorem 4.5, Feingold and Varga (1962), while the important concept of a **lower bound matrix**, also used in the proof of Theorem 4.5, seems to be due to F. Robert, in a series of papers (cf. Robert (1964), Robert (1965), and Robert (1966)). For subsequent papers on lower bound matrices, see Stoer (1968) and Bode (1968).

5.2 The material in this section again comes from Carlson and Varga (1973a). The convexity of the sets $\mathcal{G}_n^c$ or $\mathcal{G}_n$ is due to Hoffman (1969).

5.3 Minimal G-functions, which are not necessarily continuous, were first discussed in Carlson and Varga (1973a), culminating in Theorem 5.15, which is the analog of Theorem 5.12.

5.4 The material in this section, on minimal G-functions with small domains of dependence, comes from Hoffman and Varga (1970).

5.5 The material in this section is new.

6. Geršgorin-Type Theorems for Partitioned Matrices

6.1 Partitioned Matrices and Block Diagonal Dominance

The previous chapters gave a very special role to the diagonal entries of a matrix $A = [a_{i,j}] \in \mathbb{C}^{n \times n}$, in that inclusion regions for its spectrum, $\sigma(A)$, were deduced from the union of weighted Geršgorin disks in the complex plane having centers in the points $\{a_{i,i}\}_{i \in N}$, as in

$$\sigma(A) \subseteq \bigcup_{i \in N} \{z \in \mathbb{C} : |z - a_{i,i}| \leq \sum_{j \in N \backslash \{i\}} |a_{i,j}| x_j / x_i\} \quad (\text{any } \mathbf{x} > \mathbf{0} \text{ in } \mathbb{R}^n),$$

from Corollary 1.5. But these diagonal entries also were prominent in the Brauer set of (2.6), as well as in the Brualdi set of (2.40).

This special role of the diagonal entries can be generalized through the use of partitions of $\mathbb{C}^n$. By a **partition** π of $\mathbb{C}^n$, we mean a finite collection $\{W_i\}_{i=1}^{\ell}$ of pairwise disjoint linear subspaces, each having dimension at least unity, whose direct sum is $\mathbb{C}^n$:

$$\mathbb{C}^n = W_1 \dot{+} W_2 \dot{+} \cdots \dot{+} W_\ell. \tag{6.1}$$

Without essential loss of generality, this partition π is denoted by

$$\pi = \{p_j\}_{j=0}^{\ell}, \tag{6.2}$$

where the nonnegative integers $\{p_j\}_{j=0}^{\ell}$ satisfy

$$p_0 := 0 < p_1 < p_2 < \cdots < p_\ell := n,$$

and where it is assumed that

$$W_j = span\ \{\mathbf{e}_k : p_{j-1} + 1 \leq k \leq p_j\} \quad (j \in L := \{1, 2, \cdots, \ell\}); \tag{6.3}$$

here, the vectors $\{\mathbf{e}_k\}_{k=1}^{n}$ denote the standard column basis vectors in $\mathbb{C}^n$, i.e.,

$$\mathbf{e}_j = [\delta_{j,1}, \delta_{j,2}, \cdots, \delta_{j,n}]^T \quad (j \in N := (1, 2, \cdots, n)),$$

where $\delta_{i,j}$ is the familiar Kronecker delta function. It follows that

$$\dim W_j = p_j - p_{j-1} \geq 1 \quad (j \in L).$$

Next, given any $A = [a_{i,j}] \in \mathbb{C}^{n \times n}$ and given a partition $\pi = \{p_j\}_{j=0}^{\ell}$ of $\mathbb{C}^n$, the matrix A is partitioned with respect to π as

$$(6.4) \qquad A = \left[\begin{array}{c|c|c|c} A_{1,1} & A_{1,2} & \cdots & A_{1,\ell} \\ \hline A_{2,1} & A_{2,2} & \cdots & A_{2,\ell} \\ \hline \vdots & \vdots & & \vdots \\ \hline A_{\ell,1} & A_{\ell,2} & \cdots & A_{\ell,\ell} \end{array}\right] = [A_{i,j}] \quad (i, j \in L),$$

where each block submatrix $A_{i,j}$ in (6.4) represents a linear transformation from W_j to W_i. Thus from (6.3), $A_{i,j} \in \mathbb{C}^{(p_i - p_{i-1}) \times (p_j - p_{j-1})}$ for all $i, j \in L$. For additional notation, if $\pi = \{p_j\}_{j=0}^{\ell}$ is a partition of $\mathbb{C}^n$, then we define

$$\phi = (\phi_1, \phi_2, \cdots, \phi_\ell)$$

to be a **norm ℓ-tuple**, where ϕ_j is a norm on the subspace W_j, for each $j \in L$. We denote the collection of all such norm ℓ-tuples, associated with π, by Φ_π, i.e.,

$$(6.5) \; \Phi_\pi := \{\phi = (\phi_1, \phi_2, \cdots, \phi_\ell) : \phi_j \text{ is a norm on } W_j, \text{ for each } j \in L\}.$$

Given $\phi = (\phi_1, \phi_2, \cdots, \phi_\ell) \in \Phi_\pi$, the block submatrices $A_{i,j}$ of the partitioned matrix in (6.4) have the usual associated induced operator norms:

$$(6.6) \qquad \|A_{j,k}\|_\phi := \sup_{\substack{\mathbf{x} \in W_k \\ \mathbf{x} \neq \mathbf{0}}} \frac{\phi_j(A_{j,k}\mathbf{x})}{\phi_k(\mathbf{x})} = \sup_{\phi_k(\mathbf{x})=1} \phi_j(A_{j,k}\mathbf{x}) \quad (j, k \in L).$$

Now, we begin with one of the earliest generalizations, of strict diagonal dominance in Definition 1.3, to the partitioned matrix of (6.4). This was roughly simultaneously and independently considered[1] in Ostrowski (1961), Fiedler and Pták (1962a), and Feingold and Varga (1962). For each block diagonal submatrix $A_{i,i}$, they defined the numbers

$$m(A_{i,i}) := \inf_{\substack{\mathbf{x} \in W_i \\ \mathbf{x} \neq \mathbf{0}}} \frac{\phi_i(A_{i,i}\mathbf{x})}{\phi_i(\mathbf{x})} \quad (i \in L).$$

It can be verified (see Exercise 1 of this section) that if $A_{i,i}$ is nonsingular, then $m(A_{i,i}) = 1/||A_{i,i}^{-1}||_\phi$, and if $A_{i,i}$ is singular, then $m(A_{i,i}) = 0$. Thus, we can use, in place of $m(A_{i,i})$, the more suggestive notation of

[1] A perfect word to describe this phenomenon is "Zeitgeist"!

(6.7) $$\left(\|A_{i,i}^{-1}\|_\phi\right)^{-1} := \inf_{\substack{\mathbf{x}\in W_i \\ \mathbf{x}\neq \mathbf{0}}} \frac{\phi_i(A_{i,i}\mathbf{x})}{\phi_i(\mathbf{x})} \quad (i \in L),$$

and we call $(\|A_{i,i}^{-1}\|_\phi)^{-1}$ the **reciprocal norm** of $A_{i,i}$. For further notation, we set

(6.8) $$r_{i,\pi}^\phi(A) := \sum_{j\in L\backslash\{i\}} \|A_{i,j}\|_\phi \quad (i \in L),$$

(with the convention that $r_{i,\pi}^\phi(A) := 0$ if $\ell = 1$).

Definition 6.1. Given a partition π of $\mathbb{C}^n$ and given $\phi \in \Phi_\pi$, then $A = [A_{i,j}] \in C^{n\times n}$, partitioned by π, is **strictly block diagonally dominant** with respect to ϕ if

(6.9) $$\left(\|A_{i,i}^{-1}\|_\phi\right)^{-1} > r_{i,\pi}^\phi(A) \quad (\text{all } i \in L).$$

With Definition 6.1, the above-mentioned authors obtained the following generalization of Theorem 1.4:

Theorem 6.2. *Given a partition π of $\mathbb{C}^n$ and given $\phi \in \Phi_\pi$, assume that $A = [A_{i,j}] \in \mathbb{C}^{n\times n}$, partitioned by π, is strictly block diagonally dominant with respect to ϕ. Then, A is nonsingular.*

Proof. Suppose, on the contrary, that A is singular, so that $A\mathbf{x} = \mathbf{0}$ for some $\mathbf{x} \neq \mathbf{0}$ in $\mathbb{C}^n$. With P_j denoting the projection operator from $\mathbb{C}^n$ to W_j, set $P_j\mathbf{x} =: X_j$ for each $j \in L$, so that $\sum_{j\in L} A_{i,j}X_j = 0$ for each i in L. Equivalently,

(6.10) $$A_{i,i}X_i = -\sum_{j\in L\backslash\{i\}} A_{i,j}X_j \quad (i \in L).$$

As $\mathbf{x} \neq \mathbf{0}$, we may assume that $\max_{j\in L} \phi_j(X_j) = 1$, and then let k in L be such that $\phi_k(X_k) = 1$. Applying the norm ϕ_k to (6.10) in the case $i = k$, gives, with the triangle inequality and with (6.6), that

$$\phi_k(A_{k,k}X_k) \leq \sum_{j\in L\backslash\{k\}} \phi_k(A_{k,j}X_j) \leq \sum_{j\in L\backslash\{k\}} \|A_{k,j}\|_\phi \cdot \phi_j(X_j) \leq \sum_{j\in L\backslash\{k\}} \|A_{k,j}\|_\phi,$$

so that (cf. (6.8))

(6.11) $$\phi_k(A_{k,k}X_k) \leq r_{k,\pi}^\phi(A).$$

But from (6.7),

$$\phi_k(A_{k,k}X_k) \geq (\|A_{k,k}^{-1}\|_\phi)^{-1}\phi_k(X_k) = (\|A_{k,k}^{-1}\|_\phi)^{-1}.$$

Hence, it follows from (6.11) that $(\|A_{k,k}^{-1}\|_\phi)^{-1} \leq r_{k,\pi}^\phi(A)$, which contradicts the assumption of (6.9). ■

With Theorem 6.2, we immediately obtain, via our first recurring theme, the following associated block eigenvalue inclusion result of Theorem 6.3. For added notation, given a partition $\pi = \{p_j\}_{j=0}^{\ell}$ of $\mathbb{C}^n$, given $\phi \in \Phi_\pi$, and given a matrix $A = [A_{i,j}]$ in $\mathbb{C}^{n\times n}$ which is partitioned by π, then with (6.8), we define the sets

$$(6.12)\quad \begin{cases} \Gamma_{i,\pi}^{\phi}(A) := \{z \in \mathbb{C} : (\| (zI_i - A_{i,i})^{-1} \|_\phi)^{-1} \le r_{i,\pi}^{\phi}(A)\} \quad (i \in L), \\ \text{and} \\ \Gamma_{\pi}^{\phi}(A) := \bigcup_{i\in L} \Gamma_{i,\pi}^{\phi}(A), \end{cases}$$

where I_i denotes the identity matrix for the subspace W_i. We call $\Gamma_\pi^\phi(A)$the **partitioned Geršgorin set** for A, with respect to π and ϕ.

Theorem 6.3. *Given a partition π of $\mathbb{C}^n$ and given $\phi \in \Phi_\pi$, let $A = [A_{i,j}]$ be any matrix in $\mathbb{C}^{n\times n}$ which is partitioned by π. If $\lambda \in \sigma(A)$, there is an $i \in L$ such that* (cf. (6.12)) *$\lambda \in \Gamma_{i,\pi}^{\phi}(A)$. As this is true for each $\lambda \in \sigma(A)$, then*

$$(6.13)\qquad \sigma(A) \subseteq \Gamma_\pi^\phi(A).$$

We remark (see Exercise 3 of this section) that the eigenvalue inclusion of (6.13) gives, as special cases, the results of Theorem 1.1 and Corollary 1.5 of Chapter 1.

But more follows in an easy fashion. Given a matrix $B = [B_{i,j}] \in \mathbb{C}^{n\times n}$, partitioned by π, and given $\phi \in \Phi_\pi$, form the $\ell \times \ell$ nonnegative matrix $[\|B_{i,j}\|_\phi]$ in $\mathbb{R}^{\ell\times\ell}$. Then, let $\mathbb{G}_\pi(B)$ be the directed graph (see Section 1.2) for this $\ell\times\ell$ matrix. (We note that the directed graph $\mathbb{G}_\pi(B)$ is **independent** of the choice of ϕ in Φ_π; see Exercise 4 of this section.) If $\mathbb{G}_\pi(B)$ is strongly connected, then B is said to be **π-irreducible**, and **π-reducible** otherwise. This brings us to

Definition 6.4. Given a partition π of $\mathbb{C}^n$ and given $\phi \in \Phi_\pi$, then $A = [A_{i,j}] \in \mathbb{C}^{n\times n}$, partitioned by π, is **π-irreducibly block diagonally dominant** with respect to ϕ, provided that A is π-irreducible and

$$(6.14)\qquad \left(\|A_{i,i}^{-1}\|_\phi\right)^{-1} \ge r_{i,\pi}^{\phi}(A) \quad \text{for all } i \in L,$$

with strict inequality holding in (6.14) for at least one i.

With Definition 6.4, we have the following generalization, to partitioned matrices, of Taussky's Theorem 1.11. (See Exercise 5 of this section.)

Theorem 6.5. *Given a partition π of $\mathbb{C}^n$ and given $\phi \in \Phi_\pi$, assume that $A = [A_{i,j}] \in \mathbb{C}^{n\times n}$, partitioned by π, is π-irreducibly block diagonally dominant with respect to ϕ. Then, A is nonsingular.*

And with Theorem 6.5, we directly have its associated eigenvalue inclusion result of Theorem 6.6, which generalizes Taussky's Theorem 1.12.

Theorem 6.6. *Given a partition π of $\mathbb{C}^n$ and given $\phi \in \Phi_\pi$, assume that $A = [A_{i,j}] \in \mathbb{C}^{n\times n}$, partitioned by π, is π-irreducible. If $\lambda \in \sigma(A)$ is such that $\left(\|(A_{i,i} - \lambda I_i)^{-1}\|_\phi\right)^{-1} \geq r^\phi_{i,\pi}(A)$ for each $i \in L$, then*

$$\left(\|(A_{i,i} - \lambda I_i)^{-1}\|_\phi\right)^{-1} = r^\phi_{i,\pi}(A) \text{ for each } i \in L, \tag{6.15}$$

i.e., λ is on the boundary of each set $\Gamma^\phi_{i,\pi}(A)$ of (6.12).

It is of course natural to ask if there are similar extensions, of the above Geršgorin-type results, to Brauer sets and to Brualdi sets, in the partitioned matrix case. Though these topics have not been widely treated in the literature, the machinery at hand allows us to easily treat these extensions below.

First, we consider extensions of Brauer sets in the partitioned matrix case. For any matrix $A = [A_{i,j}] \in \mathbb{C}^{n\times n}$, partitioned by $\pi = \{p_j\}_{j=0}^{\ell}$ with $\ell \geq 2$, and for any $\phi \in \Phi_\pi$, we define the sets

$$(6.16)\begin{cases} K^\phi_{i,j,\pi}(A) := \{z \in \mathbb{C}: (\|(zI_i - A_{i,i})^{-1}\|_\phi)^{-1} \cdot (\|(zI_j - A_{j,j})^{-1}\|_\phi)^{-1} \leq \\ \qquad\qquad\qquad\qquad r^\phi_{i,\pi}(A) \cdot\}, \text{ for any distinct } i \text{ and } j \text{ in } L, \\ \text{and} \\ \mathcal{K}^\phi_\pi(A) := \bigcup\limits_{\substack{i,j\in L\\ i\neq j}} K^\phi_{i,j,\pi}(A). \end{cases}$$

We call $\mathcal{K}^\phi_\pi(A)$the **partitioned Brauer set** for A, with respect to π and ϕ. Then, on first generalizing Ostrowski's nonsingularity Theorem 2.1 to the partitioned case (see Exercise 6 of this section), its associated eigenvalue inclusion result is

Theorem 6.7. *Given a partition $\pi = \{p_j\}_{j=0}^{\ell}$ of $\mathbb{C}^n$ with $\ell \geq 2$, and given $\phi \in \Phi_\pi$, let $A = [A_{i,j}]$ be any matrix in $\mathbb{C}^{n\times n}$ which is partitioned by π. If $\lambda \in \sigma(A)$, there are distinct integers i and j in L such that* (cf.(6.16))

$$\lambda \in K^\phi_{i,j,\pi}(A).$$

As this is true for each $\lambda \in \sigma(A)$, then

$$\sigma(A) \subseteq \mathcal{K}^\phi_\pi(A). \tag{6.17}$$

For a similar extension to Brualdi sets, let $A = [A_{i,j}] \in \mathbb{C}^{n\times n}$ be partitioned by $\pi = \{p_j\}_{j=0}^{\ell}$, and let $\phi \in \Phi_\pi$. From the directed graph of the $\ell \times \ell$ nonnegative matrix $\left[\|A_{i,j}\|_\phi\right]$, we determine, as described in Section 2.2, its cycle set $C_\pi(A)$ of strong and weak cycles, as well as its normal reduced form (cf. (2.35)), if this matrix is reducible. As before, we remark that there is a cycle of $C_\pi(A)$ through *each* vertex of this directed graph. Then, in analogy with (2.37) - (2.40), we set

$$(6.18)\begin{cases} \mathcal{B}^{\phi}_{\gamma,\pi}(A) := \{z \in \mathbb{C} : \prod_{i\in\gamma}(\| (zI_i - A_{i,i})^{-1} \|_\phi)^{-1} \leq \prod_{i\in\gamma} \tilde{r}^{\phi}_{i,\pi}(A)\}, \\ \qquad \text{for } \gamma \in C_\pi(A), \text{ and} \\ \mathcal{B}^{\phi}_{\pi}(A) := \bigcup_{\gamma\in C_\pi(A)} \mathcal{B}^{\phi}_{\gamma,\pi}(A), \end{cases}$$

and we call $\mathcal{B}^{\phi}_{\pi}(A)$ the **partitioned Brualdi set** for A, with respect to π and ϕ. We then have (see Exercise 7 of this section) the following extension of Theorem 2.5, which has appeared, in the π-irreducible case, in Brualdi (1992).

Theorem 6.8. *Given a partition π of $\mathbb{C}^n$ and given $\phi \in \Phi_\pi$, let $A = [A_{i,j}]$ be any matrix in $\mathbb{C}^{n\times n}$ which is partitioned by π, and let $C_\pi(A)$ be its associated cycle set. If $\lambda \in \sigma(A)$, there is a $\gamma \in C_\pi(A)$ such that*

$$(6.19) \qquad \lambda \in \mathcal{B}^{\phi}_{\gamma,\pi}(A).$$

As this is true for each $\lambda \in \sigma(A)$, then

$$(6.20) \qquad \sigma(A) \subseteq \mathcal{B}^{\phi}_{\pi}(A).$$

It is also useful to include the next result, the extension of Theorem 2.8 to the partitioned case, which is thus a generalization of Taussky's original Theorem 1.12. (Its proof is left for Exercise 8 of this section.)

Theorem 6.9. *Given a partition π of $\mathbb{C}^n$ and given $\phi \in \Phi_\pi$, let $A = [A_{i,j}] \in \mathbb{C}^{n\times n}$ be π-irreducible, and let $C_\pi(A)$ be the associated cycle set. If λ, an eigenvalue of A, is such that $\lambda \notin$ int $\mathcal{B}_\gamma(A)$ for any $\gamma \in C_\pi(A)$, then* (cf.(6.18))

$$\prod_{i\in\gamma}(\| (\lambda I_i - A_{i,i})^{-1} \|_\phi)^{-1} = \prod_{i\in\gamma} r^{\phi}_{i,\pi}(A) \qquad (\text{all } \gamma \in C_\pi(A)).$$

But, how do all these eigenvalue inclusion results *compare* in the partitioned case? It turns out that these comparisons all follow *easily* from the results of Chapter 2. The analog of Theorem 2.3 in this chapter is Theorem 6.10, whose proof, given below, is based on the proof of Theorem 2.3.

Theorem 6.10. *Given a partition $\pi = \{p_j\}_{j=0}^{\ell}$ of $\mathbb{C}^n$ with $\ell \geq 2$, and given $\phi \in \Phi_\pi$, let $A = [A_{i,j}]$ be any matrix in $\mathbb{C}^{n\times n}$ which is partitioned by π. Then* (cf.(6.12) and (6.16)),

$$(6.21) \qquad \mathcal{K}^{\phi}_{\pi}(A) \subseteq \Gamma^{\phi}_{\pi}(A).$$

Proof. Let i and j be distinct elements of L, and let z be any point of $K^{\phi}_{i,j,\pi}(A)$, i.e. (cf.(6.16)),

$$(\| (zI_i - A_{i,i})^{-1} \|_{\phi})^{-1} \cdot (\| (zI_j - A_{j,j})^{-1} \|_{\phi})^{-1} \leq r^{\phi}_{i,\pi}(A) \cdot r^{\phi}_{j,\pi}(A).$$

If $r^{\phi}_{i,\pi}(A) \cdot r^{\phi}_{j,\pi}(A) = 0$, then either $(\| (zI_i - A_{i,i})^{-1} \|_{\phi})^{-1} = 0$ or $(\| (zI_j - A_{j,j})^{-1} \|_{\phi})^{-1} = 0$. Hence from (6.12), either $z \in \Gamma^{\phi}_{i,\pi}(A)$ or $z \in \Gamma^{\phi}_{j,\pi}(A)$, so that

$$z \in (\Gamma^{\phi}_{i,\pi}(A) \cup \Gamma^{\phi}_{j,\pi}(A)). \tag{6.22}$$

The case, when $r^{\phi}_{i,\pi}(A) \cdot r^{\phi}_{j,\pi}(A) > 0$, similarly implies, as in the proof of Theorem 2.3, that z again satisfies (6.22), i.e., $K^{\phi}_{i,j,\pi}(A) \subseteq (\Gamma^{\phi}_{i,\pi}(A) \cup \Gamma^{\phi}_{j,\pi}(A))$. On taking unions, as in the proof of Theorem 2.3, this gives the desired result of (6.21). ∎

With the idea in mind of the proof of the above Theorem 6.10, we similarly have the result of Theorem 6.11 (see Exercise 9 of this section), which extends Theorems 2.3 and 2.9 of Chapter 2.

Theorem 6.11. *Given a partition $\pi = \{p_j\}_{j=0}^{\ell}$ of $\mathbb{C}^n$ with $\ell \geq 2$, and given $\phi \in \Phi_{\pi}$, let $A = [A_{i,j}]$ be any matrix in $\mathbb{C}^{n \times n}$ which is partitioned by π, and let $C_{\pi}(A)$ be its associated cycle set. Then* (cf.(6.12), (6.16) and (6.18)),

$$\mathcal{B}^{\phi}_{\pi}(A) \subseteq \mathcal{K}^{\phi}_{\pi}(A) \subseteq \Gamma^{\phi}_{\pi}(A). \tag{6.23}$$

To conclude this section, we show how the above results apply to the following matrix $A = [a_{i,j}]$ in $\mathbb{C}^{4,4}$, partitioned by $\pi = \{0, 2, 4\}$:

$$A = \left[\begin{array}{cc|cc} 3 & -1 & i & 0 \\ -1 & 3 & 0 & -i \\ \hline i & 0 & 5 & -1 \\ 0 & -i & -1 & 5 \end{array}\right] = \left[\begin{array}{c|c} A_{1,1} & A_{1,2} \\ \hline A_{2,1} & A_{2,2} \end{array}\right], \tag{6.24}$$

where the spectra of A and its associated block diagonal submatrices are

$$\sigma(A) = \{2.2679,\ \ 4+i,\ \ 4-i,\ \ 5.7321\}, \text{ and}$$
$$\sigma(A_{1,1}) = \{2, 4\} \text{ and } \sigma(A_{2,2}) = \{4, 6\}.$$

(Here, we again use the convention that non-integer real or imaginary parts of eigenvalues are rounded to the first four decimal digits.) First, as is easily verified, we have from (6.24) that

$$(zI_1 - A_{1,1})^{-1} = \tfrac{1}{(z-2)(z-4)} \begin{bmatrix} z-3 & -1 \\ -1 & z-3 \end{bmatrix},$$

and

$$(zI_2 - A_{2,2})^{-1} = \tfrac{1}{(z-4)(z-6)} \begin{bmatrix} z-5 & -1 \\ -1 & z-5 \end{bmatrix},$$

for all $z \neq 2$ or 4 in the first case, and for all $z \neq 4$ or 6 in the second case. Then, if we choose $\phi = (\phi_1, \phi_2)$, where ϕ_1 and ϕ_2 are respectively the ℓ_∞-norms on each of the two-dimensional subspaces W_1 and W_2, we see in this case that (cf.(6.8))

$$r_{1,\pi}^{\phi}(A) = \left\| \begin{bmatrix} i & 0 \\ 0 & -i \end{bmatrix} \right\|_\infty = 1 = r_{2,\pi}^{\phi}(A),$$

and that

$$\left(\|(zI_1 - A_{1,1})^{-1}\|_\infty\right)^{-1} = \frac{|z-2|\cdot|z-4|}{1+|z-3|}, \text{ and}$$

$$\left(\|(zI_2 - A_{2,2})^{-1}\|_\infty\right)^{-1} = \frac{|z-4|\cdot|z-6|}{1+|z-5|}.$$

Thus, we have from (6.12) that

$$\text{(6.25)} \quad \begin{cases} \Gamma_{1,\pi}^{\phi}(A) = \{z \in \mathbb{C} : |z-2|\cdot|z-4| \leq |z-3|+1\}, \\ \quad \text{and} \\ \Gamma_{2,\pi}^{\phi}(A) = \{z \in \mathbb{C} : |z-4|\cdot|z-6| \leq |z-5|+1\}. \end{cases}$$

Further, as the directed graph of the associated 2×2 matrix $[\|A_{i,j}\|_\phi]$ is irreducible with exactly one strong cycle $\gamma = (1 \;\; 2)$, then $\mathcal{B}_\pi^\phi(A) = \mathcal{K}_\pi^\phi(A)$ in this case, and we also have (cf.(6.16) and (6.18)) that

$$\text{(6.26)} \quad \mathcal{B}_\pi^\phi(A) = \mathcal{K}_\pi^\phi(A) = \{z \in \mathbb{C} : |z-2|\cdot|z-4|^2\cdot|z-6| \leq (|z-3|+1)\cdot(|z-5|+1)\}.$$

The sets $\Gamma_\pi^\phi(A)$ and $\mathcal{B}_\pi^\phi(A) = \mathcal{K}_\pi^\phi(A)$ are shown in Fig. 6.1. That $\mathcal{B}_\pi^\phi(A)$ is a subset of $\Gamma_\pi^\phi(A)$, from (6.23) of Theorem 6.11, can be directly seen in Fig. 6.1. (The eigenvalues of A of (6.24) are shown by "x's" in Fig. 6.1.)

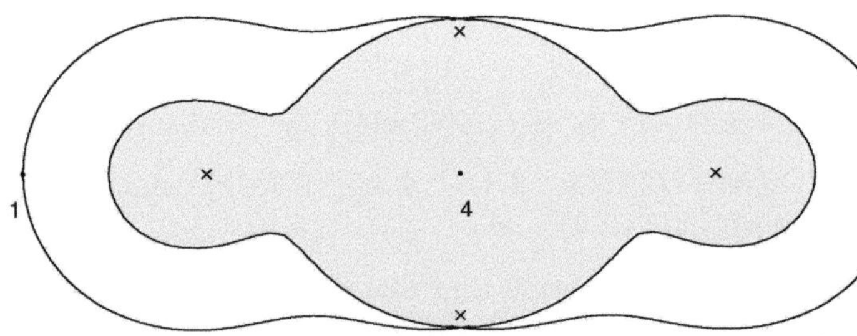

Fig. 6.1. The partitioned Geršgorin set $\Gamma_\pi^\phi(A)$ and the partitioned Brualdi set $\mathcal{B}_\pi^\phi(A)$ (shaded) for the matrix A of (6.24)

Exercises

1. Given any finite-dimensional space W and given any norm ϕ on W, let $A : W \to W$ be a linear map. Show (cf.(6.7)) that $(\| A^{-1} \|_\phi)^{-1} > 0$ if A is nonsingular, and $\left(\|A^{-1}\|_\phi\right)^{-1} = 0$if A is singular.

2. Consider the matrix $A = [a_{i,j}] \in \mathbb{C}^{4\times4}$, partitioned by $\pi = \{0, 2, 4\}$, which is given by

$$A = \left[\begin{array}{cc|cc} 0 & 1 & 0 & i/2 \\ 2/3 & 0 & 0 & 0 \\ \hline 1/4 & -1/4 & 0 & 2/3 \\ 0 & 0 & -1 & 0 \end{array}\right],$$

where $\mathbb{C}^4 = W_1 \dot{+} W_2$. Using the $\ell_\infty-$ norm on both 2-dimensional subspaces W_1 and W_2, show that A is strictly block diagonally dominant (in the sense of Definition 6.1), with respect to $\phi = (\ell_\infty, \ell_\infty)$. (Note that A is **not** strictly diagonally dominant in the sense of Definition 1.3, since all diagonal entries of A are zero.)

3. Assume that the partition π on $\mathbb{C}^n$ is given by $\pi = \{j\}_{j=0}^n$, so that each subspace W_j is one-dimensional $(1 \leq j \leq n)$.
   a. If $\phi_j(\mathbf{u}) := |\mathbf{u}|$ for any $\mathbf{u}$ in $W_j (1 \leq j \leq n)$, show that the eigenvalue inclusion of (6.13) of Theorem 6.3 reduces exactly to (1.7) of Geršgorin's Theorem 1.1.
   b. Given any positive numbers $\{x_j\}_{j=1}^n$, set $\tilde{\phi}_j(\mathbf{u}) := x_j|\mathbf{u}|$ for any $\mathbf{u} \in W_j (1 \leq j \leq n)$. Show that the eigenvalue inclusion of Theorem 6.3 reduces exactly to (1.15) of Corollary 1.5.

4. Given any partition π of $\mathbb{C}^n$ and given any $B = [B_{i,j}] \in \mathbb{C}^{n\times n}$ which is partitioned by π, show that the directed graph $\mathbb{G}_\pi(B)$ is *independent* of the choice of ϕ in Φ_π, i.e., if $B_{j,k} \not\equiv O$, then (cf.(6.6)) $\|B_{j,k}\|_\phi > 0$ for each $\phi \in \Phi_\pi$.

5. Prove Theorem 6.5. (Hint: Follow the proof of Theorem 1.11.)

6. Extend Ostrowski's nonsingularity Theorem 2.1 to the partitioned case.

7. Prove Theorem 6.8. (Hint: Follow the proof of Theorem 2.5.)

8. Prove Theorem 6.9.

9. Prove Theorem 6.11. (Hint: Follow the proof of Theorem 2.9.)

10. For the matrix A of (6.24) with the partition $\pi = \{0, 2, 4\}$, choose the $\ell_\infty-$ norm on the two associated subspaces W_1 and W_2. Verify that (6.26) is valid.

11. Given any partition π of $\mathbb{C}^n$ and given any $\phi \in \Phi_\pi$, let $A = [A_{i,j}]$ be any matrix in $\mathbb{C}^{n\times n}$ which is partitioned by π. Show that every eigenvalue of every diagonal submatrix $A_{i,i}$ is contained in $\Gamma^\phi_\pi(A)$ of (6.12), and if $\ell \geq 2$, show that the same is true for $\mathcal{K}^\phi_\pi(A)$ of (6.16) and $\mathcal{B}^\phi_\pi(A)$ of (6.18).

6.2 A Different Norm Approach

In this section, we describe an interesting variant of Householder's general norm approach, of Section 1.4, to obtain eigenvalue inclusion results for partitioned matrices. This important variant is due to Robert (1966) who derived nonsingularity results, for partitioned matrices, using the theory of M-matrices and H-matrices. But as we know (by our first recurring theme), such nonsingularity results give rise to equivalent eigenvalue inclusion sets in the complex plane, and this is what we first describe here. (Later, at the end of this section, we show the connection to the nonsingularity results of Robert, based on M-matrix and H-matrix theory, our second recurring theme.)

This work of Robert (1966) similarly makes use, from the previous section, of the partition $\pi = \{p_j\}_{j=o}^{\ell}$ of $\mathbb{C}^n$, where $p_0 = 0 < p_1 < \cdots < p_\ell = n$ and where

$$\mathbb{C}^n = W_1 \dot{+} W_2 \dot{+} \cdots \dot{+} W_\ell,$$

as well as a norm ℓ-tuple $\phi = (\phi_1, \phi_2, \cdots, \phi_\ell)$ from Φ_π, where ϕ_j is a norm on W_j, for each j with $1 \leq j \leq \ell$. What is different now is that a norm on $\mathbb{C}^n$ is needed, which is not directly provided by the norm ℓ-tuple ϕ, when $\ell > 1$. To easily rectify this, if P_j again denotes the projection operator from $\mathbb{C}^n$ to W_j for each $1 \leq j \leq \ell$, then

$$(6.27) \qquad \|\mathbf{x}\|_\phi := \max_{j \in L}\{\phi_j(P_j\mathbf{x})\} \quad (\text{any } \mathbf{x} \in \mathbb{C}^n, \text{ any } \phi \in \Phi_\pi)$$

defines a norm[2] on $\mathbb{C}^n$. (Other norm definitions in (6.27), involving $\{\phi_j(P_j\mathbf{x})\}_{j=1}^{\ell}$, are clearly possible, as shown in Exercise 2 of this section, but the ℓ_∞-type norm in (6.27) is convenient for our purposes.)

Given a partition π of $\mathbb{C}^n$, given $\phi \in \Phi_n$, and given any matrix $B = [B_{i,j}] \in \mathbb{C}^{n\times n}$ which is partitioned by π, then $\|B\|_\phi$ will denote, as usual, the induced operator norm of B with respect to the norm ϕ of (6.27):

[2] We note that the symbol ϕ can have two meanings, i.e., either a norm ℓ-tuple in Φ_π, or the related norm on $\mathbb{C}^n$ in (6.27), but this notation will cause no confusion in what follows.

$$(6.28) \qquad ||B||_\phi := \sup_{||\mathbf{x}||_\phi = 1} ||B\mathbf{x}||_\phi .$$

With P_j the projection operator from $\mathbb{C}^n$ to W_j for each $1 \leq j \leq \ell$, then on setting $X_j := P_j\mathbf{x}$, it follows from (6.27) and (6.28) that

$$(6.29) \qquad ||B||_\phi = \sup_{||\mathbf{x}||_\phi = 1} \left\{ \max_{j \in L} \left[\phi_j \left(\sum_{k \in L} B_{j,k} X_k \right) \right] \right\}.$$

But, as $||\mathbf{x}||_\phi = 1$ implies from (6.27) that $\max_{j \in L} \phi_j(X_j) = 1$, then, with the triangle inequality applied to the sum in (6.29), it follows from the definition in (6.6) (see Exercise 3 of this section) that

$$(6.30) \qquad ||B||_\phi \leq \max_{j \in L} \left[\sum_{k \in L} ||B_{j,k}||_\phi \right] =: || \left[||B_{j,k}||_\phi \right] ||_\infty .$$

Here, the right side of (6.30) is just the ℓ_∞-norm of the $\ell \times \ell$ nonnegative matrix $[||B_{j,k}||_\phi]$, so that (6.30) is just

$$||B||_\phi \leq \left\| \begin{bmatrix} ||B_{1,1}||_\phi & ||B_{1,2}||_\phi & \cdots & ||B_{1,\ell}||_\phi \\ ||B_{2,1}||_\phi & ||B_{2,2}||_\phi & \cdots & ||B_{2,\ell}||_\phi \\ \vdots & & & \vdots \\ ||B_{\ell,1}||_\phi & ||B_{\ell,2}||_\phi & \cdots & ||B_{\ell,\ell}||_\phi \end{bmatrix} \right\|_\infty .$$

This norm inequality in (6.30) will be widely used in the remainder of this chapter. We note that it is generally *easier* to compute the right side of (6.30), then the left side, namely, $||B||_\phi$ of (6.29).

Continuing, from the partitioned matrix A in (6.4), it is also convenient to let

$$(6.31) \qquad D_\pi := \text{diag}[A_{1,1}; A_{2,2}; \cdots ; A_{\ell,\ell}]$$

denote the **block-diagonal matrix of** A, with respect to the partition π. This will play a special role below.

To obtain our next eigenvalue inclusion result for a matrix $A = [A_{i,j}]$ in $\mathbb{C}^{n \times n}$, partitioned as in (6.4) with respect to the partition π, consider any eigenvalue λ of A. Then, $A\mathbf{x} = \lambda\mathbf{x}$ for some $\mathbf{x} \neq \mathbf{0}$ in $\mathbb{C}^n$. This can also be expressed as

$$(A - D_\pi)\mathbf{x} = (\lambda I_n - D_\pi)\,\mathbf{x},$$

and if $\lambda \notin \sigma(D_\pi)$, we have

$$(\lambda I_n - D_\pi)^{-1}(A - D_\pi)\mathbf{x} = \mathbf{x}.$$

Hence, assuming $||\mathbf{x}||_\phi = 1$, the above equality immediately gives, with the definition in (6.28), that

$$\|(\lambda I_n - D_\pi)^{-1}(A - D_\pi)\|_\phi \geq 1 \quad (\lambda \in \sigma(A)\backslash\sigma(D_\pi)). \tag{6.32}$$

But with the inequality of (6.30), we also see that

$$\| \left[\| \left[(\lambda I_n - D_\pi)^{-1}(A - D_\pi) \right]_{j,k} \|_\phi \right] \|_\infty \geq 1 \qquad (\lambda \in \sigma(A)\backslash\sigma(D_\pi)). \tag{6.33}$$

From this, we define the following sets in the complex plane:

$$\begin{aligned} \mathcal{H}_\pi^\phi(A) := \sigma(D_\pi) \cup \{z \in \mathbb{C} : z \notin \sigma(D_\pi) \text{ and} \\ \|(zI_n - D_\pi)^{-1}\,(A - D_\pi)\|_\phi \geq 1\}, \end{aligned} \tag{6.34}$$

and (cf. (6.30))

$$\begin{aligned} \mathcal{R}_\pi^\phi(A) := \sigma(D_\pi) \cup \{z \in \mathbb{C} : z \notin \sigma(D_\pi) \text{ and} \\ \| \left[\| \left[(zI_n - D_\pi)^{-1}\,(A - D_\pi) \right]_{j,k} \|_\phi \right] \|_\infty \geq 1\}. \end{aligned} \tag{6.35}$$

We call $\mathcal{H}_\pi^\phi(A)$, of (6.34), the **partitioned Householder set** for A, with respect to ϕ in Φ_π, because of its similarity to Householder's original set in (1.48). We likewise call $\mathcal{R}_\pi^\phi(A)$, of (6.35), the **partitioned Robert set** for A, with respect to ϕ in Φ_π, because of its similarity to constructions in Robert (1966), where nonsingularity results were derived. We remark, as in Proposition 1.23, that the sets $\mathcal{H}_\pi^\phi(A)$ and $\mathcal{R}_\pi^\phi(A)$ are (see Exercise 1 of this section) **closed and bounded sets** in $\mathbb{C}$, for any $A \in \mathbb{C}^{n\times n}$, any partition π, and any $\phi \in \Phi_\pi$.

We remind the reader that, from the triangle inequality used in deriving (6.30), the definitions in (6.34) and (6.35) directly give us that

$$\mathcal{H}_\pi^\phi(A) \subseteq \mathcal{R}_\pi^\phi(A) \qquad (\text{any } A \in \mathbb{C}^{n\times n}).$$

We also see from these definitions in (6.34) and (6.35) that

$$\mathcal{H}_\pi^\phi(A) = \mathcal{R}_\pi^\phi(A) = \sigma(D_\pi) \quad \text{ if } A = D_\pi.$$

i.e., if $A = D_\pi$, then the final sets in (6.34) and (6.35) are necessarily empty. We finally remark that since the diagonal blocks of $(zI_n - D_\pi)^{-1}(A - D_\pi)$ are all zero for any $z \notin \sigma(D_\pi)$, then

$$\mathcal{H}_\pi^\phi(A) = \mathcal{R}_\pi^\phi(A), \text{ whenever } \ell = 2 \text{ for the partition } \pi. \tag{6.36}$$

The above discussion and definitions imply that each λ in $\sigma(A)$, whether or not λ is in $\sigma(D_\pi)$, is necessarily in $\mathcal{H}_\pi^\phi(A)$ and in $\mathcal{R}_\pi^\phi(A)$. Thus, with the inequality of (6.30), we immediately have

Theorem 6.12. *Given a partition π of $\mathbb{C}^n$ and given $\phi \in \Phi_\pi$, let $A = [A_{i,j}]$ be any matrix in $\mathbb{C}^{n\times n}$ which is partitioned by π. Then,*

$$\sigma(A) \subseteq \mathcal{H}_\pi^\phi(A) \subseteq \mathcal{R}_\pi^\phi(A). \tag{6.37}$$

Given a matrix $A \in \mathbb{C}^{n\times n}$, then, on comparing the two eigenvalue inclusion sets of (6.13) of Theorem 6.3 and (6.37) of Theorem 6.12, we see that each of these sets, $\Gamma_\pi^\phi(A)$, $\mathcal{H}_\pi^\phi(A)$ and $\mathcal{R}_\pi^\phi(A)$, depend, in a different way, on the *same* basic two quantities, namely,

$$\text{the partition } \pi, \text{ and the norm } \ell\text{–tuple } \phi \text{ in } \Phi_\pi.$$

It is natural to ask which, of the associated eigenvalue inclusion sets, for A, is *smallest* in the complex plane. Using the technique of Robert (1966), this is answered in the following result.

Theorem 6.13. *Given any partition π of $\mathbb{C}^n$ and given any $\phi \in \Phi_\pi$, then for any $A = [A_{j,k}] \in \mathbb{C}^{n\times n}$, which is partitioned by π as in* (6.4), *there holds*

$$\sigma(A) \subseteq \mathcal{H}_\pi^\phi(A) \subseteq \mathcal{R}_\pi^\phi(A) \subseteq \Gamma_\pi^\phi(A). \tag{6.38}$$

Remark: This establishes that the sets $\mathcal{H}_\pi^\phi(A)$ and $\mathcal{R}_\pi^\phi(A)$, for any matrix A in $\mathbb{C}^{n\times n}$, are always smaller than or equal to its partitioned Geršgorin counterpart, $\Gamma_\pi^\phi(A)$.

Proof. Because of the inequality in (6.30) and the result of (6.37), it suffices to establish only the final inclusion of (6.38). First, assuming that the last set in $\mathcal{R}_\pi^\phi(A)$ of (6.35) is not empty, let z be any point in this set, so that $z \notin \sigma(D_\pi)$ and

$$1 \leq \left\| \left[\| \left[(zI_n - D_\pi)^{-1}(A - D_\pi) \right]_{j,k} \|_\phi \right] \right\|_\infty, \tag{6.39}$$

where $(zI_n - D_\pi)^{-1} = \text{diag}[(zI_1 - A_{1,1})^{-1}, \cdots, (zI_\ell - A_{\ell,\ell})^{-1}]$. Consider any $\mathbf{x} \in \mathbb{C}^n$ with $\|\mathbf{x}\|_\phi = 1$. With P_j the projection operator from $\mathbb{C}^n$ to W_j, set $P_j\mathbf{x} := X_j$ for $1 \leq j \leq \ell$, so that $\max_{j\in L} \phi_j(X_j) = 1$. Then with $B := (zI_n - D_\pi)^{-1}(A - D_\pi)$, it follows from (6.39) and (6.30) that

$$\begin{aligned} 1 \leq \| \left[\|B_{j,k}\|_\phi \right] \|_\infty &= \max_{j\in L} \left[\sum_{k\in L\backslash\{j\}} \|(zI_j - A_{j,j})^{-1}A_{j,k}\|_\phi \right] \\ &\leq \max_{j\in L} \left[\|(zI_j - A_{j,j})^{-1}\|_\phi \sum_{k\in L\backslash\{j\}} \|A_{j,k}\|_\phi \right] \\ &= \max_{j\in L} \left[(\|zI_j - A_{j,j})^{-1}\|_\phi \cdot r_{j,\pi}^\phi(A) \right], \end{aligned}$$

the last equality following from the definition in (6.8). Hence, with (6.39),

$$1 \leq \max_{j\in L} \left(\|(zI_j - A_{j,j})^{-1}\|_\phi \cdot r_{j,\pi}^\phi(A) \right).$$

Thus, there is an s with $1 \leq s \leq \ell$ such that $1 \leq \|(zI_s - A_{s,s})^{-1}\|_\phi \cdot r^\phi_{s,\pi}(A)$. As this last inequality implies that both $\|(zI_s - A_{s,s})^{-1}\|_\phi$ and $r^\phi_{s,\pi}(A)$ are positive, then

$$(\|(zI_s - A_{s,s})^{-1}\|_\phi)^{-1} \leq r^\phi_{s,\pi}(A).$$

But from (6.12), the above inequality gives that $z \in \Gamma^\phi_{s,\pi}(A)$, so that $z \in \Gamma^\phi_\pi(A)$.

To complete the proof, consider any $z \in \sigma(D_\pi)$, so that $z \in \sigma(A_{j,j})$ for some j with $1 \leq j \leq \ell$. From (6.35), we know that $z \in \mathcal{R}^\phi_\pi(A)$. On the other hand, as $z \in \sigma(A_{j,j})$, then $(zI_j - A_{j,j})$ is singular. Hence, from the remark preceding (6.7), $(\|(zI_j - A_{j,j})^{-1}\|_\phi)^{-1} = 0$. As the sum for $r^\phi_{j,\pi}(A)$ in (6.8) is obviously nonnegative, it follows from (6.12) that $z \in \Gamma^\phi_{j,\pi}(A)$, which gives that $z \in \Gamma^\phi_\pi(A)$. Thus, (6.38) is valid. ■

We remark that the real strength of the set $\mathcal{R}^\phi_\pi(A)$ is that it is deduced from products, such as $(zI_j - A_{j,j})^{-1}A_{j,k}$, which can be separated, as in the proof of Theorem 6.13, to obtain a comparison with the partitioned Geršgorin set of (6.12) in Section 6.1.

To illustrate the results of this section, let us consider again the partitioned matrix A of (6.24), i.e.,

$$A = \left[\begin{array}{cc|cc} 3 & -1 & i & 0 \\ -1 & 3 & 0 & -i \\ \hline i & 0 & 5 & -1 \\ 0 & -i & -1 & 5 \end{array}\right], \quad \text{where } \sigma(A_{1,1}) = \{2,4\} \text{ and } \sigma(A_{2,2}) = \{4,6\}.$$

With the choice again, as in Section 6.1, of $\phi = (\phi_1, \phi_2)$, where ϕ_1 and ϕ_2 are respectively the ℓ_∞-norms on each of the two-dimensional subspaces W_1 and W_2, it follows from (6.36) that $\mathcal{H}^\phi_\pi(A) = \mathcal{R}^\phi_\pi(A)$, which can be verified to be (see Exercise 4 of this section)

$$\mathcal{H}^\phi_\pi(A) = \sigma(D_\pi) \cup \{z \in \mathbb{C} : z \notin \sigma(D_\pi) \text{ and } \max\left[\frac{|z-3|+1}{|z-2|\cdot|z-4|}; \frac{|z-5|+1}{|z-4|\cdot|z-6|}\right] \geq 1\}.$$

This can also be expressed as

$$(6.40)\quad \mathcal{H}^\phi_\pi(A) = \{z \in \mathbb{C} : |z+3|+1 \geq |z-2|\cdot|z-4| \text{ or } |z-5|+1 \geq |z-4|\cdot|z-6|\}.$$

But, it is also interesting to see that the union of the sets in (6.25) gives exactly the *same* result, i.e.,

$$\mathcal{H}^\phi_\pi(A) = \Gamma^\phi_\pi(A).$$

Thus, the final two inclusions in (6.38) of Theorem 6.13 are one of *equality* for this particular matrix A. However, from (6.23) of Theorem 6.11 and Fig. 6.1, we have, for this matrix A of (6.24), that

$$\mathcal{B}_{\pi}^{\phi}(A) = \mathcal{K}_{\pi}^{\phi}(A) \subsetneqq \Gamma_{\pi}^{\phi}(A) = \mathcal{H}_{\pi}^{\phi}(A).$$

Hence, the set $\mathcal{H}_{\pi}^{\phi}(A)$, while in general a subset of $\Gamma_{\pi}^{\phi}(A)$ from (6.38), need *not* be a smaller set than the associated sets $\mathcal{K}_{\pi}^{\phi}(A)$ and $\mathcal{B}_{\pi}^{\phi}(A)$, in the partitioned case.

On the other hand, we give below an interesting example of a partitioned matrix for which *neither* set, $\mathcal{H}_{\pi}^{\phi}(E)$ or $\mathcal{B}_{\pi}^{\phi}(E)$, is a subset of the other. For this, we consider the following partitioned matrix of Robert (1966):

$$(6.41) \qquad E = \left[\begin{array}{cc|cc} 6 & 5 & 2.1 & 4.0 \\ 7 & 6 & 2.5 & 4.7 \\ \hline 9.6 & 4.7 & 9 & 10 \\ 8.6 & 4.2 & 8 & 9 \end{array}\right], \text{ with } D_{\pi} := \left[\begin{array}{cc|cc} 6 & 5 & 0 & 0 \\ 7 & 6 & 0 & 0 \\ \hline 0 & 0 & 9 & 10 \\ 0 & 0 & 8 & 9 \end{array}\right],$$

whose associated spectra are given by

$$(6.42) \begin{cases} \sigma(E) = \{0.0482, 0.0882, 5.1807, 24.6830\} \text{ and} \\ \sigma(E_{1,1}) = \{0.0839, 11.9161\} \text{ and } \sigma(E_{2,2}) = \{0.0557, 17.9443\}. \end{cases}$$

In this case, for $z \in \mathbb{C}\backslash\sigma(D_{\pi})$ we have

$$(6.43)\ (zI - D_{\pi})^{-1}(E - D_{\pi}) = \left[\begin{array}{cc|cc} 0 & 0 & \frac{2.1\cdot z - 0.1}{\mu} & \frac{4.0\cdot z - 0.5}{\mu} \\ 0 & 0 & \frac{2.5\cdot z - 0.3}{\mu} & \frac{4.7\cdot z - 0.2}{\mu} \\ \hline \frac{9.6\cdot z - 0.4}{\tau} & \frac{4.7\cdot z - 0.3}{\tau} & 0 & 0 \\ \frac{8.6\cdot z - 0.6}{\tau} & \frac{4.2\cdot z - 0.2}{\tau} & 0 & 0 \end{array}\right],$$

where $\mu := z^2 - 12z + 1$, and $\tau := z^2 - 18z + 1$. (We note that the four eigenvalues of D_{π} are just the zeros of the polynomials μ and τ.) Using $\ell_{\infty}-$norms on W_1 and W_2 for this example, it again follows from (6.36) that $\mathcal{H}_{\pi}^{\phi}(E) = \mathcal{R}_{\pi}^{\phi}(E)$, where

$$(6.44) \quad \begin{aligned} \mathcal{H}_{\pi}^{\phi}(E) = \{z \in \mathbb{C} : & |2.1 \cdot z - 0.1| + |4.0 \cdot z - 0.5| \geq |z^2 - 12z + 1|; \\ & |2.5 \cdot z - 0.3| + |4.7 \cdot z - 0.2| \geq |z^2 - 12z + 1|; \\ & |9.6 \cdot z - 0.4| + |4.7 \cdot z - 0.3| \geq |z^2 - 18z + 1|; \\ \text{or } & |8.6 \cdot z - 0.6| + |4.2 \cdot z - 0.2| \geq |z^2 - 18z + 1|\}, \end{aligned}$$

i.e., each point z of $\mathcal{H}_{\pi}^{\phi}(E)$ satisfies at least one of the four inequalities of (6.44). From (6.18), it can be similarly verified (see Exercise 6 of this section) that

$$(6.45) \quad \begin{aligned} \mathcal{B}_{\pi}^{\phi}(E) = \{z \in \mathbb{C} : |z^2 - 12z + 1| \cdot |z^2 - 18z + 1| \leq \\ 102.96(7 + |z - 6|) \cdot (10 + |z - 9|)\}. \end{aligned}$$

Using (6.44) and (6.45), the sets $\mathcal{H}_\pi^\phi(E)$ and $\mathcal{B}_\pi^\phi(E)$, for this matrix E of (6.41), were obtained, and they are shown in Fig. 6.2, along with the four eigenvalues of E, denoted by "x's". It is apparent from this figure that

$$\mathcal{H}_\pi^\phi(E) \not\subseteq \mathcal{B}_\pi^\phi(E) \quad \text{and} \quad \mathcal{B}_\pi^\phi(E) \not\subseteq \mathcal{H}_\pi^\phi(E).$$

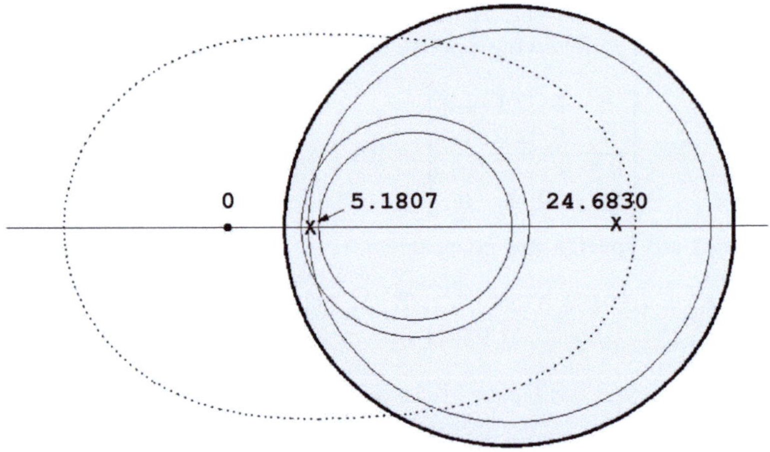

Fig. 6.2. The partitioned Householder set $\mathcal{H}_\pi^\phi(E)$ (shaded) and the partitioned Brualdi set $\mathcal{B}_\pi^\phi(E)$ (dotted boundary) for the matrix E of (6.41)

It is interesting to examine more carefully the sets in Fig. 6.2. What is most curious in this example is that the shaded set $\mathcal{H}_\pi^\phi(E)$ consists of **two** (disjoint) **components**, each being the union of four circular-like sets, where one of the components is so small, it is obscured by a dot, signifying the origin of the real axis in Fig. 6.2. This smaller component is magnified in Fig. 6.3, where one sees that this tiny component is also the union of four circular-like sets, whose boundaries are shown in Fig. 6.3. It is also the case that this tiny set actually contains two eigenvalues of E, shown as "$\times$'s" in Fig. 6.3. Thus, this smaller subset of $\mathcal{H}_\pi^\phi(E)$ gives rather sharp estimates of the two smallest eigenvalues of E (cf.(6.42)). On the other hand, the largest eigenvalue, 24.6830, of E, from Fig. 6.2, is *better* approximated by the set $\mathcal{B}_\pi^\phi(E)$, than by the set $\mathcal{H}_\pi^\phi(E)$. Finally, we remark from Fig. 6.3 that as $z = 0$ is *not* contained in $\mathcal{H}_\pi^\phi(E)$, then E is necessarily *nonsingular*, a fact which could not have been deduced from $\mathcal{B}_\pi^\phi(E)$.

Next, it is important to describe how the set $\mathcal{R}_\pi^\phi(A)$ of (6.35) of a partitioned matrix A, connects with the theory of M-matrices and H-matrices, (from Appendix C), our second recurring theme in this book. Specifically, given a partition π of $\mathbb{C}^n$, let $A = [A_{i,j}]$ be any matrix in $\mathbb{C}^{n \times n}$ which is partitioned by π. Assuming that z is an eigenvalue of A with (cf.(6.31))

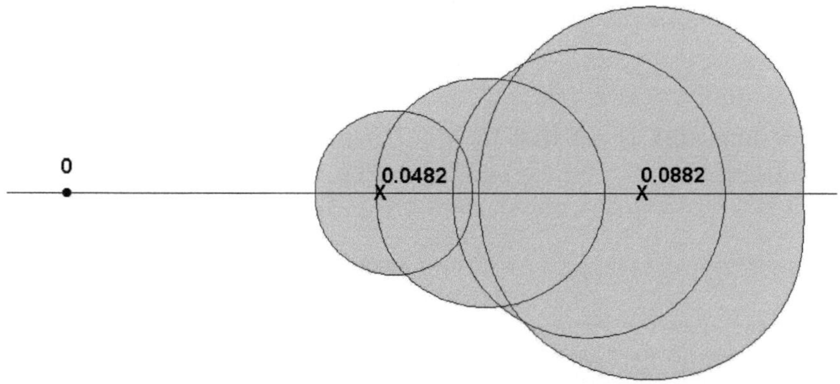

Fig. 6.3. Magnification of the component near $z = 0$ of the partitioned Householder set $\mathcal{H}_\pi^\phi(E)$ for the matrix E of (6.41)

$z \notin \sigma(D_\pi)$, then the matrix $(D_\pi - zI)^{-1}(A - zI)$ is evidently singular, and has the partitioned form (cf.(6.4))

$$(6.46) \quad \begin{aligned}(D_\pi - zI)^{-1}(A - zI) = [E_{i,j}(z)],\ 1 \le i,j \le \ell,\ \text{where } E_{i,i}(z) := I_i,\\ \text{and } E_{i,j}(z) := (A_{i,i} - zI_i)^{-1}A_{i,j}\ \text{ for } i \neq j.\end{aligned}$$

Then, given $\phi \in \Phi_\pi$, its associated partitioned **comparison matrix** $\mathcal{M}_\pi^\phi(z)$, in $\mathbb{Z}^{\ell\times\ell}$, is defined (cf. ((C.4) of Appendix C) as

$$(6.47) \quad \begin{aligned}\mathcal{M}_\pi^\phi(z) := [m_{i,j}(z)],\ \text{where }\ m_{i,i}(z) := 1 \text{ and where}\\ m_{i,j}(z) := -\|(A_{i,i} - zI_i)^{-1}A_{i,j}\|_\phi\ \text{ for } i \neq j\ (1 \le i,j \le \ell),\end{aligned}$$

which can be expressed as

$$(6.48) \qquad \mathcal{M}_\pi^\phi(z) = I_\ell - S_A(z),$$

where $S_A^\phi(z)$ is the nonnegative $\ell \times \ell$ matrix given by

$$(6.49) \quad \begin{aligned}S_A^\phi(z) := [\alpha_{i,j}(z)],\ \text{where }\ \alpha_{i,i}(z) := 0 \text{ and}\\ \alpha_{i,j}(z) := \|(A_{i,i} - zI_i)^{-1}A_{i,j}\|_\phi\ \text{ for } i \neq j\ (1 \le i,j \le \ell),\end{aligned}$$

Because of the form of $\mathcal{M}_\pi^\phi(z)$ in (6.48), it follows from Definition 5.4 that $\mathcal{M}_\pi^\phi(z)$ is a nonsingular M-matrix if and only if $\rho(S_A^\phi(z)) < 1$, where $\rho(S_A^\phi(z))$ is the spectral radius of $S_A^\phi(z)$. If $\mathcal{M}_\pi^\phi(z)$ is a nonsingular M-matrix, then by Definition C.6 of Appendix C, $(D_\pi - zI)^{-1}(A - zI)$ is a nonsingular H-matrix. But as $(D_\pi - zI)^{-1}(A - zI)$ is, by assumption, necessarily singular, it follows that

$$(6.50) \qquad 1 \le \rho(S_A^\phi(z)),$$

and as $\rho(S_A^\phi(z)) \leq \|S_A^\phi(z)\|_\infty$ is valid from (1.10), we see from (6.50) that

$$(6.51) \qquad 1 \leq \rho(S_A^\phi(z)) \leq \|S_A^\phi(z)\|_\infty.$$

Next, we immediately see that the $\ell \times \ell$ matrix $S_A^\phi(z)$ of (6.49) is just the $\ell \times \ell$ matrix $[\||(zI - D_\pi)^{-1}(A - D_\pi)\||_\phi]$, so that the inequality $1 \leq \|S_A^\phi(z)\|_\infty$, from (6.51), *precisely* determines the set $\mathcal{R}_\pi^\phi(A)$ of (6.35). In other words,

$$\mathcal{R}_\pi^\phi(A) = \sigma(D_\pi) \cup \{z \in \mathbb{C} : z \notin \sigma(D_\pi) \text{ and } \|S_A^\phi(z)\|_\infty \geq 1\},$$

which shows how the theory of M-matrices and H-matrices connects with the partitioned Robert set $\mathcal{R}_\pi^\phi(A)$.

Exercises

1. Given any partition π of $\mathbb{C}^n$, any $\phi \in \Phi_\pi$, and any matrix $A = [A_{i,j}]$ in $\mathbb{C}^{n \times n}$ which is partitioned by π, show that each of the sets $\mathcal{H}_\pi^\phi(A)$ of (6.34), or $\mathcal{R}_\pi^\phi(A)$ of (6.35), is closed and bounded in $\mathbb{C}$.

2. Given $\mathbf{x} = [x_1, x_2, \cdots, x_\ell]^T$ in $\mathbb{C}^\ell$, define $|\mathbf{x}| := [|x_1|, |x_2|, \cdots, |x_\ell|]^T$. Then $|\mathbf{x}| \leq |\mathbf{y}|$ is said to be valid if $|x_i| \leq |y_i|$, for all $i \in L$. With this notation, if ψ is a norm on $\mathbb{C}^\ell$, then ψ is a **monotone norm** (see Horn and Johnson (1985), p.285) if

$$|\mathbf{x}| \leq |\mathbf{y}| \text{ implies } \psi(\mathbf{x}) \leq \psi(\mathbf{y}) \quad (\text{any } \mathbf{x}, \text{ any } \mathbf{y} \text{ in } \mathbb{C}^\ell).$$

For any $\phi = (\phi_1, \phi_2, \cdots, \phi_\ell)$ in Φ_π and any monotone norm ψ on $\mathbb{C}^\ell$, prove, with the notation of (6.27), that

$$\|\mathbf{x}\|_{\phi,\psi} := \psi(\phi_1(P_1\mathbf{x}), \phi_2(P_2\mathbf{x}), \cdots, \phi_\ell(P_\ell\mathbf{x}))$$

is a norm on $\mathbb{C}^n$. (As a special case of this exercise, then choosing $\psi(\mathbf{x}) := \|\mathbf{x}\|_\infty$ gives that $\|\mathbf{x}\|_\phi$ of (6.27) is a norm on $\mathbb{C}^n$.)

3. Given a partition π and given $\phi \in \Phi_\pi$, let $B = [B_{j,k}]$ be any matrix in $\mathbb{C}^{n \times n}$ which is partitioned by π. Then, verify the inequality in (6.30). In addition, give an example where **strict inequality** can hold in (6.30).

4. For the partitioned matrix A of (6.24), let ϕ_1 and ϕ_2 be, respectively, ℓ_∞-norms on the subspaces W_1 and W_2. Verify that its associated set, $\mathcal{H}_\pi^\phi(A)$, is given by (6.40).

5. From (6.43), verify that the expression of (6.44) is valid.

6. Verify the expression of (6.45).

7. Consider the following 4×4 matrix with $\pi := \{0, 2, 4\}$:

$$B = \left[\begin{array}{cc|cc} 4 & -2 & -1 & 0 \\ -2 & 4 & 0 & -1 \\ \hline -1 & 0 & 4 & -2 \\ 0 & -1 & -2 & 4 \end{array}\right] = \left[\begin{array}{c|c} B_{1,1} & B_{1,2} \\ \hline B_{2,1} & B_{2,2} \end{array}\right],$$

where $\sigma(B) = \{1, 3, 5, 7\}$ and where $\sigma(B_{1,1}) = \{2, 6\} = \sigma(B_{2,2})$.

a. If $\phi := (\ell_\infty, \ell_\infty)$ is the selected norm 2-tuple in Φ_π, show that its associated set $\mathcal{H}_\pi^\phi(B)$ is given by

$$\mathcal{H}_\pi^\phi(B) = \{z \in \mathbb{C} : |z - 4| + 2 \geq |z - 2| \cdot |z - 6|\}.$$

b. If $\psi := (\ell_2, \ell_2)$ is the selected norm 2-tuple in Φ_π, show that its associated set $\mathcal{H}_\pi^\psi(B)$ is given by

$$\mathcal{H}_\pi^\psi(B) = \{z \in \mathbb{C} : |z - 2| \leq 1 \text{ or } |z - 6| \leq 1\}.$$

(Hint: Use the fact from (B.6) of Appendix B that $||A||_2 = [\rho(AA^*)]^{\frac{1}{2}}$.)

c. Show that $\mathcal{H}_\pi^\psi(B) \subsetneqq \mathcal{H}_\pi^\phi(B)$. (Hint: Show, for example, that $|z-2| = 1$, which is part of the boundary of $\mathcal{H}_\pi^\psi(B)$, is not in the boundary of $\mathcal{H}_\pi^\phi(B)$, except for $z = 1$ and $z = 3$.) In Fig. 6.4, the boundaries of $\mathcal{H}_\pi^\phi(B)$ and $\mathcal{H}_\pi^\psi(B)$ are shown, respectively, by two dashed closed curves and two closed circles. The boundary of the original Geršgorin inclusion result of Theorem 1.1 for B is the outer dotted circle in Fig. 6.4, and the crosses are the eigenvalues of B.

8. From (6.35), define the (possibly empty) component of $\mathcal{R}_\pi^\phi(A)$ by

$$\mathcal{R}_{\pi,1}^\phi(A) := \{z \in \mathbb{C} : z \notin \sigma(D_\pi) \text{ and } \\ || \left[||(zI - D_\pi)^{-1}(A - D_\pi)_{j,k}||_\phi \right] ||_\infty \geq 1\},$$

so that (cf.(6.35)) $\mathcal{R}_\pi^\phi(A) = \sigma(D_\pi) \cup \mathcal{R}_{\pi,1}^\phi(A)$. Now, $\mathcal{R}_\pi^\phi(A)$ contains $\sigma(A)$ from (6.37), and $\mathcal{R}_\pi^\phi(A)$ is also a closed and bounded set in $\mathbb{C}$. If $\overline{\mathcal{R}}_{\pi,1}^\phi(A)$ denotes the closure of the set $\mathcal{R}_{\pi,1}^\phi(A)$, find a matrix $A \in \mathbb{C}^{n \times n}$ for which the set $\overline{\mathcal{R}}_{\pi,1}^\phi(A)$ does **not** contain all eigenvalues of A. (Hint: Try some block-reducible matrix!)

9. Given a partition π of $\mathbb{C}^n$ and given $\phi \in \Phi_\pi$, let $A = [A_{i,j}]$ be any matrix in $\mathbb{C}^{n\times n}$ which is partitioned by π, and consider the associated partitioned matrix $B := (zI - D_\pi)^{-1}(A - D_\pi) =: [B_{j,k}]$, where $z \notin \sigma(D_\pi)$. If $\mathbb{G}_\pi(B)$ is the directed graph of the $\ell \times \ell$ nonnegative matrix $[||B_{j,k}||_\phi]$, (where we note that this directed graph is *independent* of ϕ and $z \notin \sigma(D_\pi)$), let $C_\pi(B)$ be the cycle set of all strong and weak cycles of $\mathbb{G}_\pi(B)$. With the set $\mathcal{R}^\phi_{\pi,1}(A)$ defined in the previous problem, show that if each cycle of $\mathbb{G}_\pi(B)$ is *strong* (i.e., $\mathbb{G}_\pi(B)$ is *weakly irreducible* (cf. Exercise 1 of Section 2.2)), then $\overline{\mathcal{R}}^\phi_{\pi,1}(A)$ contains all eigenvalues of $\sigma(A)$.

6.3 A Variation on a Theme by Brualdi

To begin, we observe that the partitioned Robert set $\mathcal{R}^\phi_\pi(A)$, in (6.35) for a partitioned matrix $A = [A_{i,j}]$ in $\mathbb{C}^{n\times n}$, had *no* dependence on a cycle set of strong and weak cycles. On the other hand, stemming from Brualdi's work, we saw in (2.48) of Theorem 2.9 that *having* the additional knowledge, of a matrix's cycle set, gave improved eigenvalue inclusions. Can the same be true in the partitioned case? The object of this section is to give new material to answer this question affirmatively!

Fixing any partition $\pi = \{p_j\}_{j=o}^{\ell}$ of $\mathbb{C}^n$ as in (6.2), assume[3] that $\ell \geq 2$, and let $A = [A_{j,k}]$ be any matrix in $\mathbb{C}^{n\times n}$ which is partitioned by π. With $D_\pi := \text{diag}[A_{1,1}; A_{2,2}; \cdots; A_{\ell,\ell}]$, let z be any complex number with $z \notin \sigma(D_\pi)$, and consider the associated partitioned $n \times n$ matrix $\tilde{A}(z)$, defined by

$$(6.52) \begin{cases} \tilde{A}(z) := (zI_n - D_\pi)^{-1}(A - D_\pi) = [\tilde{A}_{j,k}(z)], \text{where } \tilde{A}_{j,j}(z) := O \text{ and} \\ \text{where } \tilde{A}_{j,k}(z) := (zI_j - A_{j,j})^{-1} \cdot A_{j,k} \text{ for any } j \neq k \ (j,k \in L). \end{cases}$$

Next, fix any $\phi = (\phi_1, \phi_2, \cdots, \phi_\ell)$ in Φ_π and form the associated nonnegative $\ell \times \ell$ matrix $\left[||\tilde{A}_{j,k}(z)||_\phi\right]$ of norms, i.e., with the definitions in (6.6),

$$(6.53) \quad \left[||\tilde{A}_{j,k}(z)||_\phi\right] = \begin{bmatrix} 0 & ||\tilde{A}_{1,2}(z)||_\phi & \cdots & ||\tilde{A}_{1,\ell}(z)||_\phi \\ ||\tilde{A}_{2,1}(z)||_\phi & 0 & \cdots & ||\tilde{A}_{2,\ell}(z)||_\phi \\ \vdots & & & \vdots \\ ||\tilde{A}_{\ell,1}(z)||_\phi & ||\tilde{A}_{\ell,2}(z)||_\phi & \cdots & 0 \end{bmatrix}.$$

[3] The case $\ell = 1$ is trivial, as it implies from (6.31) that $A = D_\pi$, and hence, $\sigma(A) = \sigma(D_\pi)$.

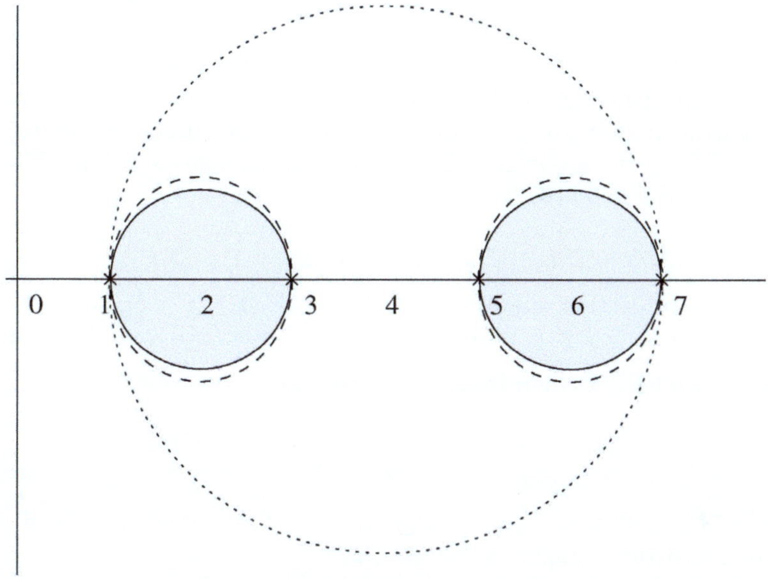

Fig. 6.4. The boundaries of $\mathcal{H}_{\pi}^{\psi}(B)$ (solid), $\mathcal{H}_{\pi}^{\phi}(B)$ (dashed) and $\Gamma(B)$ (dotted), for the matrix B of Exercise 7

With the directed graph of this nonnegative $\ell \times \ell$ matrix $\left[||\tilde{A}_{j,k}(z)||_{\phi}\right]$, determine (as described in Section 2.2) its associated cycle set $\tilde{C}_{\pi}(A)$, and its normal reduced form (cf.(2.35)), if this matrix is reducible. From the discussion relating to equations (2.35) and (2.36), we have that either the matrix $\left[||\tilde{A}_{j,k}(z)||_{\phi}\right]$ is irreducible, i.e., $m = 1$ in (2.35), in which case, there is a *strong cycle* (of some length p with $p \geq 2$) through *each* vertex of the directed graph for $\left[||\tilde{A}_{j,k}(z)||_{\phi}\right]$, or $\left[||\tilde{A}_{j,k}(z)||_{\phi}\right]$ is reducible, in which case, there is either a *strong cycle* or a *weak cycle* through *each* vertex of the directed graph of $\left[||\tilde{A}_{j,k}(z)||_{\phi}\right]$, depending, respectively, on whether that vertex is associated with a $p_j \times p_j$ irreducible matrix $R_{j,j}$ with $p_j \geq 2$, as in (2.36*i*), or this vertex is associated with a 1×1 matrix $R_{j,j}$, as in (2.36*ii*).

The next result, Lemma 6.14, states that the cycle set $\tilde{C}_{\pi}(A)$ of $\left[||\tilde{A}_{j,k}(z)||_{\phi}\right]$, for any $z \notin \sigma(D_{\pi})$, is the *same* as the cycle set $C_{\pi}(A)$ of the partitioned matrix $A = [A_{j,k}]$. Its proof is left to Exercise 1 in this section.

Lemma 6.14. *Given any partition π of $\mathbb{C}^n$ with $\ell \geq 2$, and given $\phi \in \Phi_{\pi}$, let $A = [A_{j,k}]$ be any matrix in $\mathbb{C}^{n \times n}$ which is partitioned by π, and let $C_{\pi}(A)$ be the cycle set associated with the directed graph of the $\ell \times \ell$ nonnegative matrix $[||A_{j,k}||_{\phi}]$. If $\tilde{C}_{\pi}(A)$ denotes the cycle set for the $\ell \times \ell$ matrix $\left[||\tilde{A}_{j,k}(z)||_{\phi}\right]$ of (6.53), where $z \notin \sigma(D_{\pi})$, then*

$$\tilde{C}_\pi(A) = C_\pi(A). \tag{6.54}$$

The result of Lemma 6.14 will be used below.

Assume next that λ is an eigenvalue of the given partitioned matrix $A = [A_{j,k}]$ in $\mathbb{C}^{n\times n}$, with $\lambda \notin \sigma(D_\pi)$. Then $A\mathbf{x} = \lambda\mathbf{x}$ for some $\mathbf{x} \neq \mathbf{0}$ in $\mathbb{C}^n$, which can also be written (cf.(6.52)) as

$$\mathbf{x} = (\lambda I_n - D_\pi)^{-1}(A - D_\pi)\mathbf{x} = [\tilde{A}_{j,k}(\lambda)]\mathbf{x}. \tag{6.55}$$

With P_j the projection operator from $\mathbb{C}^n$ to $W_j (1 \leq j \leq \ell)$, set $P_j\mathbf{x} =: X_j$ for any j with $1 \leq j \leq \ell$, and assume that $\max_{j\in L} \phi_j(X_j) = 1$. Recalling the hypothesis that $\ell \geq 2$, then from (6.55), we have

$$X_j = \sum_{k\in L\backslash\{j\}} \tilde{A}_{j,k}(\lambda)X_k \qquad (\text{all } j \in L). \tag{6.56}$$

Assuming $X_{j_1} \neq \mathbf{0}$, then applying the norm ϕ_{j_1} to the case j_1 of (6.56) gives, with (6.6) and the triangle inequality, that

$$0 < \phi_{j_1}(X_{j_1}) \leq \sum_{k\in L\backslash\{j_1\}} ||\tilde{A}_{j_1,k}(\lambda)||_\phi \cdot \phi_k(X_k). \tag{6.57}$$

Following along the lines of the proof of Theorem 2.5, all the products $||\tilde{A}_{j_1,k}(\lambda)||_\phi \cdot \phi_k(X_k)$ in the sum in (6.57) cannot be zero. Hence, there is a $j_2 \in L$ with $j_2 \neq j_1$ such that

$$\phi_{j_2}(X_{j_2}) = \max\{\phi_k(X_k) : k \neq j_1 \text{ and } ||\tilde{A}_{j_1,k}(\lambda)||_\phi \cdot \phi_k(X_k) > 0\}.$$

We thus have from (6.57) that

$$0<\phi_{j_1}(X_{j_1})\leq \left(\phi_{j_2}(X_{j_2})\cdot \sum_{k\in L\backslash\{j_1\}} ||\tilde{A}_{j_1,k}(\lambda)||_\phi\right)=\phi_{j_2}(X_{j_2})\cdot r^\phi_{j_1,\pi}(\tilde{A}(\lambda)), \tag{6.58}$$

where $r^\phi_{i,\pi}(\cdot)$ is defined in (6.8). This can be continued and we find, as in the proof of Theorem 2.5, that this procedure produces a **strong cycle** $\gamma = (k_1 \;\; k_2 \;\; \cdots \;\; k_p)$ in $\tilde{C}_\pi(A)$, of length $p \geq 2$, where the elements of $\{k_j\}_{j=1}^p$ are distinct with $k_{p+1} = k_1$. Taking products over these k_j's in (6.58) gives

$$\prod_{s=1}^p \phi_{k_s}(X_{k_s}) \leq \left(\prod_{s=1}^p \phi_{k_{s+1}}(X_{k_{s+1}})\right) \cdot \left(\prod_{s=1}^p r^\phi_{k_s,\pi}(\tilde{A}(\lambda))\right),$$

and as $\prod_{s=1}^p \phi_{k_s}(X_{k_s}) = \prod_{s=1}^p \phi_{k_{s+1}}(X_{k_{s+1}}) > 0$, then, on cancelling these products in the above display, we have

$$1 \leq \prod_{s=1}^{p} r^{\phi}_{k_s,\pi}(\tilde{A}(\lambda)).$$

In other words, for any $\lambda \in \sigma(A)$ with $\lambda \notin \sigma(D_\pi)$, there is a strong cycle $\gamma = (k_1 \;\; k_2 \;\; \cdots \;\; k_p)$, with $k_{p+1} = k_1$, in the directed graph of the matrix $\left[||\tilde{A}_{j,k}(\lambda)||\right]$, for which λ satisfies the above inequality. But then, with the full knowledge of the normal reduced form of the matrix $\left[||\tilde{A}_{j,k}(\lambda)||\right]$, we can replace (cf.(2.37)) $r^{\phi}_{k_s,\pi}(\tilde{A}(\lambda))$, in the above inequality, by the possibly smaller term $\tilde{r}^{\phi}_{k_s,\pi}(\tilde{A}(\lambda))$, giving

$$1 \leq \prod_{s=1}^{p} \tilde{r}^{\phi}_{k_s,\pi}(\tilde{A}(\lambda)), \tag{6.59}$$

where $\tilde{A}(\lambda) := (zI_n - D_\pi)^{-1}(A - D_\pi)$ from (6.52). In this way, we obtain the new **variation** of the partitioned Brualdi set: Given a partition π of $\mathbb{C}^n$ with $\ell \geq 2$, given $\phi \in \Phi_\pi$, and given any z with $z \notin \sigma(D_\pi)$, then determine the $\ell \times \ell$ nonnegative matrix $\left[||\tilde{A}_{i,j}(z)||_\phi\right]$ of (6.53), and its associated directed graph. Then, let $\tilde{C}_\pi(A)$ be the set of all strong and weak cycles in this directed graph. From (6.54) of Lemma 6.14, we have, for any $z \notin \sigma(D_\pi)$, that $\tilde{C}_\pi(A) = C_\pi(A)$, where $C_\pi(A)$ is the cycle set associated with the directed graph of the $\ell \times \ell$ nonnegative matrix $[||A_{j,k}||_\phi]$. Then, define the sets

$$\begin{cases} V^{\phi}_{\gamma,\pi}(A) := \{z \in \mathbb{C} : z \notin \sigma(D_\pi) \text{ and } 1 \leq \prod_{i \in \gamma} \tilde{r}^{\phi}_{i,\pi}(\tilde{A}(z)) \\ \qquad\qquad \text{(for any strong cycle } \gamma \in C_\pi(A))\}, \\ \quad \text{and} \\ V^{\phi}_{\pi}(A) := \sigma(D_\pi) \cup \bigcup_{\substack{\gamma \in C_\pi(A) \\ \gamma \text{ a strong cycle}}} V^{\phi}_{\gamma,\pi}(A). \end{cases} \tag{6.60}$$

We call $V^{\phi}_{\pi}(A)$ the **(Brualdi) variation of the partitioned Robert set** for the matrix A, with respect to π and ϕ.

As a direct consequence of our above constructions, we have the new result of

Theorem 6.15. *Given any partition π of $\mathbb{C}^n$ with $\ell \geq 2$ and given $\phi \in \Phi_\pi$, let $A = [A_{i,j}]$ be any matrix in $\mathbb{C}^{n \times n}$ which is partitioned by π. Then* (cf.(6.60)),

$$\sigma(A) \subseteq V^{\phi}_{\pi}(A). \tag{6.61}$$

We now come to the main result in this section, which compares this new eigenvalue inclusion set $V_\pi^\phi(A)$ with the previous inclusion sets $\mathcal{R}_\pi^\phi(A)$, $\mathcal{B}_\pi^\phi(A)$, $\mathcal{K}_\pi^\phi(A)$, and $\Gamma_\pi^\phi(A)$.

Theorem 6.16. *Given any partition π of $\mathbb{C}^n$ with $\ell \geq 2$, and given any $\phi \in \Phi_\pi$, then for any $A = [A_{i,j}] \in \mathbb{C}^{n \times n}$, which is partitioned by π, there holds*

$$V_\pi^\phi(A) \subseteq \mathcal{B}_\pi^\phi(A) \subseteq \mathcal{K}_\pi^\phi(A) \subseteq \Gamma_\pi^\phi(A), \tag{6.62}$$

and

$$V_\pi^\phi(A) \subseteq \mathcal{R}_\pi^\phi(A). \tag{6.63}$$

Remark. The set $V_\pi^\phi(A)$, which makes use of the directed graph of the matrix $\left[||\tilde{A}_{i,j}(z)||_\phi\right]$, is, for *any* matrix A in $\mathbb{C}^{n\times n}$, always smaller than or equal to the other sets in (6.62) and (6.63).

Proof. Because the final two inclusions of (6.62) follow directly from (6.23) of Theorem 6.11, it suffices to establish only the first inclusion of (6.62).

Given any matrix $A = [A_{i,j}] \in \mathbb{C}^{n\times n}$, which is partitioned by π, let z be any point of the set $V_\pi^\phi(A)$ which, from (6.60), contains the subset $\sigma(D_\pi)$. If $z \in \sigma(D_\pi)$, then $z \in \sigma(A_{i,i})$ for some $i \in L$, so that $(zI_i - A_{i,i})$ is singular. Thus, from the discussion preceding (6.7), $\left(||(zI_i - A_{i,i})^{-1}||_\phi\right)^{-1} = 0$. In addition, there is necessarily a (strong or weak) cycle γ of $C_\pi(A) = \tilde{C}_\pi(A)$ which passes through the vertex v_i of the directed graph of $[||A_{i,j}||_\phi]$. Then, from the definition of the associated set $\mathcal{B}_{\gamma,\pi}^\phi(A)$ of (6.18) and the fact that the $\tilde{r}_{i,\pi}^\phi(A)$'s are nonnegative, it also follows from (6.18) that $z \in \mathcal{B}_{\gamma,\pi}^\phi(A)$; whence, again from (6.18), $z \in \mathcal{B}_\pi^\phi(A)$.

Next, assuming that $V_\pi^\phi(A)\backslash\sigma(D_\pi)$ is not empty, let z be any point of $V_\pi^\phi(A)\backslash\sigma(D_\pi)$. To simplify notations, assume that the matrix $\left[||\tilde{A}_{j,k}(z)||\right]$ is irreducible, so that (cf.(2.37)) $\tilde{r}_i(\tilde{A}(z)) = r_i(\tilde{A}(z))$. Then, the construction, leading up to Theorem 6.15, gives that there is a strong cycle γ in $\tilde{C}_\pi(A) = C_\pi(A)$, such that (cf.(6.59), (6.52), and (6.8))

$$1 \leq \prod_{i\in\gamma} r_{i,\pi}^\phi(\tilde{A}(z)) = \prod_{i\in\gamma}\left\{\sum_{j\in L\backslash\{i\}} ||(zI_i - A_{i,i})^{-1}A_{i,j}||_\phi\right\}. \tag{6.64}$$

But, as $||(zI_i - A_{i,i})^{-1}A_{i,j}||_\phi \leq ||(zI_i - A_{i,i})^{-1}||_\phi \cdot ||A_{i,j}||_\phi$, we further have from (6.64) and (6.8) that

$$1 \leq \left(\prod_{i\in\gamma} ||(zI_i - A_{i,i})^{-1}||_\phi\right) \cdot \left(\prod_{i\in\gamma} r_{i,\pi}^\phi(A)\right).$$

By hypothesis, $z \notin \sigma(D_\pi)$, so this can be equivalently expressed as

$$\prod_{i\in\gamma} \left(||(zI_i - A_{i,i})^{-1}||_\phi\right)^{-1} \leq \prod_{i\in\gamma} r_{i,\pi}^\phi(A).$$

But this states, from (6.18), that $z \in \mathcal{B}_{\gamma,\pi}^\phi(A)$; whence, $z \in \mathcal{B}_\pi^\phi(A)$. Thus, the first inclusion of (6.62) is valid.

To establish the inclusion of (6.63), let z be any point of $V_\pi^\phi(A)$ of (6.60). If $z \in \sigma(D_\pi)$, which is a subset of $V_\pi^\phi(A)$, then z is, from (6.35), also in $\mathcal{R}_\pi^\phi(A)$, as $\sigma(D_\pi)$ is also a subset of $\mathcal{R}_\pi^\phi(A)$. If the set $V_\pi^\phi(A)\backslash\sigma(D_\pi)$is not empty and z is any point in this set, then there is a strong cycle γ in $\tilde{C}_\pi(A) = C_\pi(A)$ for which (cf.(6.64))

$$1 \leq \prod_{i\in\gamma} \left\{ \sum_{j\in L\backslash\{i\}} ||(zI_i - A_{i,i})^{-1}A_{i,j}||_\phi \right\}.$$

But it is evident that *all* of the sums, $\sum_{j\in L\backslash\{i\}} ||(zI_i - A_{i,i})^{-1}A_{i,j}||_\phi$ in the above product, cannot be less than unity. This implies that there is an $i \in \gamma$ for which $\sum_{j\in L\backslash\{i\}} ||(zI_i - A_{i,i})^{-1}A_{i,j}||_\phi \geq 1$, so that, with the notation of (6.52),

$$\max_{i\in L} \left(\sum_{j\in L\backslash\{i\}} ||\tilde{A}_{i,j}(z)||_\phi \right) \geq 1. \tag{6.65}$$

But, using the fact that $\tilde{A}_{i,i}(z) = O$ from (6.52), the inequality of (6.65) is *exactly* the statement (cf. (6.30)) that

$$||\left[||(zI - D_\pi)^{-1}(A - D_\pi)||_\phi\right]||_\infty \geq 1,$$

so that, from (6.35), $z \in \mathcal{R}_\pi^\phi(A)$. Thus, $V_\pi^\phi(A) \subseteq \mathcal{R}_\pi^\phi(A)$, the desired result of (6.63). ∎

It is worthwhile to again consider the Robert matrix E of (6.41), in light of the inclusions of (6.62) and (6.63) of Theorem 6.16. First, for the partitioning $\pi = \{0, 2, 4\}$ of the matrix E in (6.41), we have that the associated partitioned matrix $\tilde{E}(z) := (zI_4 - D_\pi)^{-1}(E - D_\pi) = [\tilde{E}_{j,k}(z)]$ of (6.52) is such that there is but one (strong) cycle in $C_\pi(\tilde{E})$, namely $\gamma = (1\ \ 2)$, in the directed graph of the 2×2 nonnegative matrix $\left[||\tilde{E}_{j,k}(z)||_\phi\right]$. Choosing again the ℓ_∞-norms on the subspaces W_1 and W_2 (where $\mathbb{C}^4 = W_1 \dotplus W_2$), it can be verified (see Exercise 2 of this section) that (cf.(6.60) and (6.41))

$$\begin{aligned} V_\pi^\phi(E) = \sigma(D_\pi) \cup \{&z \in \mathbb{C} : z \notin \sigma(D_\pi) \text{ and} \\ &1 \leq ||\tilde{E}_{1,2}(z)||_\infty \cdot ||\tilde{E}_{2,1}(z)||_\infty\}. \end{aligned} \tag{6.66}$$

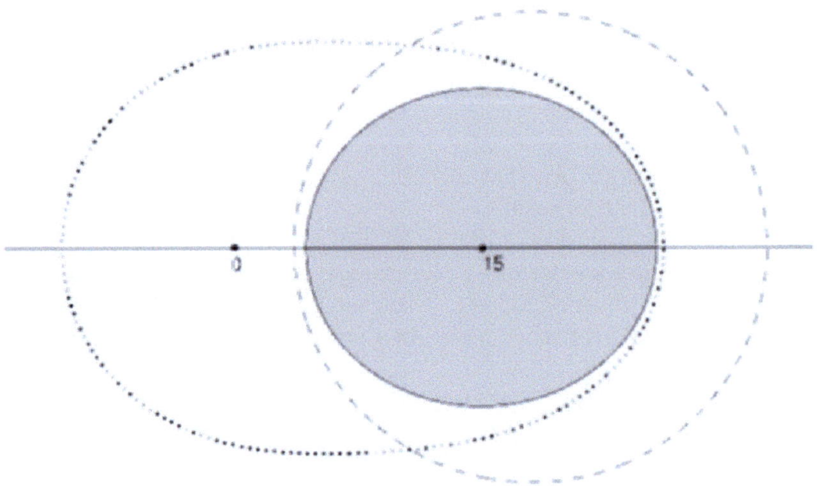

Fig. 6.5. $V_\pi^\phi(E)$ (shaded), $\mathcal{R}_\pi^\phi(E)$ (with a dashed boundary) and $\mathcal{B}_\pi^\phi(E)$ (with a dotted boundary), for the Robert matrix E of (6.41)

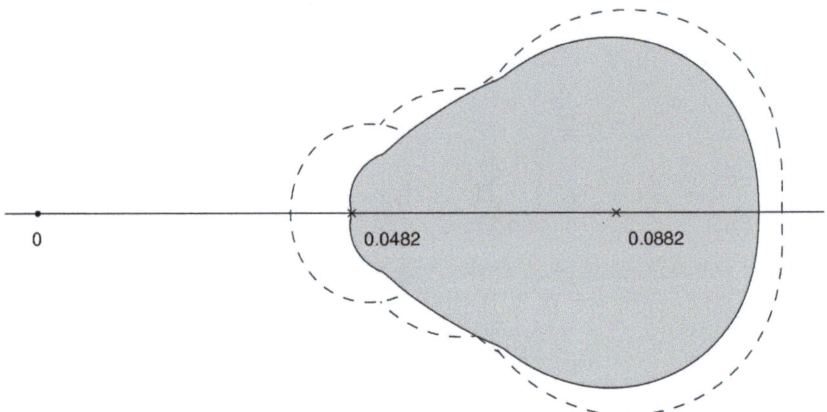

Fig. 6.6. Magnified component of $V_\pi^\phi(E)$ (shaded), and the associated component (with a dashed boundary) of $\mathcal{R}_\pi^\phi(A)$

This particular set $V_\pi^\phi(E)$ again consists of two (disjoint) components, which can be seen in Fig. 6.5, with one small component, appearing as a dot, near zero. In Fig. 6.5, the larger component of $V_\pi^\phi(E)$ is shaded with the solid boundary, while, for comparison, the larger component of $\mathcal{R}_\pi^\phi(E)$ is shown with a dashed boundary, while the boundary of $\mathcal{B}_\pi^\phi(A)$ is shown with a dotted boundary. In Fig. 6.6, the smaller component of $V_\pi^\phi(E)$ is magnified and is the shaded set with a solid boundary, while, for comparison, the smaller component of $\mathcal{R}_\pi^\phi(A)$ is shown with the dotted boundary. The inclusions of Theorem 6.16 are clearly seen to be true in these two figures!

Exercises

1. Give a proof of Lemma 6.14.

2. For the partitioned matrix E of (6.41), show, from (6.43) and (6.52), that

$$||\tilde{E}_{1,2}(z)||_\infty = \max\left\{\frac{|2.1\cdot z-0.1|+|4.0\cdot z-0.5|}{|z^2-12z+1|};\frac{|2.5\cdot z-0.3|+|4.7\cdot z-0.2|}{|z^2-12z+1|}\right\},$$
$$||\tilde{E}_{2,1}(z)||_\infty = \max\left\{\frac{|9.6z-0.4|+|4.7\cdot z-0.3|}{|z^2-18z+1|};\frac{|8.6\cdot z-0.6|+|4.2\cdot z-0.2|}{|z^2-18z+1|}\right\};$$

which are used in (6.66).

3. From (6.60), consider the set $V^{\phi}_{\gamma,\pi}(A)$, where $\ell \geq 2$ and where γ is a strong cycle in $C_\pi(A)$. As this set depends on $\tilde{A}(z)$ in (6.52), which in turn depends on the assumption that $z \notin \sigma(D_\pi)$, then $V^{\phi}_{\gamma,\pi}(A)$ is not necessarily a closed set in $\mathbb{C}$. Let $\overline{V}^{\phi}_{\gamma,\pi}(A)$ denote its **closure** in $\mathbb{C}$. Next, if $\gamma = (j)$ is a weak cycle in $C_\pi(A)$, then define $V^{\phi}_{\gamma,\pi}(A) := \sigma(A_{j,j})$. Show that the Brualdi variation of the partitioned Robert set, $V^{\phi}_{\pi}(A)$ of (6.60), can also be expressed as

$$V^{\phi}_{\pi}(A) = \left(\bigcup_{\substack{\gamma \text{ a strong cycle}\\ \text{of } C_\pi(A)}} \overline{V}^{\phi}_{\gamma,\pi}(A)\right) \cup \left(\bigcup_{\substack{\gamma=(j) \text{ a weak}\\ \text{cycle of } C_\pi(A)}} \sigma(A_{j,j})\right).$$

6.4 G-Functions in the Partitioned Case

The focus, thus far in this chapter, has been on the extensions of Geršgorin sets, Brauer sets, and Brualdi sets to the partitioned matrix case. In this section, we similarly extend results of Chapter 5, on G-functions, to partitioned matrices, with results from Carlson and Varga (1973b) and Johnston (1965) and (1971).

With the notations of Section 6.1, let $\pi = \{p_j\}_{j=0}^{\ell}$ be a partition of $\mathbb{C}^n$, as in (6.1), let $\phi = (\phi_1, \phi_2, \cdots, \phi_\ell)$ be a norm ℓ-tuple in Φ_π, and let $A = [A_{i,j}] \in \mathbb{C}^{n\times n}$ denote, as in (6.4), the partitioning of A, with respect to π. Then, $\mathbb{C}^{n\times n}_\pi$ is defined as the subset of $\mathbb{C}^{n\times n}$ of all matrices $A = [A_{i,j}]$, partitioned by π, for which each block diagonal submatrix $A_{i,i}$ is *nonsingular*

$(1 \le i \le \ell)$. (For our purposes here, the partition π and ϕ will be fixed, as are the projection operators $P_j : \mathbb{C}^n \to W_j$, for each j $(j \in L := \{1, 2, \cdots, j\})$.

Definition 6.17. Let $\mathcal{F}_\pi$ be the collection of all functions $F = [F_1, F_2, \cdots, F_\ell]$ for which

i) $F_i : \mathbb{C}_\pi^{n\times n} \to \mathbb{R}_+$, i.e., $0 \le F_i(A) < +\infty$ for any $i \in L$ and any $A \in \mathbb{C}_\pi^{n\times n}$;

ii) for each $B = [B_{i,j}] \in \mathbb{C}_\pi^{n\times n}, F_i(B)$ depends only on the products

$$\left(||B_{k,k}^{-1}||_\phi\right)^{-1} \cdot ||B_{k,k}^{-1} \cdot B_{k,j}||_\phi \qquad \text{(all } k, j \in L \text{ with } k \ne j), \tag{6.67}$$

where the norms used here are defined in (6.6) and (6.7).
In other words, if the matrix $C = [C_{k,j}] \in \mathbb{C}_\pi^{n\times n}$ satisfies

$$(||C_{k,k}^{-1}||_\phi)^{-1} \cdot ||C_{k,k}^{-1} C_{k,j}||_\phi = (||B_{k,k}^{-1}||_\phi)^{-1} \cdot ||B_{k,k}^{-1} B_{k,j}||_\phi$$

for all $k \ne j$ in L, then $F_i(C) = F_i(B)$ for all $i \in L$.

We remark that part *ii)* of Definition 6.17, involving the dependence of the products in (6.67), does effectively reduce to *ii)* of Definition 5.4, when the subspaces W_i are all one-dimensional. (See Exercise 1 of this section.)
In analogy with Definition 5.2, we make the following

Definition 6.18. Let $F = [F_1, F_2, \cdots, F_\ell] \in \mathcal{F}_\pi$. Then, F is a **G-function** (relative to the partition π and $\phi \in \Phi_\pi$) if, for any $B = [B_{i,j}] \in \mathbb{C}_\pi^{n\times n}$, the relations

$$(||B_{i,i}^{-1}||_\phi)^{-1} > F_i(B) \qquad \text{(for all } i \in L) \tag{6.68}$$

imply that B is nonsingular. The collection of all **G-functions**, relative to π and ϕ, is denoted by $\mathcal{G}_\pi^\phi$.

For additional notation, if $F \in \mathcal{F}_\pi$ and if $B = [B_{i,j}] \in \mathbb{C}_\pi^{n\times n}$, then we set

$$\begin{cases} \mathcal{M}^F(B) := [\alpha_{i,j}] \in \mathbb{R}^{\ell\times\ell}, \text{ where} \\ \alpha_{i,i} := F_i(B), \text{ and } \alpha_{i,j} := -(||B_{i,i}^{-1}||_\phi)^{-1} \cdot ||B_{i,i}^{-1} B_{i,j}||_\phi, i \ne j \, (i, j \in L), \end{cases} \tag{6.69}$$

which is the partitioned analog of (5.7).
As an extension of Theorem 5.5 to the partitioned case, we have

Theorem 6.19. *Let $F = [F_1, F_2, \cdots, F_\ell] \in \mathcal{F}_\pi$. Then, $F \in \mathcal{G}_\pi^\phi$ if and only if $\mathcal{M}^F(B)$ is an M-Matrix for every $B \in \mathbb{C}_\pi^{n\times n}$.*

Remark. As in Chapter 5, an M-matrix here *can* be singular.

Proof. First, for any $B = [B_{i,j}]$ in $\mathbb{C}_\pi^{n\times n}$, assume that $\mathcal{M}^F(B)$ of (6.69) is an M-matrix, and assume further that B satisfies (6.68). In this part of the proof, we are to show that B is nonsingular, so that $F \in \mathcal{G}_\pi^\phi$. Then with (6.68), set

(6.70) $$d_i := (||B_{i,i}^{-1}||_\phi)^{-1} - F_i(B) > 0 \qquad (\text{each } i \in L),$$

so that $\text{diag}[d_1, d_2, \cdots, d_\ell]$ is a positive diagonal matrix in $\mathbb{C}^{\ell\times\ell}$. But then, it follows from (6.69) and (6.70) that

(6.71) $$\begin{cases} \mathcal{E} := \mathcal{M}^F(B) + \text{diag}[d_1, d_2, \cdots, d_\ell] = [e_{i,j}] \in \mathbb{R}^{\ell\times\ell}, \text{ where} \\ e_{i,i} = (||B_{i,i}^{-1}||_\phi)^{-1} \text{and } e_{i,j} = -(||B_{i,i}^{-1}||_\phi)^{-1} \cdot (||B_{i,j}||_\phi), i \neq j (i,j \in L). \end{cases}$$

Moreover, as $\mathcal{E}$ is the sum of an M-matrix and a positive diagonal matrix, then (cf. Propositions C.4 and C.5 of Appendix C) $\mathcal{E}$ is a nonsingular M-matrix with $\mathcal{E}^{-1} \geq O$. This can be used as follows.

With the projection operators $P_j : \mathbb{C}^n \to W_j$ $(1 \leq j \leq \ell)$ and with $\phi = (\phi_1, \phi_2, \cdots, \phi_\ell) \in \Phi_\pi$, we define the associated **vectorial norm**:

(6.72) $$\mathbf{p}_\phi(\mathbf{x}) := [\phi_1(P_1\mathbf{x}), \phi_2(P_2\mathbf{x}), \cdots, \phi_\ell(P_\ell\mathbf{x})]^T \qquad (\text{any } \mathbf{x} \in \mathbb{C}^n).$$

Next, with the given matrix $B = [B_{i,j}] \in \mathbb{C}_\pi^{n\times n}$, set (cf.(6.31))

$$D_\pi := \text{diag}[B_{1,1}; B_{2,2}; \cdots; B_{\ell,\ell}].$$

From (6.68), $(||B_{i,i}^{-1}||_\phi)^{-1} > 0$ for all $i \in L$, so that D_π is necessarily nonsingular, and we write

(6.73) $$B = D_\pi(I - \Gamma), \text{ where } \Gamma := (I - D_\pi^{-1}B).$$

(It is important to note that the partitioned matrix Γ of (6.73) has zero block-diagonal submatrices.) Using the definition in (6.72), then for any $\mathbf{x} \in \mathbb{C}^n$, it can first be verified from (6.7) that

$$\begin{aligned} \mathbf{p}_\phi(B\mathbf{x}) &= \mathbf{p}_\phi\left(D_\pi(I-\Gamma)\mathbf{x}\right) \\ &\geq \text{diag}\left[(||B_{1,1}^{-1}||_\phi)^{-1}, \cdots, (||B_{\ell,\ell}^{-1}||_\phi)^{-1}\right] \mathbf{p}_\phi\left((I-\Gamma)\mathbf{x}\right), \end{aligned}$$

and with (6.6), that

$$\mathbf{p}_\phi((I-\Gamma)\mathbf{x}) \geq \begin{bmatrix} 1 & -||B_{1,1}^{-1}B_{1,2}||_\phi & \cdots & -||B_{1,1}^{-1}B_{1,\ell}||_\phi \\ -||B_{2,2}^{-1}B_{2,1}||_\phi & 1 & \cdots & -||B_{2,2}^{-1}B_{2,\ell}||_\phi \\ \vdots & & & \vdots \\ -||B_{\ell,\ell}^{-1}B_{\ell,1}||_\phi & -||B_{\ell,\ell}^{-1}B_{\ell,2}||_\phi & \cdots & 1 \end{bmatrix} \cdot \mathbf{p}_\phi(\mathbf{x}).$$

Putting these inequalities together gives, with the definition of the matrix $\mathcal{E}$ in (6.71), that

(6.74) $$\mathbf{p}_\phi(B\mathbf{x}) \geq \mathcal{E}\mathbf{p}_\phi(\mathbf{x}) \qquad (\text{any } \mathbf{x} \in \mathbb{C}^n),$$

so that the $\ell \times \ell$ matrix $\mathcal{E}$ of (6.71) is a **lower bound matrix** for B, as in (C.8) of Appendix C. Moreover, since $\mathcal{E}^{-1} \geq O$, then multiplying on the left by $\mathcal{E}^{-1}$, in (6.74), preserves these inequalities, giving

$$\mathcal{E}^{-1}\mathbf{p}_\phi(B\mathbf{x}) \geq \mathbf{p}_\phi(\mathbf{x}) \qquad (\text{any } \mathbf{x} \in \mathbb{C}^n), \tag{6.75}$$

which, as in (5.12), gives that B is nonsingular, the first part of the desired result.

Conversely, assume that $F \in \mathcal{G}_\pi^\phi$, and consider the $\ell \times \ell$ matrix $\mathcal{M}^F(B)$ of (6.69), for any $B \in \mathbb{C}_\pi^{n\times n}$. Here, we are to show that $\mathcal{M}^F(B)$ is an M-matrix. From (6.69), we see (cf.(C.3) of Appendix C) that $\mathcal{M}^F(B) \in \mathbb{Z}^{\ell\times\ell}$, so that it has the proper sign-pattern to be an M-matrix. If, on the contrary, $\mathcal{M}^F(B)$ is not an M-matrix, there necessarily exist constants $\delta_i > 0$ (for all $i \in L$) such that

$$\mathcal{T} := \mathcal{M}^F(B) + \text{diag}[\delta_1, \delta_2, \cdots, \delta_\ell]$$

is singular; whence, there is a $\mathbf{y} = [y_1, y_2, \cdots, y_\ell]^T \neq \mathbf{0}$ in $\mathbb{C}^\ell$ with $\mathcal{T}\mathbf{y} = \mathbf{0}$. Equivalently, $\mathcal{T}\mathbf{y} = \mathbf{0}$ can be expressed from (6.69) as

$$(F_i(B)+\delta_i)y_i - \sum_{j\in L\backslash\{i\}} (||B_{i,i}^{-1}||_\phi)^{-1} \cdot ||B_{i,i}^{-1}B_{i,j}||_\phi y_j = 0 \text{ (all } i \in L). \tag{6.76}$$

Now, let $\{\boldsymbol{\xi}_i\}_{i\in L}$ be fixed vectors such that $\boldsymbol{\xi}_i \in W_i$ with $\phi_i(\boldsymbol{\xi}_i) = 1$ for each $i \in L$, and set $\mathbf{z} := \sum_{i\in L} y_i\boldsymbol{\xi}_i$, so that $\mathbf{z} \in \mathbb{C}^n$ with $\mathbf{z} \neq \mathbf{0}$. We now construct a partitioned matrix $C = [C_{i,j}] \in \mathbb{C}_\pi^{n\times n}$ such that $C\mathbf{z} = \mathbf{0}$, and such that

$$(||C_{i,i}^{-1}||_\phi)^{-1} \cdot ||C_{i,i}^{-1}C_{i,j}||_\phi = (||B_{i,i}^{-1}||_\phi)^{-1} \cdot ||B_{i,i}^{-1}B_{i,j}||_\phi \text{ (all } i \neq j; i,j \in L). \tag{6.77}$$

For the diagonal blocks $C_{i,i}$, set

$$C_{i,i} := (F_i(B) + \delta_i)I_i \qquad (\text{all } i \in L), \tag{6.78}$$

where I_i is the identity operator on W_i, so that each $C_{i,i}$ is nonsingular, as $F_i(B) + \delta_i > 0$. Thus, $C = [C_{i,j}] \in \mathbb{C}_\pi^{n\times n}$. With these choices for $C_{i,i}$, it follows that (6.77) reduces to

$$||C_{i,j}||_\phi = (||B_{i,i}^{-1}||_\phi)^{-1} \cdot ||B_{i,i}^{-1}B_{i,j}||_\phi \qquad (\text{all } i \neq j; i,j \in L). \tag{6.79}$$

Next, we verify from (6.76) and (6.78) that $C\mathbf{z} = \mathbf{0}$ if

$$C_{ij}\boldsymbol{\xi}_j = -(||B_{i,i}^{-1}||_\phi)^{-1} \cdot ||B_{i,i}^{-1}B_{i,j}||_\phi\boldsymbol{\xi}_i \qquad (\text{all } i \neq j; i,j \in L). \tag{6.80}$$

Our problem then reduces to constructing the nondiagonal submatrices $C_{i,j} : W_j \to W_i$, for all $i,j \in L$ with $i \neq j$, which *simultaneously* satisfy (6.79) and (6.80). Following Johnston (1965) and Johnston (1971), let ϕ_i^D be the **dual norm** to ϕ_i on W_i, for each $i \in L$. As a well-known consequence of the Hahn-Banach Theorem (see Horn and Johnson (1985), p.288) there is a vector $\boldsymbol{\sigma}_i$ in W_i for which

$$(6.81) \qquad \phi_i^D(\boldsymbol{\sigma}_i) = 1 \text{ and } \boldsymbol{\sigma}_i^* \boldsymbol{\xi}_i = 1 \quad (\text{all } i \in L),$$

i.e., if $\boldsymbol{\xi}_i = [s_1, s_2, \cdots, s_{d_i}]^T$, with the notation that $\dim W_i := d_i = p_i - p_{i-1}$ from (6.3), and if $\boldsymbol{\sigma}_i = [t_1, t_2, \cdots, t_{d_j}]^T$, then $\boldsymbol{\sigma}_j^* := [\bar{t}_1, \bar{t}_2, \cdots, \bar{t}_{d_j}]$. On defining the submatrices $C_{i,j} : W_j \to W_i$, for $i \neq j$, by

$$C_{i,j} := -(||B_{i,i}^{-1}||_\phi)^{-1} \cdot ||B_{i,i}^{-1} B_{i,j}||_\phi \boldsymbol{\xi}_i \cdot \boldsymbol{\sigma}_j^* \qquad (\text{all } i \neq j; i, j \in L),$$

we see that both (6.79) and (6.80) are satisfied. In summary, the above construction gives us from (6.79) that $C\mathbf{z} = \mathbf{0}$ where $\mathbf{z} \neq \mathbf{0}$, so that C is singular, a contradiction. Thus, $\mathcal{M}^F(B)$ is an M-matrix for each $B \in \mathbb{C}_\pi^{n \times n}$, which was the final item to be proved. ∎

We remark that extensions of Theorems 5.12 and 5.14, to the partitioned case, can be found in Carlson and Varga (1973b), where the wordings of these extensions are exactly the *same* as those of Theorems 5.12 and 5.14, except for the change of $\mathcal{G}_n^c$ into $\mathcal{G}_\pi^{\phi,c}$, where $\mathcal{G}_\pi^{\phi,c}$ denotes the subset of *continuous functions* in $\mathcal{G}_\pi^\phi$ of Definition 6.18.

Exercises

1. With $\ell = n$, so that all subspace W_j in (6.3) are one-dimensional, define the norm n-tuple $\phi = (\phi_1, \phi_2, \cdots, \phi_n)$, in Φ_π, simply by $\phi_j(\mathbf{x}) := |x_j|$ for all $1 \leq j \leq n$, where $\mathbf{x} = [x_1, x_2, \cdots, x_n]^T \in \mathbb{C}^n$. Show, for any matrix $B = [b_{i,j}] \in \mathbb{C}^{n \times n}$ which has nonzero diagonal entries, that *ii)* of Definition 6.18 reduces to that of Definition 5.1.

2. Verify the steps leading up to (6.74).

Bibliography and Discussion

6.1 As mentioned in the text, the first generalization of the nonsingularity of strictly diagonally dominant matrices to partitioned matrices were roughly simultaneously and independently considered in Ostrowski (1961), Fiedler and Pták (1962a), and Feingold and Varga (1962). From these results came their associated eigenvalue inclusion results, which we call in this section **partitioned Geršgorin sets**.

The extensions of these results, in the partitioned case, to **partitioned Brauer sets** were briefly mentioned in the last of the above papers, but with no comparison results, as in Theorem 6.10, with partitioned Geršgorin sets. There was a lengthy hiatus of many years in further developments in this area of partitioned matrices. Though partitioned matrices were not considered in the seminal paper of Brualdi (1982), this paper served as the basis of results for partitioned matrices in Brualdi (1992), where partitioned Brualdi sets are considered, but with no comparison results on the associated eigenvalue inclusion with partitioned Brauer sets or partitioned Geršgorin sets. In this sense, the results of Theorem 6.11 are new.

We remark that Theorem 6.9 gives a **sufficient** condition for a partitioned π-irreducible matrix to have an eigenvalue on the boundary of its associated Brualdi lemniscates. Necessary **and** sufficient conditions for this have been nicely studied in Kolotolina (2003a), where the norm on each subspace W_i is chosen to be the ℓ_2-norm.

6.2 The norm, defined in (6.27), is generalized in Exercise 1 of this section. Such general compound norms are treated at length in Ostrowski (1961), but it turns out that these more general compound norms do *not* improve the results derived, from (6.27), on partitioned Householder and Robert sets.

As mentioned in this section, the new term, **partitioned Householder set** of (6.34) for a partitioned matrix A in $\mathbb{C}^{n\times n}$, is used here because of its similarity to Householder's original set in (1.48) from his publications in 1956 and 1964, though Householder did not directly consider partitioned matrices. The new term, **partitioned Robert set** of (6.35), was suggested by constructions in Robert (1966) and Robert (1969), associated with partitions, norms, and M-matrix theory, though Robert did not directly consider eigenvalue inclusion results. The interesting 4×4 matrix of (6.41) comes from Robert (1966). The connections with M-matrices and H-matrices, at the end of this section, is directly related to the work of Robert (1966) and Robert (1969).

6.3 The new (Brualdi) **variation** of the partitioned Robert set gives rise to a new eigenvalue inclusion set for a partitioned matrix which compares favorably, in **all** cases from Theorem 6.16, with partitioned Geršgorin, partitioned Brauer, and partitioned Brualdi sets, where we note that this variant depends on the additional knowledge of the cycle set, derived from the directed graph of the associate $\ell \times \ell$ nonnegative matrix in (6.53).

6.4 The material in this section extends results of Chapter 5 to G-functions in the partitioned case, and draws upon works of Carlson and Varga (1973b) and Johnston (1965) and (1971). It is interesting to see in (6.74) the partitioned analog of a **lower bound matrix** used in (5.11). Further extensions of G-functions in the partitioned case appear in Carlson and Varga (1973b).

Appendix A. Geršgorin's Paper from 1931, and Comments on His Life and Research.

Appendix A.1. Semen Aronovich Geršgorin

It is interesting to first comment on the contents of Geršgorin's original paper from 1931 (in German), on estimating the eigenvalues of a given $n \times n$ complex matrix, which is *reproduced*, for the reader's convenience, at the end of this appendix. There, one can see the originality of Geršgorin pouring forth in this paper! His Satz II corresponds exactly to our Theorem 1.1, his Satz III corresponds to our Theorem 1.6, and his Satz IV, on separated Geršgorin disks, appears in Exercise 4 of Section 1.1. In his final result of Satz V, he uses a positive diagonal similarity transformation, as in our (1.14), which is dependent on a single parameter α, with $0 < \alpha < 1$, to obtain better eigenvalue inclusion results. This approach was subsequently used by Olga Taussky in Taussky (1947) in the practical estimation of eigenvalues in the flutter of airplane wings! However, we must mention that his Satz I is **incorrect**. His statement in Satz I is that if $A = [a_{i,j}] \in \mathbb{C}^{n \times n}$ satisfies

$$|a_{i,i}| \geq r_i(A) := \sum_{j \in N \backslash \{i\}} |a_{i,j}|, \quad \text{for all } i \in N,$$

with strict inequality for at least one i, then A is nonsingular. But, as we have seen in Section 1.2, the matrix $A = \begin{bmatrix} 1 & 0 \\ 0 & 0 \end{bmatrix}$ is a *counterexample*, as A satisfies the above conditions, but is singular. (Olga Taussky was certainly aware of this error, but she was probably just too polite to mention this in print!) As we now know, her assumption of **irreducibility** in Taussky (1949), (cf. Theorem 1.9 in Chapter 1) clears this up nicely, but see also Exercise 1 of Sec. 1.2.

We also mention here the important contribution of Fujino and Fischer (1998) (in German) which provided us with the biographical data below on

Geršgorin, as well as a list of his significant publications. This paper of Fujino and Fischer (1998) also contains pictures, from the Deutsches Museum in Munich, of **ellipsographs**, a mechanical device to draw ellipses, which were built by Geršgorin. There is a very new contribution on the life and works of Geršgorin by Garry Tee (see Tee (2004)).

Semen Aronovich Geršgorin

- Born: 24 August 1901 in Pruzhany (Brest region), Belorussia
- Died: 30 May 1933 in St. Petersburg
- Education: St. Petersburg Technological Institute, 1923
- Professional Experience: Professor 1930-1933, St. Petersburg Machine-Construction Institute

SIGNIFICANT PUBLICATIONS

1. Instrument for the integration of the Laplace equation, Zh. Priklad. Fiz. 2 (1925), 161-7.
2. On a method of integration of ordinary differential equations, Zh. Russkogo Fiz-Khimi. O-va. 27 (1925), 171-178.
3. On the description of an instrument for the integration of the Laplace equation, Zh. Priklad. Fiz. 3(1926), 271-274.
4. On mechanisms for the construction of functions of a complex variable, Zh. Fiz.- Matem. O-va 1 (1926), 102-113.
5. On the approximate integration of the equations of Laplace and Poisson, Izv. Leningrad Polytech. Inst. 20 (1927), 75-95.
6. On the number of zeros of a function and its derivative, Zh. Fiz.- Matem. O-va 1(1927), 248-256.
7. On the mean values of functions on hyper-spheres in n-dimensional space, Mat. Sb. 35 (1928), 123-132.
8. A mechanism for the construction of the function $\xi = \frac{1}{2}(z - \frac{r^2}{z})$, Izv. Leningrad Polytech. Inst. 2 (26) (1928), 17-24.
9. On the electric nets for the approximate solution of the Laplace equation, Zh. Priklad. Fiz. 6 (3-4) (1929), 3-30.
10. Fehlerabschätzung für das Differenzverfahren zur Lösung partieller Differentialgleichungen, J. Angew. Math. Mech. 10 (1930).
11. Über die Abgrenzung der Eigenwerte einer Matrix. Dokl. Akad. Nauk (A), Otd. Fiz.-Mat. Nauk (1931), 749-754.
12. Über einen allgemeinen Mittelwertsatz der mathematischen Physik, Dokl. Akad. Nauk. (A) (1932), 50-53.
13. On the conformal map of a simply connected domain onto a circle, Mat. Sb. 40 (1933), 48-58.

Of the above papers, three papers, 10, 11, and 13, stand out as **seminal contributions**. Paper 10 was the first paper to treat the important topic of the **convergence** of finite-difference approximations to the solution of

Laplace-type equations, and it is quoted in the book by Forsythe and Wasow (1960). Paper 11 was Geršgorin's original result on estimating the eigenvalues of a complex $n \times n$ matrix, from which the material of this book has grown. Paper 13, on numerical conformal maps, is quoted in the book by Gaier (1964). But what is most impressive is that these three papers of Geršgorin are still being referred today in research circles, after more than 70 years!

Next, we have been given permission to give below a translation, from Russian to English, of the following obituary of Geršgorin's passing, as recorded in the journal, Applied Mathematics and Mechanics 1 (1933), no.1, page 4. Then, after this obituary, Geršgorin's original paper (in German) is given in full.

APPLIED MATHEMATICS AND MECHANICS
Volume 1, 1933, No.1

Semen Aronovich Geršgorin has passed away. This news will cause great anguish in everybody who knew the deceased.

The death of a great scientist is always hard to bear, as it always causes a feeling of emptiness that cannot be filled; it is especially sad when a young scientist's life ends suddenly, with his talent in its full strength, when he is still full of unfulfilled research potential.

Semen Aronovich died at the age of 32. Having graduated from the Technological Institute and having defended a brilliant thesis in the Division of Mechanics, he quickly became one of the leading figures in Soviet Mechanics and Applied Mathematics. Numerous works of S.A., in the theory of Elasticity, Theory of Vibrations, Theory of Mechanisms, Methods of Approximate Numerical Integration of Differential Equations and in other parts of Mechanics and Applied Mathematics, attracted attention and brought universal recognition to the author. Already the first works showed him to be a very gifted young scientist; in the last years his talent matured and blossomed. The main features of Geršgorin's individuality are his methods of approach, combined with the power and clarity of analysis. These features are already apparent in his early works (for example, in a very clever idea for constructing the profiles of aeroplane wings), as well as in his last brilliant (and not yet completely published) works in elasticity theory and in theory of vibrations.

S.A. Geršgorin combined a vigorous and active research schedule which, in his last years, centered around the Mathematical and Mechanical Institute at Leningrad State University, as well as around the Turbine Research Institute (NII Kotlo-Turbiny) with wide-ranging teaching activities.

In 1930 he became a Professor at the Institute of Mechanical Engineering (Mashinostroitelnyi); he then became head of the Division of Mechanics at the Turbine Institute. He also taught very important courses at Leningrad State University and at the Physical-Mechanical Institute of Physics and Mechanics.

A vigorous, stressful job weakened S.A.'s health; he succumbed to an accidental illness, and a brilliant and successful young life has ended abruptly.

S.A. Geršgorin's death is a great and irreplaceable loss to Soviet Science. He occupied a unique place in the Soviet science - this place is now empty.

A careful collection and examination of everything S.A. has done, has been made, so that none of his ideas are lost - this is the duty of Soviet science in honor of one of its best representatives.

ИЗВЕСТИЯ АКАДЕМИИ НАУК СССР. 1931

BULLETIN DE L'ACADÉMIE DES SCIENCES DE L'URSS

Classe des sciences mathématiques et naturelles

Отделение математических и естественных наук

ÜBER DIE ABGRENZUNG DER EIGENWERTE EINER MATRIX

Von S. GERSCHGORIN

(Présenté par A. Krylov, membre de l'Académie des Sciences)

§ **1.** Haben wir eine Matrix

$$A = \begin{Vmatrix} a_{11}, & a_{12}, & \dots & a_{1n} \\ a_{21}, & a_{22}, & \dots & a_{2n} \\ \dots & \dots & \dots & \dots \\ a_{n1}, & a_{n2}, & \dots & a_{nn} \end{Vmatrix} \tag{1}$$

wo die Elemente a_{ik} beliebige komplexe Zahlen sein dürfen, und bezeichnen wir durch s_k $(k = 1, 2, \dots n)$ ihre Eigenwerte, d. h. die Wurzeln der Gleichung

$$\begin{Vmatrix} a_{11} - s, & a_{12}, & \dots & a_{1n} \\ a_{21}, & a_{22} - s, & \dots & a_{2n} \\ \dots & \dots & \dots & \dots \\ a_{n1}, & a_{n2}, & \dots & a_{nn} - s \end{Vmatrix} = 0, \tag{2}$$

so gilt nach Bendixson und Hirsch * die Ungleichung

$$|s_k| \leq na,$$

wo a den Maximalwert aller Zahlen $|a_{ik}|$ bedeutet.

Wir wollen im folgenden zeigen, dass man im allgemeinen viel schärfere Aussagen über die Lage der Eigenwerte machen kann.

* Sur les racines d'une équation fondamentale. Acta Mathematica, t. 25 (1900).

Wir beweisen zunächst den folgenden Satz, der einem von L. Lévy * über Matrizen mit reellen Elementen ausgesprochenen völlig analog ist.

Satz I. Sind in der Matrix (1) die Bedingungen

$$|a_{ii}| \geq \sum_k{}' |a_{ik}|,^{**} \qquad (i = 1, \ldots n \tag{3}$$

erfüllt (wobei das Ungleichheitszeichen mindestens für einen Wert von i gilt), so ist die Determinante Δ dieser Matrix gewiss von 0 verschieden.

Zum Beweis betrachten wir das zu der Matrix (1) zugehörige homogene Gleichungssystem

$$\begin{cases} a_{11} x_1 + a_{12} x_2 + \cdots + a_{1n} x_n = 0, \\ a_{21} x_1 + a_{22} x_2 + \cdots + a_{2n} x_n = 0, \\ \cdots\cdots\cdots\cdots\cdots\cdots \\ a_{n1} x_1 + a_{n2} x_2 + \cdots + a_{nn} x_n = 0. \end{cases} \tag{4}$$

Sollte entgegen der gemachten Annahme $\Delta = 0$ sein, so hat das System (4) eine nichtverschwindende Lösung $x_1^0, x_2^0, \ldots x_n^0$ (wobei diese Werte auch nicht alle einander gleich sein können). Sei $|x_\mu^0|$ die grösste unter den Zahlen $|x_i^0|$, so dass

$$|x_i^0| \leq |x_\mu^0| \qquad (i = 1, \ldots, n). \tag{5}$$

Wir betrachten nun die μ-te der Gleichungen (4), welche lautet

$$a_{\mu\mu} x_\mu^0 = -\sum_k{}' a_{\mu k} x_k^0. \tag{6}$$

Aus den Ungleichungen (3) und (5) folgt aber

$$|a_{\mu\mu}|\,|x_\mu^0| > \sum_k{}' |a_{\mu k}|\,|x_k^0|,$$

was mit der Gleichung (6) unvereinbar ist. Damit ist der Satz bewiesen.***

* Sur la possibilité de l'équilibre électrique. C. R. de l'Académie des Sciences, t. XCIII (1881).

** $\sum_k{}'$ bedeutet die Summation über alle Werte von k, ausser $k = i$.

*** Eine analoge Überlegung wurde schon früher von R. Kusmin zum Beweis des L. Lévy'schen Satzes verwendet.

§ 2. Verwenden wir den oben gefundenen Satz zur Matrix

$$
(7) \qquad \begin{Vmatrix} a_{11}-z, & a_{12}, & \dots & a_{1n} \\ a_{21}, & a_{22}-z, & \dots & a_{2n} \\ \dots & \dots & \dots & \dots \\ a_{n1}, & a_{n2}, & \dots & a_{nn}-z \end{Vmatrix}
$$

so finden wir, dass die zugehörige Determinante von 0 verschieden ist, falls die Bedingungen

$$
(8) \qquad |a_{ii}-z| \geq \sum_k{}' |a_{ik}| \qquad (i=1,\dots n)
$$

(wo das Ungleichheitszeichen mindestens für ein i gilt) erfüllt sind.

Die geometrische Interpretation dieses Resultates führt uns auf den folgenden Satz.

Satz II. Die Eigenwerte $z_1, \dots z_n$ der Matrix (1) liegen nur innerhalb des abgeschlossenen Gebietes G, das aus allen Kreisen K_i $(i=1,\dots n)$ der z-Ebene mit den Mittelpunkten a_{ii} und zugehörigen Radien

$$
R_i = \sum_k{}' |a_{ik}|
$$

besteht.

Es kann vorkommen, dass m von den Kreisen K_i $(m=1,\dots n)$ zu einem zusammenhängenden Gebiet $H_{(m)}$ zusammenfallen, wobei alle übrigen Kreise ausserhalb dieses Gebietes liegen. Über die Verteilung der Eigenwerte unter verschiedenen so definierten Gebieten $H_{(m)}$ kann der folgende Satz ausgesprochen werden.

Satz III. In jedem Gebiet $H_{(m)}$ liegen genau m Eigenwerte der Matrix (1).

Es sei $H_{(m)}$ aus den Kreisen

$$
K_{i_1}, K_{i_2}, \dots K_{i_m}
$$

gebildet. Wir betrachten neben der Matrix A eine andere Matrix A', bei welcher alle nicht in der Diagonale stehende Elemente der Zeilen

$$
i_1, i_2, \dots i_m
$$

verschwinden, die übrigen aber denjenigen der Matrix A gleich sind. Die Matrix A' hat sicher die Eigenwerte

$$a_{i_1 i_1}, a_{i_2 i_2}, \ldots a_{i_m i_m}.$$

Nun fangen wir an die oben erwähnten verschwindenden Elemente der Matrix A' von 0 bis zu ihren Werten in der Matrix A so stetig zu verändern, dass ihre absoluten Beträge monoton wachsen. Die Kreise

$$K_{i_1}, K_{i_2}, \ldots K_{i_m}$$

wachsen dabei stetig, bleiben jedoch immer von den übrigen festen Kreisen K_i der z-Ebene getrennt. Da die Eigenwerte der Matrix stetig von ihren Elementen abhängen, folgt daraus, dass in den Kreisen

$$K_{i_1}, K_{i_2}, \ldots K_{i_m}$$

immer m Eigenwerte liegen müssen. Die Zahl der Eigenwerte in $H_{(m)}$ kann nicht m überschreiten, da ihre gesamte Anzahl in allen Gebieten $H_{(m)}$ genau n gleich sein muss. Damit ist unser Satz bewiesen.*

Liegen alle Kreise K_i getrennt voneinander, was durch die Bedingungen

$$(9) \qquad |a_{ii} - a_{jj}| \geq \sum_k{}' |a_{ik}| + \sum_k{}' |a_{jk}| \quad (i = 1, \ldots, n;\ j = 2, \ldots, n;\ j > i)$$

ausgedrückt werden kann, so sind alle Eigenwerte voneinander abgegrenzt. Da eine Gleichung mit reellen Koeffizienten nur paarweise konjugierte komplexe Wurzeln besitzen kann, folgt daraus unter anderen der folgende Satz.

Satz IV. Sind alle Elemente der Matrix (1) reel und bestehen die Relationen (9), so sind die sämtlichen Eigenwerte dieser Matrix reel.

§ 3. In allen vorstehenden Sätzen kann man statt der Zeilen die Spalten heranziehen. Wir gelangen in dieser Weise im allgemeinen zu einem neuen System G' von Kreisen K_i', welche auch zur Abgrenzung der Wurzeln dienen können. Wir können auch mehrere solche Kreissysteme bekommen, indem wir unsere Matrix verschiedenen Transformationen unterwerfen, bei

* Der Satz bleibt auch dann richtig, wenn sich $H_{(m)}$ mit den übrigen Kreisen von aussen berührt, so dass man bei Bestimmung der Gebiete $H_{(m)}$ solche Berührungen ausser acht lassen kann.

denen das Spektrum sich nicht ändert. Man gelangt dabei im allgemeinen zu einer besseren Abgrenzung der Eigenwerte, da die letzteren nur in denjenigen Punkten liegen dürfen, welche sämtlichen Kreissystemen gehören. Genauer: es seien die Kreissysteme G_λ $(\lambda = 1, \ldots l)$ vorhanden, von denen jedes aus den Kreisen $K_i^{(\lambda)}$ $(i = 1, \ldots n)$ besteht. Wir stellen uns vor, dass die Kreise von G_λ in n_λ $(n_\lambda \leq n)$ voneinander getrennte zusammenhängende Gebiete

$$H_1^{(\lambda)}, H_2^{(\lambda)}, \ldots H_{n_\lambda}^{(\lambda)}$$

zerfallen. Zu jedem Gebiet $H_j^{(\lambda)}$ $(j = 1, \ldots, n_\lambda)$ soll $m_j^{(\lambda)}$ von den Kreisen $K_i^{(\lambda)}$ gehören. Wir bezeichnen weiter durch $S_{j_1, \ldots j_l}$ ein Gebiet, welches allen Gebieten

$$H_{j_1}^{(1)}, H_{j_2}^{(2)}, \ldots H_{j_l}^{(l)}$$

gemeinsam ist (wo j_λ bestimmte Zahlen $\leq n_\lambda$ bedeuten). Dann liegen im Gebiet $S_{j_1, \ldots j_l}$ (es kann auch nicht zusammenhängend sein) genau $m_{j_1, \ldots j_l}$ Eigenwerte, wo $m_{j_1 \ldots j_l}$ die kleinste der Zahlen

$$m_{j_1}^{(1)}, m_{j_2}^{(2)}, \ldots m_{j_l}^{(l)}$$

ist.

Wir können diese Überlegung in folgender Weise verwenden. Es sei $H_{(m)}$ ein aus den Kreisen

$$K_{i_1}, K_{i_2}, \ldots K_{i_m} \quad (m < n)$$

bestehendes zusammenhängendes Gebiet, welches von den anderen Kreisen K_i getrennt liegt. Wir unterwerfen unsere Matrix A einer Transformation mit Hilfe der Matrix $S = ||s_{ik}||$, wo

$$s_{ik} = 0 \qquad (i \neq k)$$

$$s_{ii} = \begin{cases} \alpha & (i = i_1, i_2, \ldots i_m) \\ 1 & (i \neq i_1, i_2, \ldots i_m). \end{cases}$$

Die Zahl $0 < \alpha < 1$ ist noch später genauer zu definieren. Die transformierte Matrix $B = SAS^{-1}$ entsteht aus A durch Multiplikation der Reihen $i_1, i_2, \ldots i_m$ mit α und Division der entsprechenden Spalten durch α. Wir können α so wählen, dass die Kreise

$$K_{i_1}, K_{i_2}, \ldots K_{i_m}$$

des Bereiches $H_{(m)}$ verkleinert werden, ohne die übrigen Kreise K_i, welche sich dabei vergrössern, zu schneiden (es darf höchstens eine Berührung von aussen eintreten). Damit erreichen wir eine bessere Abgrenzung der in $H_{(m)}$ liegenden Eigenwerte.

Wir wollen näher auf den Fall $m = 1$ eingehen. Es sei K_i ein isoliert liegender Kreis. Die Bedingungen für α lauten dann

$$(10) \qquad |a_{ii} - a_{jj}| \geq \alpha \sum' |a_{ik}| + \frac{1}{\alpha} |a_{ji}| + \sum_k'' |a_{jk}|, \quad (j = 1, \dots n;\ j \neq i)$$

wobei $\sum\limits_k''$ die Summation über alle k mit Ausnahme $k = i$ und $k = j$ bedeutet. Man kann, wie leicht zu ersehen ist, allen über α gestellten Bedingungen genügen, indem wir setzen*

$$\alpha = \max \frac{|a_{ii} - a_{jj}| - \sum\limits_k'' |a_{jk}| - \sqrt{(|a_{ii} - a_{jj}| - \sum\limits_k'' |a_{jk}|)^2 - 4 |a_{ji}| \sum\limits_k' |a_{ik}|}}{2 \sum\limits_k' |a_{ik}|}.$$

Wir kommen damit zum folgenden Resultat.

Satz V. Ist K_i ein isoliert liegender Kreis des Gebietes G, so liegt der zugehörige Eigenwert innerhalb des zu K_i konzentrischen kleineren Kreises K_i' mit dem Radius

$$R_i' = \alpha R_i =$$

$$= \max \frac{1}{2} \left[|a_{ii} - a_{jj}| - \sum_k'' |a_{jk}| - \sqrt{\left(|a_{ii} - a_{jj}| - \sum_k'' |a_{jk}|\right)^2 - 4 |a_{ji}| \sum_k' |a_{ik}|} \right].$$

---

* Das Zeichen max bedeutet das Maximum der nachstehenden Grösse für alle Werte von j ausser $j = i$.

Appendix B. Vector Norms and Induced Operator Norms.

With $\mathbb{C}^n$ denoting, for any positive integer n, the complex n-dimensional vector space of all column vectors $\mathbf{v} = [v_1, v_2, \cdots, v_n]^T$, where each v_i is a complex number, we have

Definition B.1. Let $\varphi : \mathbb{C}^n \to \mathbb{R}$. Then, φ is a **norm** on $\mathbb{C}^n$ if

$$\text{(B.1)} \quad \begin{array}{rl} i) & \varphi(\mathbf{x}) \geq 0 \quad (\text{all } \mathbf{x} \in \mathbb{C}^n); \\ ii) & \varphi(\mathbf{x}) = 0 \text{ if and only if } \mathbf{x} = \mathbf{0}; \\ iii) & \varphi(\gamma\mathbf{x}) = |\gamma|\varphi(\mathbf{x}) \quad (\text{any scalar } \gamma, \text{ any } \mathbf{x} \in \mathbb{C}^n); \\ iv) & \varphi(\mathbf{x}+\mathbf{y}) \leq \varphi(\mathbf{x}) + \varphi(\mathbf{y}) \qquad (\text{all } \mathbf{x}, \mathbf{y} \in \mathbb{C}^n). \end{array}$$

Next, given a norm φ on $\mathbb{C}^n$, consider any matrix $B = [b_{i,j}] \in \mathbb{C}^{n \times n}$, so that B maps $\mathbb{C}^n$ into $\mathbb{C}^n$. Then,

$$\text{(B.2)} \qquad ||B||_{\varphi} := \sup_{\mathbf{x} \neq \mathbf{0}} \frac{\varphi(B\mathbf{x})}{\varphi(\mathbf{x})} = \sup_{\varphi(\mathbf{x})=1} \varphi(B\mathbf{x})$$

is called the **induced operator norm** of B, with respect to φ.

Proposition B.2. *Given any $A = [a_{i,j}] \in \mathbb{C}^{n \times n}$, let $\sigma(A)$ denote its* **spectrum**, *i.e.,*

$$\sigma(A) := \{\lambda \in \mathbb{C} : \det(\lambda I - A) = 0\},$$

and let $\rho(A)$ denote its **spectral radius**, *i.e.,*

$$\rho(A) := \max\{|\lambda| : \lambda \in \sigma(A)\}.$$

Then, for any norm φ on $\mathbb{C}^n$,

$$\text{(B.3)} \qquad \rho(A) \leq ||A||_{\phi}.$$

Proof. For any $\lambda \in \sigma(A)$, there is an $\mathbf{x} \neq \mathbf{0}$ in $\mathbb{C}^n$ with $\lambda\mathbf{x} = A\mathbf{x}$. Then, given any norm φ on $\mathbb{C}^n$, we normalize $\mathbf{x}$ so that $\varphi(\mathbf{x}) = 1$. Thus, from (B.1*iii*), (B.2), and our normalization, we have

$$\varphi(\lambda\mathbf{x}) = |\lambda|\varphi(\mathbf{x}) = |\lambda| = \varphi(A\mathbf{x}) \leq ||A||_{\phi} \cdot \varphi(\mathbf{x}) = ||A||_{\varphi},$$

i.e., $|\lambda| \leq ||A||_{\varphi}$. As this is true for each $\lambda \in \sigma(A)$, then $\rho(A) \leq ||A||_{\varphi}$. ■

Proposition B.3. *Let A and B be any matrices in $\mathbb{C}^{n\times n}$, and let φ be any norm on $\mathbb{C}^n$. Then, the induced operator norms of $A+B$, and $A\cdot B$ satisfy*

$$||A+B||_\varphi \leq ||A||_\varphi + ||B||_\varphi \text{ and } ||A\cdot B||_\varphi \leq ||A||_\varphi \cdot ||B||_\varphi. \tag{B.4}$$

Proof. From (B.1) and (B.2), we have

$$\begin{aligned} ||A+B||_\varphi &= \sup_{\varphi(\mathbf{x})=1} \varphi((A+B)\mathbf{x}) = \sup_{\varphi(\mathbf{x})=1} \varphi(A\mathbf{x}+B\mathbf{x}) \\ &\leq \sup_{\varphi(\mathbf{x})=1} \{\varphi(A\mathbf{x}) + \varphi(B\mathbf{x})\} \\ &\leq \sup_{\varphi(\mathbf{x})=1} \varphi(A\mathbf{x}) + \sup_{\varphi(\mathbf{x})=1} \varphi(B\mathbf{x}) \\ &= ||A||_\varphi + ||B||_\varphi. \end{aligned}$$

Similarly, from (B.2)

$$||A\cdot B||_\varphi = \sup_{\mathbf{x}\neq\mathbf{0}} \frac{\varphi(A(B\mathbf{x}))}{\varphi(\mathbf{x})} \leq \sup_{\mathbf{x}\neq\mathbf{0}} \left\{ ||A||_\varphi \cdot \frac{\varphi(B\mathbf{x})}{\varphi(\mathbf{x})} \right\} \leq ||A||\cdot||B||.$$

■

For $\mathbf{x} := [x_1, x_2, \cdots, x_n]^T \in \mathbb{C}^n$, perhaps the three most widely used norms on $\mathbb{C}^n$ are $\ell_1, \ell_2,$ *and* ℓ_∞, where

$$\begin{cases} ||\mathbf{x}||_{\ell_1} := \sum_{j=1}^n |x_j|, \quad ||\mathbf{x}||_{\ell_2} := \left(\sum_{i=1}^n |x_i|^2 \right)^{\frac{1}{2}}, \\ \text{and} \\ ||\mathbf{x}||_{\ell_\infty} := \max_{1\leq i\leq n} |x_i|. \end{cases} \tag{B.5}$$

Given any matrix $C = [c_{i,j}] \in \mathbb{C}^{n\times n}$, the associated induced operator norms of C for the norms of (B.5) are easily shown to be

$$\begin{cases} ||C||_{\ell_1} = \max_{1\leq j\leq n} \left(\sum_{i=1}^n |a_{i,j}| \right); ||C||_{\ell_2} = [\rho(CC^*)]^{\frac{1}{2}}, \\ \text{and} \\ ||C||_{\ell_\infty} = \max_{1\leq i\leq n} \left(\sum_{j=1}^n |a_{i,j}| \right), \end{cases} \tag{B.6}$$

where $C^* := [\bar{c}_{j,i}] \in \mathbb{C}^{n\times n}$.

Appendix C. The Perron-Frobenius Theory of Nonnegative Matrices, M-Matrices, and H-Matrices.

To begin, if $B = [b_{i,j}] \in \mathbb{R}^{n \times n}$ is such that $b_{i,j} \geq 0$ for all $1 \leq i, j \leq n$, we write $B \geq O$. Similarly, if $\mathbf{x} = [x_1, x_2, \cdots, x_n]^T \in \mathbb{R}^n$ is such that $x_i > 0$ ($x_i \geq 0$) for all $1 \leq i \leq n$, we write $\mathbf{x} > \mathbf{0}$ ($\mathbf{x} \geq \mathbf{0}$). We also recall Definition 1.7 from Chapter 1, where **irreducible** and **reducible** matrices in $\mathbb{C}^{n \times n}$ are defined. Then, we state the following strong form of the **Perron-Frobenius Theorem** for irreducible matrices $A \geq O$ in $\mathbb{C}^{n \times n}$. Its complete proof can be found, for example, in Horn and Johnson (1985), Section 8.4, Meyer (2000), Chapter 8, or Varga (2000), Chapter 2. For notation, we again have $N := \{1, 2, \cdots, n\}$.

Theorem C.1. *(Perron-Frobenius Theorem) Given any $A = [a_{i,j}] \in \mathbb{R}^{n \times n}$, with $A \geq O$ and with A irreducible, then:*

- *i)* *A has a positive real eigenvalue equal to its spectral radius $\rho(A)$;*
- *ii)* *to $\rho(A)$, there corresponds an eigenvector $\mathbf{x} = [x_1, x_2, \cdots, x_n]^T > \mathbf{0}$;*
- *iii)* *$\rho(A)$ increases when any entry of A increases;*
- *iv)* *$\rho(A)$ is a simple eigenvalue of A;*
- *v)* *the eigenvalue $\rho(A)$ of A satisfies*

$$\text{(C.1)} \quad \sup_{\mathbf{x}>\mathbf{0}} \left\{ \min_{i \in N} \left[\frac{\sum_{j \in N} a_{i,j} x_j}{x_i} \right] \right\} = \rho(A) = \inf_{\mathbf{x}>\mathbf{0}} \left\{ \max_{i \in N} \left[\frac{\sum_{j \in N} a_{i,j} x_j}{x_i} \right] \right\}.$$

In the case that $A \geq O$ but is not necessarily irreducible, then the analog of Theorem C.1 is

Theorem C.2. *Given any $A = [a_{i,j}] \in \mathbb{R}^{n \times n}$ with $A \geq O$, then:*

- *i)* *A has a nonnegative eigenvalue equal to its spectral radius $\rho(A)$;*
- *ii)* *to $\rho(A)$, there corresponds an eigenvector $\mathbf{x} \geq \mathbf{0}$ with $\mathbf{x} \neq \mathbf{0}$;*
- *iii)* *$\rho(A)$ does not decrease when any entry of A increases;*
- *iv)* *$\rho(A)$ may be a multiple eigenvalue of A;*
- *v)* *the eigenvalue of $\rho(A)$ of A satisfies*

$$\text{(C.2)} \qquad \rho(A) = \inf_{\mathbf{x}>\mathbf{0}} \left\{ \max_{i\in N} \left[\frac{\sum_{j\in N} a_{i,j}x_j}{x_i} \right] \right\}.$$

Next, given $A = [a_{i,j}] \in \mathbb{R}^{n\times n}$, then A is said (cf. Birkhoff and Varga (1958)) to be **essentially nonnegative** if $a_{i,j} \geq 0$ for all $i \neq j$, $(i, j \in N)$, and **essentially positive** if, in addition, A is irreducible. Similarly, we use the notation

$$\text{(C.3)} \quad \mathbb{Z}^{n\times n} := \left\{ A = [a_{i,j}] \in \mathbb{R}^{n\times n} : a_{i,j} \leq 0 \text{ for all } i \neq j \ (i, j \in N) \right\},$$

which also is given in equation (5.5) of Chapter 5. We see immediately that A is essentially nonnegative if and only if $-A \in \mathbb{Z}^{n\times n}$.

For additional notation, consider any $A = [a_{i,j}] \in \mathbb{C}^{n\times n}$. We say that $\mathcal{M}(A) := [\alpha_{i,j}] \in \mathbb{R}^{n\times n}$ is the **comparison matrix** of A if $\alpha_{i,i} := |a_{i,i}|$, and $\alpha_{i,j} := -|a_{i,j}|$ for $i \neq j$ $(i, j \in N)$, i.e.,

$$\text{(C.4)} \qquad \mathcal{M}(A) := \begin{bmatrix} +|a_{1,1}| & -|a_{1,2}| & \cdots & -|a_{1,n}| \\ -|a_{2,1}| & +|a_{2,2}| & \cdots & -|a_{2,n}| \\ \vdots & & & \vdots \\ -|a_{n,1}| & -|a_{n,2}| & \cdots & +|a_{n,n}| \end{bmatrix},$$

where we note that $\mathcal{M}(A) \in \mathbb{Z}^{n\times n}$, for any $A \in \mathbb{C}^{n\times n}$. This brings us to our next important topic of M-matrices.

Given any $A = [a_{i,j}] \in \mathbb{Z}^{n\times n}$, let $\mu := \max_{i\in N} a_{i,i}$, so that $A = \mu I - B$, where the entries of $B = [b_{i,j}] \in \mathbb{R}^{n\times n}$ satisfy $b_{i,i} = \mu - a_{i,i} \geq 0$ and $b_{i,j} = -a_{i,j} \geq 0$ for all $i \neq j$. Thus, $b_{i,j} \geq 0$ for all $1 \leq i, j \leq n$, i.e., $B \geq O$. Then, as in Definition 5.4, we have

Definition C.3. Given any $A = [a_{i.j}] \in \mathbb{Z}^{n\times n}$, let $A = \mu I - B$ be as described above, where $B \geq O$. Then, A is an M**-matrix** if $\mu \geq \rho(B)$. More precisely, A is a **nonsingular** M-matrix if $\mu > \rho(B)$, and a singular M-matrix if $\mu = \rho(B)$.

With Definition C.3, we come to

Proposition C.4. *Given any $A = [a_{i,j}] \in \mathbb{R}^{n\times n}$ which is a nonsingular M-matrix (i.e., $A = \mu I - B$ where $B \geq O$ with $\mu > \rho(B)$), then $A^{-1} \geq O$.*

Proof. Since $A = \mu I - B$ where $B \geq O$ with $\mu > \rho(B)$, we can write that $A = \mu\{I - (B/\mu)\}$, where $\rho(B/\mu) < 1$. Then $I - (B/\mu)$ is also nonsingular, with its known convergent matrix expansion of

$$\text{(C.5)} \qquad \{I - (B/\mu)\}^{-1} = I + (B/\mu) + (B/\mu)^2 + \cdots.$$

Since B/μ is a nonnegative matrix, so are all powers of (B/μ), and it follows from (C.5) that

$$\{I - (B/\mu)\}^{-1} \geq O; \text{ whence, } A^{-1} = \frac{1}{\mu}\{I - (B/\mu)\}^{-1} \geq O.$$

■

In a similar way (cf. Berman and Plemmons (1994), (A_3) of 4.6 Theorem), Proposition C.4 can be extended to

Proposition C.5. *Given any $A = [a_{i,j}] \in \mathbb{R}^{n \times n}$ which is a (possible singular) M-matrix (i.e., $A = \mu I - B$ with $B \geq O$ and $\mu \geq \rho(B)$), then, for any $\mathbf{x} = [x_1, x_2, \cdots, x_n]^T > \mathbf{0}$, $A+$ diag$[x_1, \cdots, x_n]$ is a nonsingular M-matrix.*

Now, we come to the associated topic of H-matrices. Given $A = [a_{i,j}] \in \mathbb{C}^{n \times n}$, let $\mathcal{M}(A)$ be its comparison matrix of (C.4).

Definition C.6. Given $A = [a_{i,j}] \in \mathbb{C}^{n \times n}$, then A is an **H-matrix** if $\mathcal{M}(A)$ of (C.4) is an M-matrix.

Proposition C.7. *Given any $A = [a_{i,j}] \in \mathbb{C}^{n \times n}$ for which $\mathcal{M}(A)$ is a nonsingular M-matrix, then A is a nonsingular H-matrix.*

Proof. By Definition C.6, A is certainly an H-matrix, so it remains to show that A is nonsingular. As in the proof of Theorem 5.5 in Chapter 5, given any $\mathbf{u} = [u_i, u_2, \cdots, u_n]^T \in \mathbb{C}^n$, then the particular **vectorial norm $\mathbf{p}(\mathbf{u})$** on $\mathbb{C}^n$ is defined by

$$\text{(C.6)} \quad \mathbf{p}(\mathbf{u}) := [|u_1|, |u_2|, \cdots, |u_n|]^T \quad (\text{any } \mathbf{u} = [u_1, u_2, \cdots, u_n]^T \in \mathbb{C}^n).$$

Now, it follows by the reverse triangle inequality that, for any $\mathbf{y} = [y_1, y_2, \cdots, y_n]^T$ in $\mathbb{C}^n$,

$$\text{(C.7)} \quad |(A\mathbf{y})_i| = \left| \sum_{j \in N} a_{i,j} y_i \right| \geq |a_{i,i}| \cdot |y_i| - \sum_{j \in N \backslash \{i\}} |a_{i,j}| \cdot |y_j| \qquad (\text{any } i \in N).$$

Recalling the definitions of $\mathcal{M}(A)$ of (C.4) and $\mathbf{p}(\mathbf{u})$ in (C.6), the inequalities of (C.7) nicely reduce to

$$\text{(C.8)} \qquad \mathbf{p}(A\mathbf{y}) \geq \mathcal{M}(A)\mathbf{p}(\mathbf{y}) \quad (\text{any } \mathbf{y} \in \mathbb{C}^n),$$

and we say that $\mathcal{M}(A)$ is a **lower bound matrix** for A. But as $\mathcal{M}(A)$ is, by hypothesis, a nonsingular M-matrix, then $(\mathcal{M}(A))^{-1} \geq O$, from Proposition C.4. As multiplying (on the left) by $(\mathcal{M}(A))^{-1}$ preserves the inequalities of (C.8), we have

$$\text{(C.9)} \qquad (\mathcal{M}(A))^{-1}\mathbf{p}(A\mathbf{y}) \geq \mathbf{p}(\mathbf{y}) \qquad (\text{any } \mathbf{y} \in \mathbb{C}^n).$$

But, the inequalities of (C.9) give us that A is nonsingular, for if A were singular, we could find a $\mathbf{y} \neq \mathbf{0}$ in $\mathbb{C}^n$ with $A\mathbf{y} = \mathbf{0}$, so that $\mathbf{p}(\mathbf{y}) \neq \mathbf{0}$ and $\mathbf{p}(A\mathbf{y}) = \mathbf{0}$. But this contradicts the inequalities of (C.9). ■

It is important to mention that the terminology of H- and M- matrices was introduced in the seminal paper of Ostrowski (1937b). Here, A.M. Ostrowski paid homage to his teacher, H. Minkowski, and to J. Hadamard, men who had inspired Ostrowski's work in this area. By naming these two classes of matrices after them, their names are forever honored and remembered in mathematics.

The theory of M-matrices and H-matrices has proved to be an incredibly useful tool in linear algebra, and it is as fundamental to linear algebra as topology is to analysis. For example, one finds 50 *equivalent* formulations of a nonsingular M-matrix in Berman and Plemmons (1994). Some additional equivalent formulations can be found in Varga (1976), and it is plausible that there are now over 70 such equivalent formulations of a nonsingular M-matrix.

Appendix D. Matlab 6 Programs.

In this appendix, Professor Arden Ruttan of Kent State University has kindly gathered several of the various Matlab 6 programs for figures generated in this book, so the interested readers can study these programs and alter them, as needed, for their own purposes.

Programs are listed on the following pages according to their figure numbers.

Fig. 2.1

```
x=[-2.5:0.05:2.5];
y=[-2.5:0.05:2.5];;
[X,Y]=meshgrid(x,y);
hold on
plot([1],[0],'Marker','o','MarkerSize',2)
plot([-1],[0],'Marker','o','MarkerSize',2)
axis equal
colormap([.7,.7,.7;1,1,1])
caxis([-1 1])
Z=abs(X+i*Y-1).*abs(X+i*Y+1)-2.0^2;
contourf(X,Y,-Z-1,[-1 -1],'k')
Z=abs(X+i*Y-1).*abs(X+i*Y+1)-1.41^2;
contourf(X,Y,-Z-1,[-1 -1],'k')
Z=abs(X+i*Y-1).*abs(X+i*Y+1)-1.2^2;
contourf(X,Y,-Z-1,[-1 -1],'k')
Z=abs(X+i*Y-1).*abs(X+i*Y+1)-1.0^2;
contourf(X,Y,-Z,[0 0],'k')
Z=abs(X+i*Y-1).*abs(X+i*Y+1)-0.9^2;
contourf(X,Y,-Z,[0 0],'k')
Z=abs(X+i*Y-1).*abs(X+i*Y+1)-0.5^2;
contourf(X,Y,-Z-1,[-1 -1],'k')
plot([-1],[0],'.k')
plot([1],[0],'.k')

title('Figure 2.1')
```

Fig. 2.2

```
hold on
x=[-.5:0.05:2.5];
y=[-1.5:0.05:1.5];;
[X,Y]=meshgrid(x,y);
Z=abs(X+i*Y-1)-1;
contour(X,Y,-Z,[0 0],'k')
y=[-.5:0.05:2.5];
x=[-1.5:0.05:1.5];;
[X,Y]=meshgrid(x,y);
Z=abs(X+i*Y-i)-1;
contourf(X,Y,-Z,[0 0],'k')
x=[-2.5:0.05:0.5];
y=[-1.5:0.05:1.5];;
[X,Y]=meshgrid(x,y);
Z=abs(X+i*Y+1)-1;
contourf(X,Y,-Z,[0 0],'k')
y=[-2.5:0.05:0.5];
x=[-1.5:0.05:1.5];;
[X,Y]=meshgrid(x,y);
Z=abs(X+i*Y+i)-1;
contourf(X,Y,-Z,[0 0],'k')

x=[-2.5:0.05:2.5];
y=[-2.5:0.05:2.5];;
[X,Y]=meshgrid(x,y);
Z=abs(X+i*Y-1).*abs(X+i*Y-i)-1;
contourf(X,Y,-Z-1,[-1 -1],'k')
axis equal
colormap([.7,.7,.7;1,1,1])
axis([-2.2,2.2,-2.2,2.2])
Z=abs(X+i*Y-1).*abs(X+i*Y+1)-1;
contourf(X,Y,-Z-1,[-1 -1],'k')
Z=abs(X+i*Y-1).*abs(X+i*Y+i)-1;
contourf(X,Y,-Z-1,[-1 -1],'k')
Z=abs(X+i*Y+1).*abs(X+i*Y-i)-1;
contourf(X,Y,-Z-1,[-1 -1],'k')
Z=abs(X+i*Y-i).*abs(X+i*Y+i)-1;
contourf(X,Y,-Z-1,[-1 -1],'k')
```

```
Z=abs(X+i*Y+1).*abs(X+i*Y+i)-1;
contourf(X,Y,-Z-1,[-1 -1],'k')
plot([0],[1],'.k')
plot([0],[-1],'.k')
plot([1],[0],'.k')
plot([-1],[0],'.k')
text(0,.8,'i')
text(0,-1.2,'i')
text(1,-.2,'1')
text(-1,-.2,'-1')

title('Figure 2.2')
a='Set Transparency of grey part to .5'
```

Fig. 2.7

```
x=[-2.5:0.05:2.5];
y=[-2.5:0.05:2.5];;
[X,Y]=meshgrid(x,y);
hold on
axis equal
caxis([-1,0])
colormap([.7,.7,.7;1,1,1])
axis([-2,2,-2,2])
Z=abs((X+i*Y).^2-1)-1;
contourf(X,Y,-Z,[0 0],'k')
Z=(abs(X+i*Y-1).^2).*abs(X+i*Y+1)-1/2.0;
contourf(X,Y,-Z-1,[-1 -1],'k')
plot([1],[0],'.k','MarkerSize',10)
plot([-1],[0],'.k','MarkerSize',10)
text(-1.08,-.075,'-1')
text(1,-.1,'1')
text(0,-.2,'0')
text(-.3,.5,'|z-1| |z+1|=1/2')
text(-.3,-.6,'|z -1|=1')

title('Figure 2.7')
```

Fig. 2.9

```
x=[-2.5:0.05:2.5];
y=[-2.5:0.05:2.5];;
[X,Y]=meshgrid(x,y);
hold on
axis equal
caxis([-1,0])
colormap([.7,.7,.7;1,1,1])
axis([-2,2,-2,2])
Z=abs((X+i*Y).^4-1)-1;
contourf(X,Y,-Z-1,[-1 -1],'k')
Z=abs(X+i*Y-1).*abs(X+i*Y-i)-1.0;
contourf(X,Y,-Z-1,[-1 -1],'k')
%plot([1],[0],'Marker','+','MarkerSize',10)
%plot([-1],[0],'Marker','+','MarkerSize',10)
a='Set transparency to 0.5'
title('Figure 2.9')
```

Fig. 3.2

```
x=[-.5:0.05:2.5];
y=[-1.5:0.05:1.5];;
[X,Y]=meshgrid(x,y);
Z=abs(X+i*Y-1)-1;
contour(X,Y,Z,[0 0],'k')
hold on
axis equal
colormap([.7,.7,.7;1,1,1])
caxis([-1 0])
axis([-2.2,2.2,-2.2,2.2])
y=[-.5:0.05:2.5];
x=[-1.5:0.05:1.5];;
[X,Y]=meshgrid(x,y);
Z=abs(X+i*Y-i)-1;
contour(X,Y,Z,[0 0],'k')
x=[-2.5:0.05:0.5];
y=[-1.5:0.05:1.5];;
[X,Y]=meshgrid(x,y);
Z=abs(X+i*Y+1)-1;
contour(X,Y,Z,[0 0],'k')
y=[-2.5:0.05:0.5];
x=[-1.5:0.05:1.5];;
[X,Y]=meshgrid(x,y);
Z=abs(X+i*Y+i)-1;
contour(X,Y,Z,[0 0],'k')

x=[-2.5:0.05:2.5];
y=[-2.5:0.05:2.5];;
[X,Y]=meshgrid(x,y);
Z=abs((X+i*Y).^4-1)-1;
contourf(X,Y,-Z-1,[-1 -1],'k')
%plot([1],[0],'Marker','+','MarkerSize',10)
%plot([2],[0],'Marker','+','MarkerSize',10)
title('Figure 3.2')
```

Fig. 3.4

```
x=[0:0.05:5];
y=[-2:0.05:2];
[X,Y]=meshgrid(x,y);
caxis([-1 0])
colormap([.7,.7,.7;1,1,1])
Z=(abs(X+i*Y-2).^2).*abs(X+i*Y-1)
   -abs(X+i*Y-1)-abs(X+i*Y-2);
contourf(X,Y,-Z-1,[-1 -1],'k')
axis equal
hold on
Z=(abs(X+i*Y-2).^2).*abs(X+i*Y-1)
   -abs(X+i*Y-1)+abs(X+i*Y-2);
contourf(X,Y,-Z,[0 0],'k')
Z=(abs(X+i*Y-2).^2).*abs(X+i*Y-1)
   +abs(X+i*Y-1)-abs(X+i*Y-2);
contourf(X,Y,-Z,[0 0],'k')
text(2,.4,'(13)(2)')
text(.7,.25,'(1)(23)')
text(1.5,1.5,'(1)(2)(3)')
plot([1],[0],'.k')
plot([2],[0],'.k')
title('Figure 3.4')
text(1,-.2,'1')
text(2,-.2,'2')
text(2,-.2,'0')
```

Fig. 6.1

```
x=[-20:0.1:40];
y=[-20:0.1:20];
[X,Y]=meshgrid(x,y);
hold on
axis equal
axis([.5 7.5 -2 2])
colormap bone
brighten(.9)
Z=-100*bc1(X,Y);% 0.059759, 5.831406
contourf(X,Y,Z,[0 0],'k.')
Z=-bc2(X,Y); % 0.063666, 4.693469
contour(X,Y,Z,[0 0],'k')
     Z=-bc3(X,Y); %3.617060, 32.247282
contour(X,Y,Z,[0 0],'k')
plot([2.2679],[0],'kx')
plot([4],[-1],'kx')
plot([4],[1],'kx')
plot([5.7321],[0],'kx')

with files bc1, bc2, and bc3, respectively:
function mm=bc1(x,y)
z=x+i*y;
mm=abs(z-2).*(abs(z-4).^2).*abs(z-6)
   -(abs(z-3)+1).*(abs(z-5)+1);

function mm=bc2(x,y)
z=x+i*y;
mm=abs(z-2).*abs(z-4)-(abs(z-3)+1);

function mm=bc3(x,y)
z=x+i*y;
mm=abs(z-4).*abs(z-6)-(abs(z-5)+1);
```

Fig. 6.2, and 6.3

```
x=[-20:0.1:40];
y=[-20:0.1:20];
[X,Y]=meshgrid(x,y);
hold on
Z=bb1(X,Y);
contour(X,Y,Z,[0 0], 'b--')
Z=bb2(X,Y);
axis equal
axis([-15 35 -16 16])
contour(X,Y,Z,[0 0], 'b--')
     Z=bb3(X,Y);
contour(X,Y,Z,[0 0], 'b')
     Z=bb4(X,Y);
contour(X,Y,Z,[0 0], 'b--')
W=bb(X,Y);
contour(X,Y,W,[102.96 102.96])
x=[0.03:.0005:0.12];
y=[-.04:.0005:0.04];
[X,Y]=meshgrid(x,y);
Z=bb1(X,Y);
contour(X,Y,Z,[0 0])
Z=bb2(X,Y);
contour(X,Y,Z,[0 0])
Z=bb3(X,Y);
contour(X,Y,Z,[0 0])
Z=bb4(X,Y);
contour(X,Y,Z,[0 0])
figure
hold on
Z=bb1(X,Y);
contourf(X,Y,Z,[0 0])
Z=bb2(X,Y);
contourf(X,Y,Z,[0 0],'k-')
Z=bb3(X,Y);
contourf(X,Y,Z,[0 0],'k-')
Z=bb4(X,Y);
contourf(X,Y,Z,[0 0],'k-')
```

Fig. 6.5

```
x=[-20:0.1:40];
y=[-20:0.1:20];
hold on
axis equal
axis([-15 35 -15 15])
caxis([-1 0])
colormap([.7,.7,.7;1 1 1])
[X,Y]=meshgrid(x,y);
Z=b1(X,Y)-1;
contour(X,Y,Z,[0 0],'k')
Z=b2(X,Y)-1;
contour(X,Y,Z,[0 0],'k')
Z=b3(X,Y)-1;
contour(X,Y,Z,[0 0],'k')
Z=b4(X,Y)-1;
contour(X,Y,Z,[0 0],'k')
Z=bb(X,Y)-102.96;
contour(X,Y,Z,[0 0],'k')
Z=bb1(X,Y).*bb3(X,Y)-1;
contourf(X,Y,Z-1,[-1 -1],'k')
plot([15],[0],'.k')
plot([-14 35],[0 0],'-k')
text(0,-1,'0')
text(15,-1,'15')
```

Fig. 6.6

```
x=[0.03:0.001:.12];
y=[-0.04:0.001:0.04];
[X,Y]=meshgrid(x,y);
hold on
axis equal
axis([-.005 .12 -.04 .04])
colormap([1,1,1;.7,.7,.7])
Z=b1(X,Y)-1;
contourf(X,Y,Z-1,[-1 -1],'k')
Z=b2(X,Y)-1;
contourf(X,Y,Z-1,[-1 -1],'k')
Z=b3(X,Y)-1;
contourf(X,Y,Z-1,[-1 -1],'k')
Z=b4(X,Y)-1;
contourf(X,Y,Z-1,[-1 -1],'k')
Z=bb1(X,Y).*bb3(X,Y)-1;
contourf(X,Y,Z,[0 0],'k')
plot([-.005 .12],[0 0],'-k')
plot([.0482],[0],'.k')
plot([.0882],[0],'.k')
plot([0],[0],'.k')
text(.09,-.004,'0.0882')
text(.05,-.004,'0.0482')
text(0,-.004,'0')
```

References

Bauer, F. L. (1962) On the field of values subordinate to a norm, Numer. Math. 4, 103-113. *[96]*

Bauer, F. L. (1968) Fields of values and Gershgorin disks, Numer. Math. 12, 91-95. *[96]*

Beauwens, R. (1976) Semistrict diagonal dominance, SIAM J. Numer. Anal. 13, 109-112. *[17]*

Beckenbach, E.F. and Bellman, R. (1961) Inequalities, Springer-Verlag, Berlin. *[136, 140]*

Berman, A. and Plemmons, R.J. (1994) Nonnegative Matrices in the Mathematical Sciences, Classics in Applied Mathematics 9, SIAM, Philadelphia. *[12, 108, 133, 203, 204]*

Birkhoff, G. and MacLane, S (1960) A Survey of Modern Algebra, third edition, MacMillan Co., New York. *[45]*

Birkhoff, G. and Varga, R.S. (1958) Reactor criticality and non-negative matrices, J. Soc. Indust. Appl. Math. 6, 354-377. *[202]*

Bode, A. (1968) Matrizielle untere Schranken linearer Abbildungen und M-Matrizen, Numer. Math. 11, 405-412. *[153]*

Brauer, A. (1946) Limits for the characteristic roots of a matrix, Duke Math. J. 13, 387-395. *[31]*

Brauer, A. (1947) Limits for the characteristic roots of a matrix II, Duke Math. J. 14, 21-26. *[36, 37]*

Brauer, A. (1952) Limits for the characteristic roots of a matrix V, Duke Math. J. 19, 553-562. *[72]*

Brualdi, R. (1982) Matrices, eigenvalues and directed graphs, Linear Multilinear Algebra 11, 143-165. *[45, 47, 49, 51, 71, 186]*

Brualdi, R. (1992) The symbiotic relationship of combinatorics and matrix theory, Linear Algebra Appl. 162/164, 65-105. *[160]*

Brualdi, R. and Mellendorf, S. (1994) Regions in the complex plane containing the eigenvalues of a matrix, Amer. Math. Monthly 101, 975-985. *[92, 94, 96]*

Camion, P. and Hoffman, A. J. (1966) On the nonsingularity of complex matrices, Pacific J. Math. 17, 211-214. *[115, 125]*

Carlson, D. H. and Varga, R. S. (1973a) Minimal G-functions, Linear Algebra and Appl. 6, 97-117. *[127, 153]*

Carlson, D. H. and Varga, R. S. (1973b) Minimal G-functions II, Linear Algebra and Appl. 7, 233-242. *[181, 185, 187]*

Cvetkovic, L. (2001). Private communication. *[17]*

Cvetkovic, L., Kostic, V., and Varga, R. S. (2004). A new Geršgoirn-type eigenvalue inclusion set, ETNA (Electronic Transactions on Numerical Analysis) (to appear). *[84, 85, 87, 96]*

Dashnic, L. S. and Zusmanovich, M. S. (1970) O nekotoryh kriteriyah regulyarnosti matric i lokalizacii ih spectra, Zh. vychisl. matem. i matem. fiz 5, 1092-1097. *[84, 86, 87, 96]*

Desplanques, J. (1887) Théorèm d'algébre, J. de Math. Spec. 9, 12-13. *[6, 31]*

Deutsch, E. and Zenger, C. (1975) On Bauer's generalized Gershgorin disks. Numer. Math. 24, 63-70. *[96]*

Eiermann, M. and Niethammer, W. (1997) Numerische Lösung von Gleichungssystemen, Studientext der Fernuniversität Hagen. *[96]*

Elsner, L. (1968) Minimale Gerschgorin-Kreise, Z. Angew. Math. Mech. 48, 51-55. *[125]*

Engel, G. M. (1973) Regular equimodular sets of matrices for generalized matrix functions, Linear Algebra and Appl. 7, 243-274. *[40, 70, 125]*

Fan, K. and Hoffman, A. J. (1954) Lower bounds for the rank and location of the eigenvalues of a matrix. Contributions to the Solution of Systems of Linear Equations and the Determination of Eigenvalues (O. Taussky, ed.), pp. 117-130. National Bureau of Standards Applied Mathematics Series 39, U.S. Government Printing Office. *[22, 25, 33]*

Fan, K. (1958) Note on circular disks containing the eigenvalues of a matrix, Duke Math. J. 25, 441-445. *[23, 33, 127, 132, 153]*

Feingold, D. G. and Varga, R. S. (1962) Block diagonally dominant matrices and generalizations of the Gerschgorin circle theorem, Pacific J. Math. 12, 1241-1250. *[72, 153, 156, 186]*

Fiedler, M. and Pták, V. (1962a) Generalized norms of matrices and the location of the spectrum, Czechoslovak Math. J. 12(87), 558-571. *[153, 156, 186]*

Forsythe, G.E. and Wasow, W.R. (1960) Finite-difference Methods for Partial Differential Equations, Wiley and Sons, New York. *[191]*

Fujino, S. and Fischer, J. (1998) Über S.A. Gerschgorin, GAMM-Mitteilungen, Heft. 1, 15-19. *[189, 190]*

Gaier, D. (1964) Konstruktive Methoden der Konformen Abbildung, Springer-Verlag, Berlin. *[191]*

Geršgorin, S. (1931) Über die Abgrenzung der Eigenwerte einer Matrix, Izv. Akad. Nauk SSSR Ser. Mat. 1, 749-754. *[1, 3, 6, 8, 10, 31]*

Hadamard, J. (1903) Leçons sur la propagation des ondes, Hermann et fils, Paris, reprinted in (1949) by Chelsea, New York. *[6, 31]*

Hoffman, A. J. (1969) Generalization of Gersgorin's Theorem: G-Generating Families, Lecture notes, University of California at Santa Barbara, 46 pp. *[127, 135, 141, 153]*

Hoffman, A. J. (1971) Combinatorial aspects of Gerschgorin's theorem, Recent Trends in Graph Theory, (Proc. Conf., New York, 1970), pp. 173-179. Lecture Notes in Mathematics 186, Springer-Verlag, New York. 1971. *[153]*

Hoffman, A.J. (1975) Linear G-functions., Linear and Multilinear Alg. 3, 45-72. *[153]*

Hoffman, A. J. (2000) Geršgorin variations. I. On a theme of Pupkov and Solov'ev, Linear Algebra and Appl. 304 (2000), 173-177. *[92, 96]*

Hoffman, A. J. (2003) Selected Papers of Alan Hoffman, with Commentary. (edited by C.A. Micchelli), World Scientific, New Jersey. *[153]*

Hoffman, A. J. and Varga, R. S. (1970) Patterns of dependence in generalizations of Gerschgorin's theorem, SIAM J. Numer. Anal. 7, 571-574. *[145]*

Horn, R. A. and Johnson, C. R. (1985) Matrix Analysis, Cambridge University Press, Cambridge. *[9, 44, 45, 83, 172, 185, 201]*

Horn, R. A. and Johnson, C. R. (1991) Topics in Matrix Analysis, Cambridge University Press, Cambridge. *[79, 96]*

Householder, A. S. (1956) On the convergence of matrix iterations, J. Assoc. Comput. Mach. 3, 314-324. *[26, 33]*

Householder, A. S. (1964) The Theory of Matrices in Numerical Analysis. Blaisdell Publ. Co., New York, 257 pp. *[26, 33]*

Householder, A. S., Varga, R. S. and Wilkinson, J. H. (1972) A note on Gerschgorin's inclusion theorem for eigenvalues of matrices, Numer. Math. 16, 141-144. *[30]*

Huang, T. Z. (1995) A note on generalized diagonally dominant matrices, Linear Algebra Appl., 225, 237-242. *[84, 86, 96]*

Huang, T. Z. and You, Z. Y. (1993) G-block diagonal dominance of matrices (Chinese) Gongcheng Shuxue Xuebao 10, 75-85. MR 96b:15043. *[153]*

Huang, T. Z. and Zhong, S. M. (1999) G-functions and eigenvalues of block matrices, Acta Math. Sci. (Chinese) 19, 62-66. MR 2000m: 15007. *[153]*

Johnson, C. R. (1973) A Geršgorin inclusion set for the field of values of a finite matrix, Proc. Amer. Math. Soc. 41, 57-60.. *[82, 95]*

Johnston, R. L. (1965) Block generalizations of some Gerschgorin-type theorems, Ph.D. Thesis, Case Institute of Technology, 43 pp. *[181, 184, 187]*

Johnston, R. L. (1971) Gerschgorin theorems for partitioned matrices, Linear Algebra and Appl. 4, 205-220. *[184]*

Karow, M. (2003) Geometry of Spectral Value Sets, Ph.D. Thesis, Universität Bremen, Bremen, Germany. *[58, 71]*

Kolotolina, L. Yu. (2001) On Brualdi's theorem, Notes of the LOMI-Seminars 284, 1-17 (in Russian). *[71]*

Kolotolina, L. Yu. (2003a) Nonsingularity/singularity criteria for nonstrictly block diagonally dominant matrices, Linear Algebra and Appl. 359, 133-159. *[33, 186]*

Kolotolina, L. Yu. (2003b) Generalizations of the Ostrowski-Brauer Theorem, Linear Algebra and Appl. 364, 65-80. *[33]*

Korganoff, A. (1961) Calcul Numérique, Tome 1, Dunod, Paris. *[70]*

Levinger, B. W. and Varga, R. S. (1966a) Minimal Gerschgorin sets II, Pacific J. Math. 17, 199-210. *[110, 125]*

Lévy, L. (1881) Sur le possibilité du l'equibre électrique, C.R. Acad. Sci. Paris 93, 706-708. *[6, 31]*

Li, B. and Tsatsomeros, M.J. (1997) Doubly diagonally domninant matrices, Linear Algebra and Appl. 261, 221-235. *[71]*

Loewy, R. (1971) On a theorem about the location of eigenvalues of matrices, Linear Algebra and Appl. 4, 233-242. *[121]*

Marcus, M. and Minc, H. (1964) A Survey of Matrix Theory and Matrix Inequalities, Allyn and Bacon, Boston, 1964. *[44]*

Meyer, C.D. (2000) Matrix Analysis and Applied Linear Algebra, SIAM, Philadelphia. *[201]*

Minkowski, H. (1900) Zur Theorie der Einheiten in den algebraischen Zahlkörpern, Nachr. Königlichen Ges. Wiss. Göttingen Math. - Phys. Kl., 90-93. Gesamelte Abh. 1, 316-319. *[6, 31]*

Nirschl, N. and Schneider, H. (1964) The Bauer fields of values of a matrix, Numer. Math. 6, 355-365. *[96]*

Nowosad, P. (1965) On the functional (x^{-1}, Ax) and some of its applications, An. Acad. Brasil. Ci. 37, 163-165. *[127, 153]*

Nowosad, P. and Tover (1980) Spectral inequalities and G-functions, Linear Algebra Appl. 31, 179-197.. *[153]*

Ostrowski, A. M. (1937b) Über die Determinanten mit überwiegender Hauptdiagonale, Comment. Math. Helv. 10, 69-96. *[35, 70, 132, 153, 204]*

Ostrowski, A. M. (1951a) Über das Nichtverschwinden einer Klasse von Determinanten und die Lokalisierung der charakteristischen Wurzeln von Matrizen, Compositio Math. 9, 209-226. *[20, 31, 33, 140]*

Ostrowski, A.M. (1951b) Sur les conditions générales pour la régularité des matrices, Univ. Roma. Ist. Naz. Alta Mat. Rend. Mat. e. Appl. (5), 10, 156-168. *[22, 25, 31, 33]*

Ostrowski, A. M. (1960) Solution of Equations and Systems of Equations, Academic Press, New York. *[8, 27]*

Ostrowski, A. M. (1961) On some metrical properties of operator matrices and matrices partitioned into blocks, J. Math. Anal. Appl. 2, 161-209. *[153, 156, 186]*

Parodi, M. (1952) Sur quelques propriétés des valeurs caractéristiques des matrices carrées, Mémor. Sci. Math. 118, Gauthier-Villars, Paris. *[73, 95, 110]*

Parodi, M. (1959) La localisation des valeurs caratéristiques des matrices et ses applications, Gauthier-Villars, Paris. *[95]*

Pupkov, V. A. (1984) Some sufficient conditions for the non-degeneracy of matrices, U.S.S.R. Comput. Math. and Math. Phys. 24, 86-89. *[92, 96]*

Rein, H.J. (1967) Bemerkung zu einem Satz von A. Brauer, Kleine Mitteilungen, Z. Angew. Math. Mech. 47, 475-476. *[52, 72]*

Robert, F. (1964) Normes vectorielles de vecteurs et de matrices, Rev. Française Traitement Information Chiffres 7, 261-299. *[153]*

Robert, F. (1965) Sur les normes vectorielles régulieères sur un espace de dimension finie, C.R. Acad. Sci. Paris Ser. A-B 260, 5193-5176. *[153]*

Robert, F. (1966) Recherche d'une M-matrice parmi les minorantes d'un opérator lineaire, Numer. Math. 9, 189-199. *[153, 164, 166, 167, 169, 186]*

Robert, F. (1969) Blocs-H-matrices et convergence des methodes iteratives classiques par blocks, Linear Algebra and Appl. 2, 223-265. *[186]*

Rohrbach, H. (1931) Bemerkungen zu einem Determinantensatz von Minkowski, Jahresber. Deutsch. Math. Verein. 40, 49-53. *[31]*

Royden, H.L. (1988) Real Analysis, Macmillan, Third Edition, New York. *[79]*

Schneider, H. (1954) Regions of exclusion for the latent roots of a matrix, Proc. Amer. Soc. 5, 320-322. *[73, 95, 110, 125]*

Solov'ev, V. N. (1984) A generalization of Geršgorin's theorem. Math. U.S.S.R. Izvestiya 23, 545-559. *[92, 96]*

Stoer, J. (1968) Lower bounds of matrices, Numer. Math. 12, 146-158. *[153]*

Taussky, O. (1947) A method for obtaining bounds for characteristic roots of matrices with applications to flutter calculations, Aero. Res. Council (Great Britain), Report 10. 508, 19 pp. *[189]*

Taussky, O. (1948) Bounds for characteristic roots of matrices, Duke Math. J. 15, 1043-1044. *[15]*

Taussky, O. (1949) A recurring theorem on determinants, Amer. Math. Monthly 56, 672-676. *[14, 15, 32, 189]*

Taussky, O. (1988)How I became a torch bearer for Matrix Theory, Amer. Math. Monthly 95, 801-812. *[32]*

Tee, G. J. (2004) Semyon Aronovich Gerschgorin, Image (to appear). *[190]*

Varga, R. S. (1965) Minimal Gerschgorin sets, Pacific J. Math. 15, 719-729. *[125]*

Varga, R. S. (1976) On recurring theorems on diagonal dominance, Linear Algebra and Appl. 13, 1-9. *[204]*

Varga, R. S. (2000) Matrix Iterative Analysis, Second revised and expanded edition, Springer-Verlag, Berlin. *[201]*

Varga, R. S. (2001a) Gerschgorin-type eigenvalue inclusion theorems and their sharpness, ETNA(Electronic Transactions on Numerical Analysis). 12, 113-133. *[62, 72, 122, 125]*

Varga, R. S. (2001b) Gerschgorin disks, Brauer ovals of Cassini (a vindication), and Brualdi sets, Information 4, 171-178. *[71]*

Varga, R. S. and Kraustengl, A. (1999) On Geršgorin-type problems and ovals of Cassini, ETNA (Electronic Transactions on Numerical Analysis) 8, 15-20. *[38, 40, 70]*

Walsh, J. L. (1969) Interpolation and Approximation by Rational Functions in the Complex Domain, Colloquium Publications, vol. 20, fifth edition, Amer. Math. Soc., Providence. *[43]*

Zenger, C. (1968) On convexity properties of the Bauer field of values of a matrix, Numer. Math. 12, 96-105. *[96]*

Zenger, C. (1984) Positivity in complex spaces and its application to Gershgorin disks, Numer. Math. 44, 67-73. *[96]*

Zhang, Xian and Gu, Dun He (1994) A note on A. Brauer's theorem. Linear Algebra and Appl. 196, 163-174. *[52, 72]*

Index

Symbol Index

MIX
Papier aus verantwortungsvollen Quellen
Paper from responsible sources
FSC® C105338

If you have any concerns about our products,
you can contact us on
ProductSafety@springernature.com

In case Publisher is established outside the EU,
the EU authorized representative is:
Springer Nature Customer Service Center GmbH
Europaplatz 3, 69115 Heidelberg, Germany

Printed by Libri Plureos GmbH
in Hamburg, Germany